NONLINEAR SIGNAL AND IMAGE ANALYSIS

ANNALS OF THE NEW YORK ACADEMY OF SCIENCES
Volume 808

NONLINEAR SIGNAL AND IMAGE ANALYSIS

Edited by J. Robert Buchler and Henry Kandrup

The New York Academy of Sciences
New York, New York
1997

Library of Congress Cataloging-in-Publication Data

Nonlinear signal and image analysis/edited by J. Robert Buchler and Henry Kandrup.
p. cm.—(Annals of the New York Academy of Sciences, ISSN 0077-8923; v. 808)
Includes bibliographic references and index.
ISBN 1-57331-044-1 (cloth: alk. paper).—ISBN 1-57331-045-X (paper: alk. paper).
1. Chaotic behavior in systems. 2. Astrophysics. 3. Time-series analysis. 4. Image processing. I. Buchler, J. R. (J. Robert). II. Kandrup, Henry E. III. Series.
Q11.N5 vol. 808
[QB466.C45]
500 s—dc21
[523.01]

96-29561
CIP

S/PCP
Printed in the United States of America
ISBN 1-57331-044-1 (cloth)
ISBN 1-57331-045-X (paper)
ISSN 0077-8923

ANNALS OF THE NEW YORK ACADEMY OF SCIENCES

Volume 808
January 30, 1997

NONLINEAR SIGNAL AND IMAGE ANALYSIS[a]

Editors and Conference Organizers
J. ROBERT BUCHLER and HENRY KANDRUP

CONTENTS

[a]This volume is the result of the Eleventh Annual Florida Workshop in Nonlinear Astronomy and Physics, entitled Nonlinear Signal and Image Analysis, which was held on November 30 to December 2, 1995, at the University of Florida at Gainesville.

Financial assistance was received from:

- COLLEGE OF LIBERAL ARTS AND SCIENCES, UNIVERSITY OF FLORIDA
- DEPARTMENT OF ASTRONOMY, UNIVERSITY OF FLORIDA
- DEPARTMENT OF PHYSICS, UNIVERSITY OF FLORIDA
- OFFICE OF RESEARCH, TECHNOLOGY, AND GRADUATE EDUCATION, UNIVERSITY OF FLORIDA

Preface

J. ROBERT BUCHLER[a] AND HENRY KANDRUP[b]

[a] *Department of Physics*
[b] *Department of Astronomy*
University of Florida
Gainesville, Florida 32611

The Eleventh Florida Workshop in Nonlinear Astronomy and Physics was devoted to recent developments in time series and image processing techniques and the interesting physics that can be extracted using these techniques. It has been recognized for some time that more conventional linear techniques are inadequate for analyzing many physical systems, especially those where chaos plays an important role. For this reason, many scientists in a variety of different areas have been developing alternatives. In the last few years, several of these novel alternatives have matured to the point where they have started to yield interesting practical results. A workshop in this general area thus seemed especially appropriate. The specific aim of this workshop was to assemble a broad range of practitioners of these modern arts of extracting useful information about the physics that underlies apparently random temporal time series and spatial structures.

This workshop was made possible through the financial support of a number of people at the University of Florida, including Karen Holbrook, Vice President of the Division of Sponsored Research, W. W. Harrison, Dean of the College of Liberal Arts and Sciences, Neil Sullivan and Stan Dermott, Chairs of the Physics and Astronomy Departments, and Pierre Ramond, Director of the Institute for Fundamental Theory.

The Second through the Tenth Florida Workshops have also appeared in the *Annals* of the New York Academy of Sciences:

- BUCHLER, J. R. & H. EICHHORN, Eds. 1987. Chaotic Phenomena in Astrophysics: Proceedings of the Second Florida Workshop in Nonlinear Astronomy. Ann. N.Y. Acad. Sci. 497.
- BUCHLER, J. R., J. R. IPSER & C. WILLIAMS, Eds. 1988. Integrability in Dynamical Systems: Proceedings of the Third Florida Workshop in Nonlinear Astronomy. Ann. N.Y. Acad. Sci. **536**.
- BUCHLER, J. R., S. GOTTESMAN & J. H. HUNTER, Eds. 1989. Galactic Models: Proceedings of the Fourth Florida Workshop in Nonlinear Astronomy. Ann. N.Y. Acad. Sci. **596**.
- BUCHLER, J. R. & S. GOTTESMAN, Eds. 1990. Nonlinear Astrophysical Fluid Dynamics: Proceedings of the Fifth Florida Workshop in Nonlinear Astronomy. Ann. N.Y. Acad. Sci. **617**.
- BUCHLER, J. R., S. L. DETWEILER & J. R. IPSER, Eds. 1991. Nonlinear Problems in Relativity and Cosmology: Proceedings of the Sixth Florida Workshop in Nonlinear Astronomy. Ann. N.Y. Acad. Sci. **631**.
- DERMOTT, S. F., J. H. HUNTER & R. E. WILSON, Eds. 1992. Astrophysical Disks: Proceedings of the Seventh Florida Workshop in Nonlinear Astronomy. Ann. N.Y. Acad. Sci. **675**.
- BUCHLER, J. R. & H. KANDRUP, Eds. 1993. Stochastic Processes in Astrophysics: Proceedings of the Eighth Florida Workshop in Nonlinear Astronomy. Ann. N.Y. Acad. Sci. **706**.

- KANDRUP, H., S. GOTTESMAN & J. R. IPSER, Eds. 1995. Three-Dimensional Systems: Proceedings of the Ninth Florida Workshop in Nonlinear Astronomy. Ann. N.Y. Acad. Sci. **751**.
- HUNTER, J. H. & R. E. WILSON, Eds. 1995. Waves in Astrophysics: Proceedings of the Tenth Florida Workshop in Nonlinear Astronomy. Ann. N.Y. Acad. Sci. **773**.

Tools for the Analysis of Chaotic Data

HENRY D. I. ABARBANEL[a]

Department of Physics
and
Marine Physical Laboratory
Scripps Institution of Oceanography
University of California, San Diego
La Jolla, California 92093-0402

RECONSTRUCTING PHASE SPACE

The analysis of time series has long been focused on regular, periodic or quasi-periodic time dependence—with all other time courses considered as "noise". Chaotic motions lie between these extremes as they possess *structure in state space* and have *limited predictability* while they possess continuous broad Fourier spectra, are nonperiodic, and are unstable everywhere in state space. These characteristics mean chaotic signals are not amenable to the enormous body of signal analysis tools built up over the past sixty or more years which are tuned to globally linear signal sources and rest heavily on Fourier analysis to simplify the tasks at hand.

We have developed a set of analysis tools for the study of time series[1,2] which rest on geometric and dynamical lessons from the study of nonlinear systems—the only systems in which chaos can appear. These are time domain tools. They are also useful for regular signals from linear sources, but probably they do not compete well with the vast selection of methods developed in frequency domain or based on newer linear ideas such as the use of wavelets. They are indispensable for the analysis of chaotic signals.

The analysis tools rest on the idea of *phase space reconstruction*[1] which takes scalar observations made on some system at intervals τ_s $s(n) = s(t_0 + n\tau_s)$ and creates a d-dimensional multivariate space from $s(n)$ and its time lags at intervals $T\tau_s$. The observed scalar data are replaced by the d-dimensional vectors

$$\mathbf{y}(n) = [s(n), s(n + T), s(n + 2T), \ldots, s(n + (d - 1)T)], \tag{1}$$

which are points in a phase space or state space which is dynamically identical to the original, but unknown, space in which the signal source operates as long as d is large enough. The geometric theorem underlying this construction tells us that if we have an infinite amount of infinitely accurate data the value of the lag T is not important and when the dimension of the system attractor is d_A, which could be fractional, then choosing the integer d such that $d > 2d_A$, we have a *sufficient* condition to capture the properties of the underlying dynamics. $d > 2d_A$ is not always necessary. We never have an infinite amount of data—if we did, then accounting for

[a]Institute for Nonlinear Science.

the time to collect it we would have little time left for its analysis. We never have infinitely accurate data. Clearly if we wish to move beyond entertaining, but useless mathematics, we need other assistance.

To use the vectors $\mathbf{y}(n)$ we need to know "good" values for the lag T and the necessary dimension $d = d_E$—the embedding dimension. If we have selected a value for T on the basis of some prescription, then we need to choose d_E so that strands of the system orbit in d-dimensional space do not overlap in an inappropriately chosen space having too few dimensions. This overlap due to projection from the correct higher dimensional space is in dimension $d - 1$, then $d - 2$, ... until the last overlaps occur as points. When the global dimension of the vectors $\mathbf{y}(n)$ is large enough to remove overlaps in points, any dimension larger than this will be perfectly acceptable for the analysis of the observed data. Additional coordinates will not serve to remove any further overlaps—there are none—so using the dimension where overlaps cease for d_E is the necessary dimension for the unfolding of the attractor.

Now the criterion of necessity is dictated by the goals one has in mind. We have *prediction* in mind, and if two strands of the attractor overlap there are regions or points in state space where it is ambiguous as to where the orbit should go, and this can only be resolved by viewing the orbits in higher dimensional space. If one wishes to compute only the fractal dimension of the attractor, $d > d_A$ will do[1].

The analysis of chaotic signals in the space in which they are observed is sure to fail. Chaos only occurs when the dynamics takes place in three or more dimensions if the system is governed by differential equations; that is, continuous time, or in two or more dimensions if the time label is discrete. Using $s(n)$ directly can only lead to erroneous conclusions about prediction or control or classification or system identification; that is, it is generically the wrong way to proceed to answer the usual questions of signal processing for nonlinear systems. The methods we present here and elaborate on in our papers[1,2] may be replaced by newer, more direct and computationally more efficient methods over the years, but they address directly the multivariate nature of the state space for nonlinear dynamical systems. If one wishes to know the answer to the usual class of questions about the source of observed signals and that source is nonlinear—that is, it is remotely connected to properties of the real physical or biological world—methods such as the ones we discuss are required.

Very little serious analysis of the methods we discuss have been performed by statisticians who wish to answer questions bearing on the number of data one requires for this or that accuracy in each of the analyses we shall present. Also the answers to questions bearing on the distribution of errors expected in making this or that kind of analysis discussed below are absent. In other words, the field is rather wide open and cries out for the application of quantitative statistical analyses of the methods one utilizes to understand nonlinear dynamics when presented with observed data. I have no doubt that the road is rich with remarkable results, and I compliment the organizers of this meeting in bringing together some persons who have the statistical skills and interests to address these unaddressed and unanswered matters and those of us who have struggled to make do without theorems to guide us in applying these ideas to real world data of direct physical or biological interest.

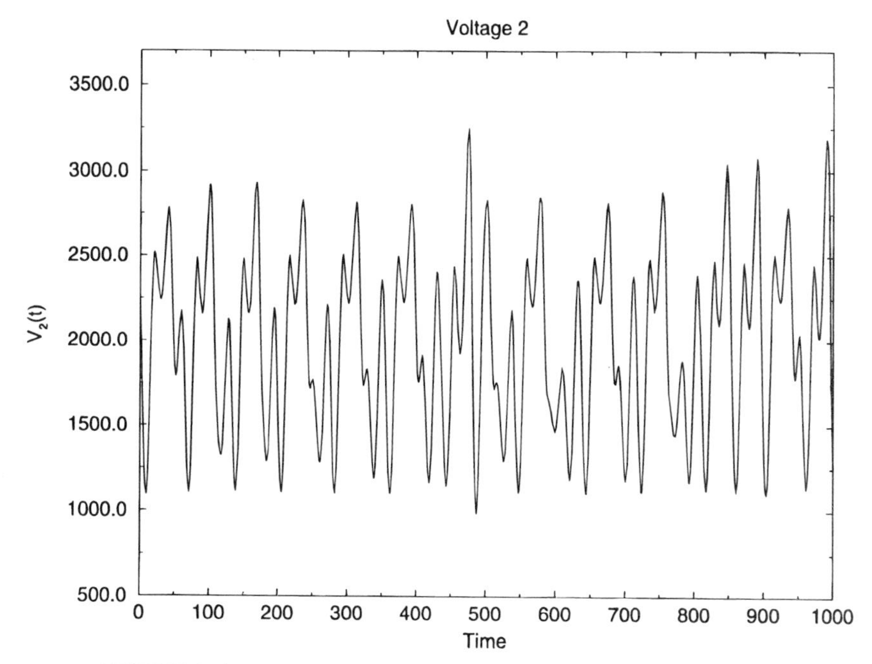

FIGURE 1. A sample of the original time series of the false nearest neighbors.

FALSE NEAREST NEIGHBORS

To determine d_E for the data vectors $\mathbf{y}(n)$ we employ a statistic about these data known as false nearest neighbors.[3] The construction of the d_E dimensional data vectors is purely geometric once a time delay T has been chosen. One is seeking a multivariate space in which the orbits $\mathbf{y}(n)$ do not overlap because of projection from a higher dimension. The main idea is that if the data come from a system governed by autonomous differential equations or discrete time maps with time-independent parameters, then orbits cannot cross each other. If they appear to cross, it is because we are viewing them in a dimension which is too small or which is smaller than the number of active degrees of freedom in the source of our observations.

To establish whether the space has become big enough we look at *each* point $\mathbf{y}(n)$ and its nearest neighbor $\mathbf{y}^{NN}(n)$ in dimension d. The time index k associated with $\mathbf{y}^{NN}(n) = \mathbf{y}(k)$ bears little resemblance to the time index n of the data point we are examining. As data folds back on itself in dissipative systems, the neighborhood of $\mathbf{y}(n)$ may be populated by points such as $\mathbf{y}^{NN}(n)$ of quite different time index. Now we ask what happens to the distance between $\mathbf{y}(n)$ as seen in dimension $d + 1$ where it is the vector

$$\mathbf{y}(n) = [s(n), s(n + T), \ldots, s(n + Td)], \tag{2}$$

and the vector $\mathbf{y}^{NN}(n)$ in dimension $d + 1$, where it becomes

$$\mathbf{y}^{NN}(n) = \mathbf{y}(k) = [s(k), s(k + T), \ldots, s(k + Td)]. \qquad (3)$$

If the distance in dimension $d + 1$ is large, we label $\mathbf{y}^{NN}(n)$ a *false nearest neighbor* of $\mathbf{y}(n)$. When d becomes large enough, the number of false neighbors will go to zero, for the overlaps due to projections will be removed. At that dimension the attractor is unfolded in the coordinate system defined by our choice of time delay T. Now the idea of "large" distances requires some threshold for deciding that a nearest neighbor $\mathbf{y}^{NN}$ is false. The choice of false neighbor turns out to be independent of this threshold over a wide range of choices for its value.[3,4]

An example of the false nearest neighbors method is seen in FIGURES 1, 2 and 3. In FIGURE 1 we see a sample of the original time series from a nonlinear hysteretic circuit operated at the US Naval Research Laboratory by Tom Carroll and Lou Pecora. $\tau_s = 0.1$ ms. FIGURE 2 is the percentage of false nearest neighbors when the vectors $\mathbf{y}(n)$ are constructed with a time lag $T = 6$; clearly $d_E = 3$ for these data. In FIGURE 3 we show the attractor reconstructed in the space $\mathbf{y}(n) = [s(n), s(n + 6), s(n + 12)]$ for the hysteretic circuit. The structure in state space is clearly revealed only in the last view of the data. The time series in FIGURE 1 and the Fourier spectrum of these data are of little value in characterizing the source of these signals.

For a statistician (if the members of this audience will permit my presuming to suggest problems of interest—which I do with humility as I make no (false) claims to have expertise or experience in statistics!) it may be of interest to determine what

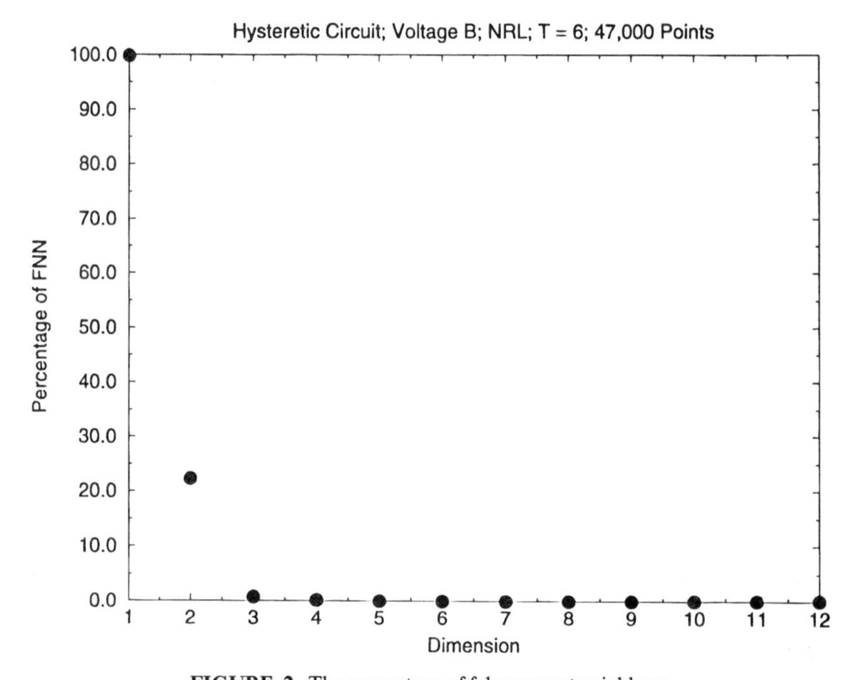

FIGURE 2. The percentage of false nearest neighbors.

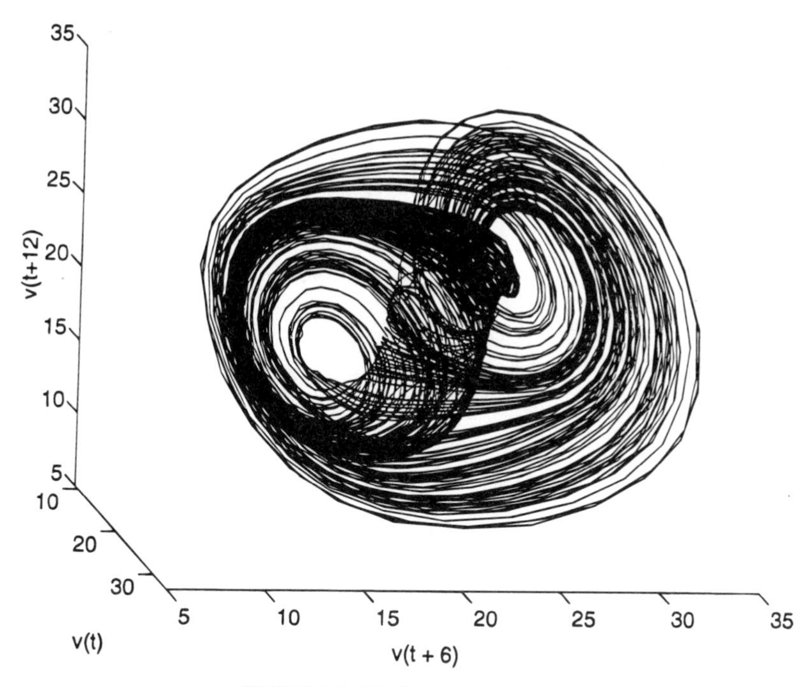

FIGURE 3. The hysteretic attractor.

is the distribution of the false nearest neighbor statistic as a function of the number of data, the amount of the attractor visited by the orbit (similar to the number of data issue), the signal to noise ratio of the observations, the time lag T, and other questions of interest in the practical use of this statistic for the determination of d_E.

LOCAL FALSE NEAREST NEIGHBORS

The integer dimension d_E just discussed tells us the global dimension required to unfold the attractor from its projection on the observation axis $s(n)$. The local dimension d_L of the dynamics may be less than or equal to d_E. If $d_E > d_L$, it means that the particular coordinate system, namely that of the time delay vectors $\mathbf{y}(n)$, twists the attractor about so that unfolding the attractor from the observations requires additional dimensions to "untwist it". The local or dynamical dimension is the same in any coordinate system of $\mathbf{y}(n)$ or other "proxy" multivariate vectors representing the original coordinates as long as the connections among these coordinate systems are smooth enough. d_L represents the integer number of differential equations (or maps, if time is discrete) required to describe the local evolution of the system.

We use the method of local false nearest neighbors[5] which starts with a working dimension $d_W \geq d_E$ with d_E as determined by the global false neighbors algorithm described above. All distances between points $\mathbf{y}(n)$ are computed in d_W. Now we choose the local dimension d starting with $d = 1$ and proceeding to $d = d_W$, asking at each dimension how well we can *predict* the evolution of a cluster of neighbors about $\mathbf{y}(n)$ for each $n = 1, 2, \ldots, N$ on the attractor. In a dimension d which is lower than the true local dimension d_L, there will be points in a neighborhood which got there by projection and not by a dynamical rule. The ability to predict where these points go will be impaired for $d < d_L$. As we increase d up to d_L our ability to predict will improve until it levels off and becomes independent of the estimated local dimension d as well as the number of neighbors we use to define a neighborhood whose evolution we predict. In dimensions $d_L \leq d \leq d_W$ predictability should become independent of d as we have enough dynamical degrees of freedom already to capture the evolution of the data. The quality of prediction is determined by how far ahead we wish to predict, and by how large an error we tolerate until we say the prediction has failed.

To be more precise about the details of the method of local false nearest neighbors we note that what is involved in the prediction is a local map from one neighborhood of the attractor to another neighborhood of the attractor. The members of a neighborhood are determined by distances evaluated in a dimension $d_W \geq d_E$ with d_E the global embedding dimension determined by the false nearest neighbors method. This assures that all neighbors are true neighbors. Then in a neighborhood, we ask if a local model in dimension $d \leq d_W$ accurately relating the neighbors of the point $\mathbf{y}(n)$ to the neighbors of $\mathbf{y}(n + 1)$ can be made. This local model is a rule which gives $\mathbf{y}(n + 1)$ in terms of a polynomial constructed out of the vectors $\mathbf{y}(n)$. The coefficients in this polynomial rule are determined by a least squares fit minimizing the residual errors in the rule which takes $\mathbf{y}(n) \rightarrow \mathbf{y}(n + 1)$. This rule is used to predict ahead a time T equal to the time lag used before. If in predicting ahead this amount of time, the error between the prediction and the known observed data point is larger than a certain fraction β of the attractor size, we deem it a bad prediction. The fraction of bad predictions P_K is collected as we move over the whole attractor, and this is displayed against the dimension d of the local model for various choices of the number of neighbors N_B of the points $\mathbf{y}(n)$ on the orbit. We seek a dimension where the model, represented by its fraction of bad predictions, becomes independent of the local dimension and the number of neighbors used to establish the model. The d at which this occurs determines the dimension d_L of a good local model. As one changes the fraction β of the attractor size which defines the allowed error sphere or changes the time which one predicts ahead, the fraction of bad predictions P_K will move up and down in absolute value. The dimension d_L at which P_K becomes independent of d and N_B will remain the same. As it is d_L which one wishes to extract from the local false nearest neighbors statistic, so the absolute size of P_K is of no special significance.

We illustrate the method of local false nearest neighbors with data from the volume of the Great Salt Lake (GSL) in Utah. The GSL, located at approximately 40° to 42° N and 110° to 112° W, is the fourth largest, perennial, closed basin, saline lake in the world. It drains an area of 90,000 km^2. The GSL's immediate atmospheric catchment is the Pacific Ocean, and its fluctuations exhibit high spectral coherence with the Pacific North American Circulation Index, and the Southern Oscillation Index at interannual time scales.[6] These indices measure the fluctuations in the Northern Pacific atmospheric flow and tropical Pacific atmospheric variability related to the El Niño Southern Oscillation, respectively. Its fluctuations have been related[7] to those in the hemispheric surface temperature and sea level pressure records.

The levels of the GSL have been recorded by various methods since 1847. From these data a biweekly time series of GSL volumes has been compiled by Sangoyomi[8] which shows substantial variation both on an annual time scale as well as on interannual and interdecadal time scales. Lall and Mann[9] speculate that the low frequency oscillations found on spectral analysis of this and related hydroclimatic data sets may represent unstable oscillations that arise from the nonlinearity of the climate dynamics, rather than stable periodic motion.

FIGURE 4 shows the time series for the GSL data, and the false nearest neighbors for $T = 12$ are seen in FIGURE 5. $d_E = 4$ is clearly selected. FIGURE 6 shows the local false nearest neighbors, and we also see quite clearly that $d_L = 4$ is required by these data.

In the statistic we call local false nearest neighbors we have added dynamics to the geometrical considerations underlying global false nearest neighbors. The dynamics is added in the form of local predictions from neighborhood to neighborhood of the attractor in multidimensional space. This prediction assumes that underlying the data is a discrete time map (for finite τ_s all dynamics is discrete or a

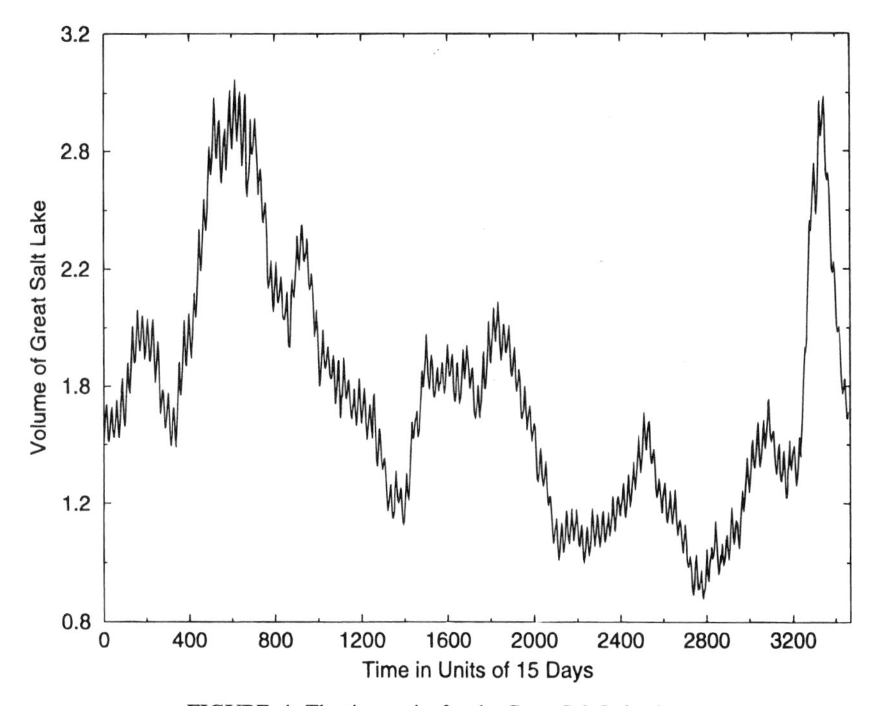

FIGURE 4. The time series for the Great Salt Lake data.

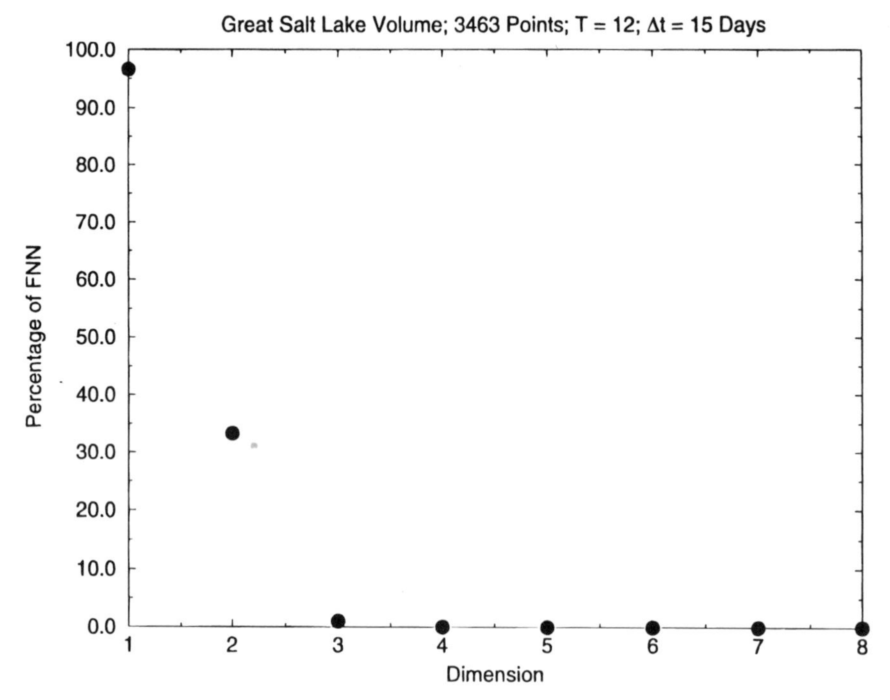

FIGURE 5. The false nearest neighbors for the GSL data.

discrete approximation to a flow)

$$\mathbf{y}(n) \rightarrow \mathbf{F}(\mathbf{y}(n)) = \mathbf{y}(n+1), \tag{4}$$

and the main job is to estimate $\mathbf{F}(\mathbf{x})$ accurately. The estimation is made in terms of some local basis function set $\phi_k(\mathbf{x})$ as

$$\mathbf{F}(\mathbf{x}) = \sum_{k=1}^{M} \mathbf{c}(k)\phi_k(\mathbf{x}). \tag{5}$$

We want to evaluate the coefficients $\mathbf{c}(k)$ locally in phase space by selecting N_B neighbors of the point $\mathbf{y}(n)$ and observing how they evolve into the points near $\mathbf{y}(n+1)$. So we use phase spatial information to capture temporal evolution. The determination of the coefficients is made using a least squares fit to the points $\mathbf{y}(n+1)$ compared to

$$\sum_{k=1}^{M} \mathbf{c}(k)\phi_k(\mathbf{y}(n)). \tag{6}$$

There are a number of clear statistical issues:

- what is the "best" choice of basis functions $\phi_k(\mathbf{x})$ for each problem?

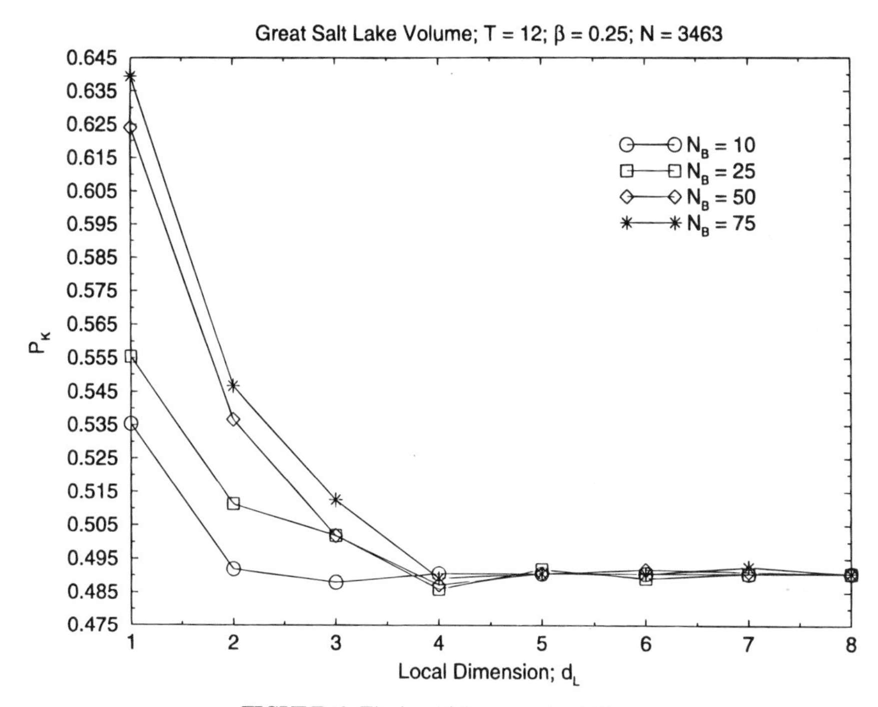

FIGURE 6. The local false nearest neighbors.

- how many neighbors N_B do we need in dimension d_E with local dimension d_L for N data points?
- what is the convergence of the approximation as a function of these parameters?
- what order do we require for what quality of approximation to $\mathbf{F}(\mathbf{x})$?
- what is the distribution of errors as a function of the properties of the attractor $(d_E, d_L, d_A, \lambda_a(\mathbf{x}, L), \ldots)$ and the given data set $(N, N_B, \ldots)$?
- etc.

It seems fair to assume that as these questions are addressed we shall see new algorithms for each of the questions. Their implementation and analysis will be most valuable to the full development of the kind of nonlinear dynamics analysis toolkit discussed here.

LOCAL AND GLOBAL LYAPUNOV EXPONENTS

The evolution of small perturbations to an observed orbit $\mathbf{y}(n)$ is governed by the linearized equations of motion, whatever they may be. As we do not know them from looking at the data alone, we assume there is an underlying nonlinear process $\mathbf{y}(n) \rightarrow \mathbf{y}(n+1) = \mathbf{F}(\mathbf{y}(n))$ which moves the system ahead one sampling time τ_s. A small perturbation $\mathbf{w}(n)$ to the orbit $\mathbf{y}(n)$ satisfies

$$\mathbf{y}(n+1) + \mathbf{w}(n+1) = \mathbf{F}(\mathbf{y}(n) + \mathbf{w}(n))$$
$$\mathbf{w}(n+1) = \mathbf{DF}(\mathbf{y}(n)) \cdot \mathbf{w}(n) + \text{Order } \mathbf{w}(n)^2, \tag{7}$$

where the Jacobian matrix is

$$\mathbf{DF}(\mathbf{x})_{ab} = \frac{\partial F_a(\mathbf{x})}{\partial x_b}, \tag{8}$$

and $a, b = 1, 2, \ldots, d_L$.

The eventual growth or shrinkage of the perturbation under this linear evolution rule is determined by the eigenvalues $e^{\lambda_a(\mathbf{x}, L)}$ of the Oseledec matrix[1,2,10]

$$\mathbf{OSL}(\mathbf{x}, L) = [(\mathbf{DF}^L(\mathbf{x}))^T \cdot \mathbf{DF}^L(\mathbf{x})]^{1/2L}. \tag{9}$$

The matrix $\mathbf{DF}^L(\mathbf{x})$ is the composition of L Jacobian matrices along the observed orbit $\mathbf{y}(n)$ starting at location $\mathbf{x}$. As $L \to \infty$, the $\lambda_a(\mathbf{x}, L) \to \lambda_a$, which are the usual global Lyapunov exponents. The λ_a are independent of $\mathbf{x}$ in the basin of attraction of the attractor. They are invariants of the dynamics and characterize it. They are also independent of the coordinate system in which they are evaluated. To determine the λ_a reliably we need to know the value of d_L as this is the dimension of the dynamics. If we work in a space with $d > d_L$, then $d - d_L$ of the eigenvalues of the Oseledec matrix will be false, and we need a reliable rule to establish which are true and which are not. Similarly if we work in $d < d_L$, we will not have unfolded the local dynamics in such a way which would allow the correct evaluation of the $\mathbf{DF}(\mathbf{x})$, and thus the λ_a would be in error.

Using $d_E = d_L = 4$, we computed the $\lambda_a(\mathbf{x}, L)$ for a large number of starting locations $\mathbf{x}$ on the GSL attractor, then determined the value of these quantities as a function of the number of steps we look ahead of these starting points. This allows us to define an average local Lyapunov exponent

$$\bar{\lambda}_a(L) = \frac{1}{N_s} \sum_{k=1}^{N_s} \lambda_a(\mathbf{y}(k), L), \tag{10}$$

for N_S starting locations $\mathbf{y}(k)$.

For the GSL data we estimated the local Jacobian matrices by making local cubic neighborhood-to-neighborhood maps in $d_E = d_L = 4$ state space and picking off the linear part of these local maps. These maps can be thought of as regressions of the GSL volume $V(n+1)$ on the vector defined by the embedding, $[V(n), V(n+T), \ldots, V(n+(d-1)T)]$. These estimated local Jacobians were then multiplied to give the Oseledec matrix and diagonalized as indicated. In FIGURE 7 we display the average local Lyapunov exponents as a function of number of steps $2^{(L-1)}$ along the orbit for $L = 0, 1, \ldots, 12$. The average exponents have been averaged over 750 different starting locations on the attractor. These locations are spaced approximately equally in time from the center of the measured data set. Had one chosen them randomly from the data set or essentially any other way, when the number is large enough, the results will be the same as implied by the multiplicative ergodic theorem.[10] The exponents converge to the approximate global values: $\lambda_1 = 0.17$, $\lambda_2 = 0$, $\lambda_3 = -0.13$, and $\lambda_4 = -0.68$. These estimates used a reconstructed state space with time lag $T = 17$. In principle, they are independent of the lag T.

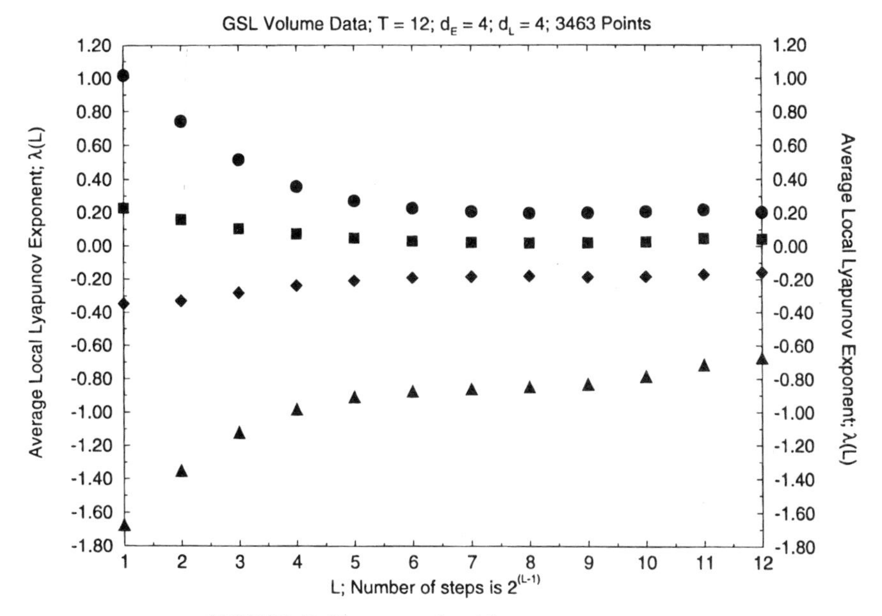

FIGURE 7. The average local Lyapunov exponents.

We conclude two important things from these values for the λ_a. First, one of the exponents is zero within numerical accuracy. This means that the underlying dynamics comes from a differential equation as a zero global exponent is always associated with a flow in state space. Secondly, the value of the largest exponent is $\lambda_1 \approx 1/5.5$, in units of $1/\tau_s$. Errors along the orbit or in initial conditions grow as $\exp[t/5.5\tau_s]$, so when the time after some perturbation or initial state is order a few times $5.5\tau_s$ or a few hundreds of days or a year or so, we lose the ability to predict the behavior of the volume of the GSL.

These estimates for the Lyapunov exponents can be used to estimate the fractal dimension of the attractor using the idea of Lyapunov dimension[11] which has been conjectured to be the same as the information dimension. This Lyapunov dimension is evaluated by finding at which K the sum

$$\sum_{a=1}^{K} \lambda_a \tag{11}$$

is positive, but the sum

$$\sum_{a=1}^{K+1} \lambda_a \tag{12}$$

is negative. The Lyapunov dimension is defined as

$$D_L = K + \frac{\sum_{a=1}^{K} \lambda_a}{|\lambda_{K+1}|}. \tag{13}$$

For the GSL data we find $D_L \approx 3.05$. This means that we should be able to get a good idea of the structure of the attractor by looking at it in three dimensions with $T = 12$. This is displayed in FIGURE 8. The three-dimensional view of the attractor suggests that the annual cycle is approximately motion around the smaller radius of the "spool" on which the attractor lies, while the longer term motion which has larger amplitude moves the orbits along the longer axis of this "spool". Such motions may correspond to the spectral power identified by Lall and Mann[9] at certain interannual and interdecadal frequencies. Interestingly, such structure persists even if larger delays (T) that obscure the annual cycle are used.

The issues requiring statistical analysis here are initially those mentioned earlier with some fascinating twists. One clearly needs to make a good local map

$$\mathbf{y}(n) \rightarrow \sum_{k=1}^{M} \mathbf{c}(k)\phi_k(\mathbf{y}(n)), \tag{14}$$

as above, but then one must very accurately pick off from this map the local Jacobian with components

$$\mathbf{DF}(\mathbf{y}(n))_{ab} = \sum_{k=1}^{M} c_a(k) \frac{\partial \phi_k(\mathbf{y}(n))}{\partial y_b(n)}, \tag{15}$$

and one must then combine many of these local Jacobians to construct the very ill-conditioned Oseledec matrix whose eigenvalues we seek. How well this construction can be expected to work as a function of the usual dynamical quantities d_E, d_L,

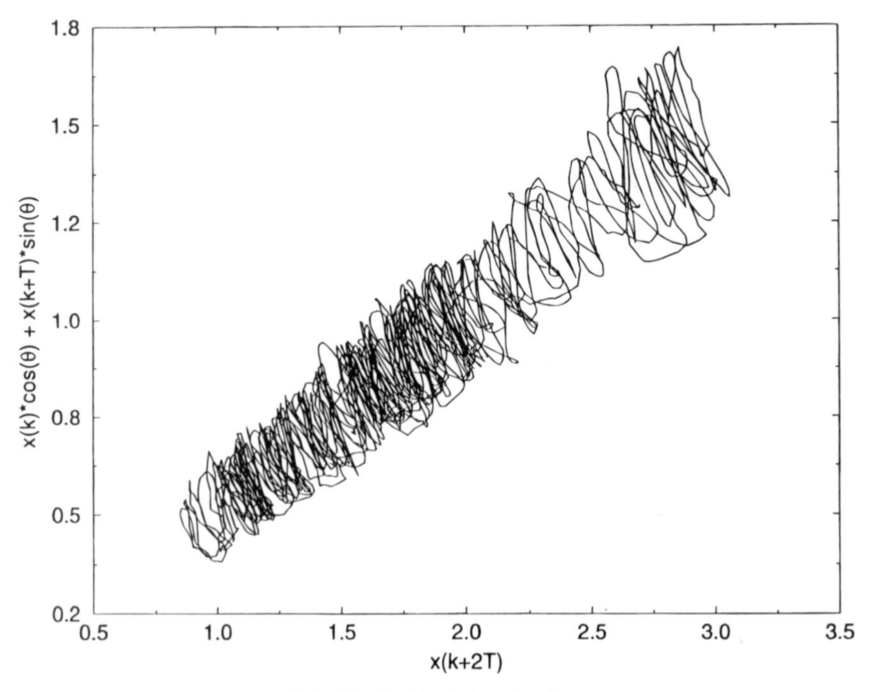

FIGURE 8. The GSL attractor.

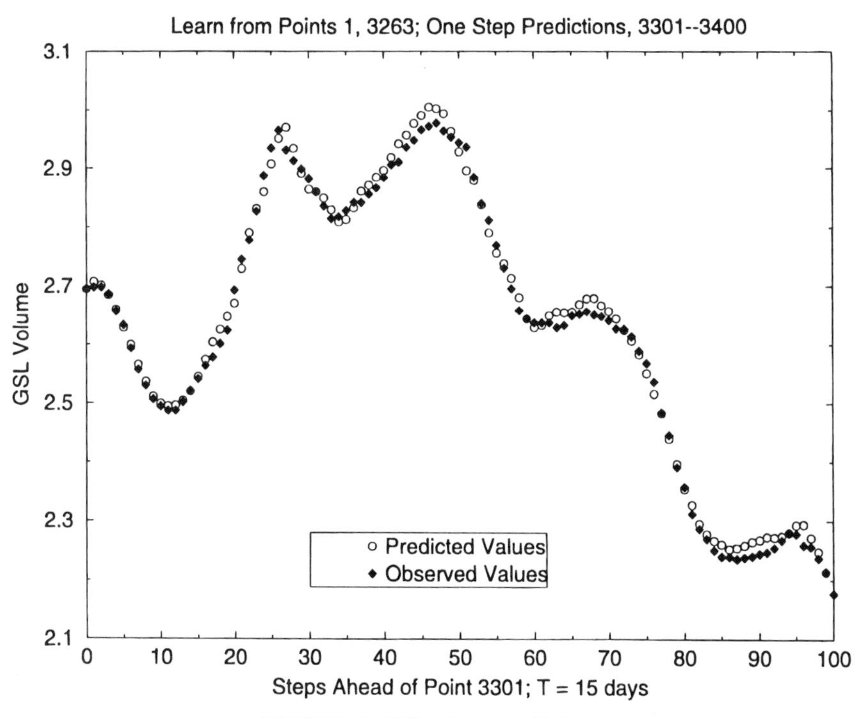

FIGURE 9. GSL volume predictions.

... and the number of data is quite important. Similarly, how the errors in these estimations of the eigenvalues of the Oseledec matrix are distributed as a function of the number of data, the size of neighborhoods, etc., is well worth investigating.

PREDICTING IN RECONSTRUCTED PHASE SPACE

Even without knowledge of the dynamical equations for the GSL system we can use the information we have acquired so far to make predictive models for the GSL volume evolution. The method[2] utilizes the compactness of the attractor in $\mathbf{y}(n)$ space by noting that we have knowledge of the evolution of whole phase space neighborhoods into whole neighborhoods later in time. We can use this to make local models of this evolution and then use these models as interpolating rules for the evolution of new phase space points near the attractor. We use only local polynomial models, although basis sets other than polynomials are certainly quite useful. Indeed, in this work we use only local linear models as we have substantial amounts of data and thus good coverage of the attractor; namely, every neighborhood is rather well populated.

The idea is that locally on the attractor we find the N_B neighbors $\mathbf{y}^{(r)}(n)$; $r = 1, 2, \ldots, N_B$ of each point $\mathbf{y}(n)$ and make the local linear model $\mathbf{y}(r; n + 1) = \mathbf{A}_n + \mathbf{B}_n \cdot$

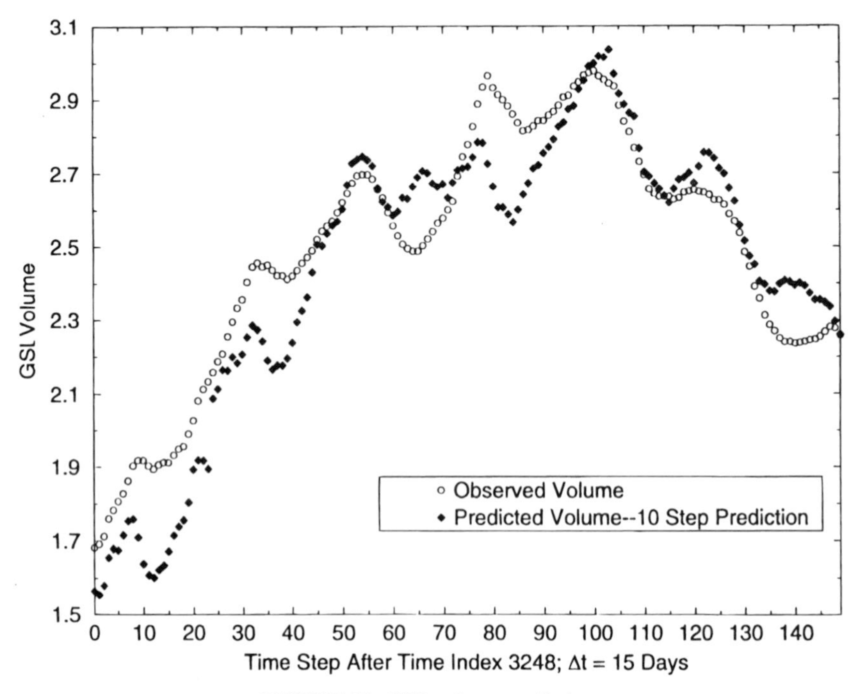

FIGURE 10. GSL volume predictions.

$\mathbf{y}^{(r)}(n)$, where $\mathbf{y}(r; n + 1)$ is the point to which $\mathbf{y}^{(r)}(n)$ goes in one time step. The coefficients $\mathbf{A}_n$ and $\mathbf{B}_n$ are determined by minimizing at each time location n

$$\sum_{r=0}^{N_B} |\mathbf{y}(r; n + 1) - \mathbf{A}_n - \mathbf{B}_n \cdot \mathbf{y}^{(r)}(n)|^2, \tag{16}$$

where $r = 0$ means $\mathbf{y}(n)$ itself and $\mathbf{y}(n + 1)$. When we have a new point $\mathbf{z}(k)$ on or near the attractor, we seek the nearest neighbor $\mathbf{y}(Q)$ among all the data in the set we used to determine the $\mathbf{A}_n$ and the $\mathbf{B}_n$. The predicted point $\mathbf{z}(k + 1)$ is then

$$\mathbf{z}(k + 1) \approx \mathbf{A}_Q + \mathbf{B}_Q \cdot \mathbf{z}(k). \tag{17}$$

This works remarkably accurately within the limits of prediction dictated by the largest Lyapunov exponent λ_1. When we try to predict beyond the instability horizon, that is for times much greater than

$$\frac{\tau_s}{\lambda_1}, \tag{18}$$

our predictions should rapidly lose accuracy.

Prediction on the GSL Attractor

These methods were used to predict the volume of the GSL in $d_E = d_L = 4$. In FIGURE 9 we display the result of predicting one step ahead for the 100 time points after time index 3301; April 24, 1985 to June 2, 1989. The data used to determine the local linear maps were taken from the beginning of the time series up to point 3263. It is clear that the ability to predict one step (15 days) ahead is very high. Next we determined the local linear maps from data starting at the beginning of the data set and using points up to index 3050. In FIGURE 10 we display the result of predicting 10 time steps ahead, that is 150 days or about a half-year, starting with time index 3248. In FIGURE 11 we learned local linear maps from the first 3200 points of the data set, and then predicted ahead 10 steps from time index 3348. These predictions are actually rather good, and, as we can see, the prediction ability degrades with the number of steps ahead one wishes to predict, as it is expected to, given that the largest Lyapunov exponent is positive, implying that the system is chaotic. The main message in these predictions is that we can accurately follow the features of the data without elaborate computations by working in dimension four as our systematic analysis would suggest. Of course, it is possible to devise even better predictors using more elaborate methods.

Many of the potentially interesting statistical issues associated with prediction on the attractor are extensions of well-studied matters in conventional statistics with the very interesting twist that the focus is now on the *local* predictability of a system which is everywhere unstable in the state space in which one is predicting. Many of

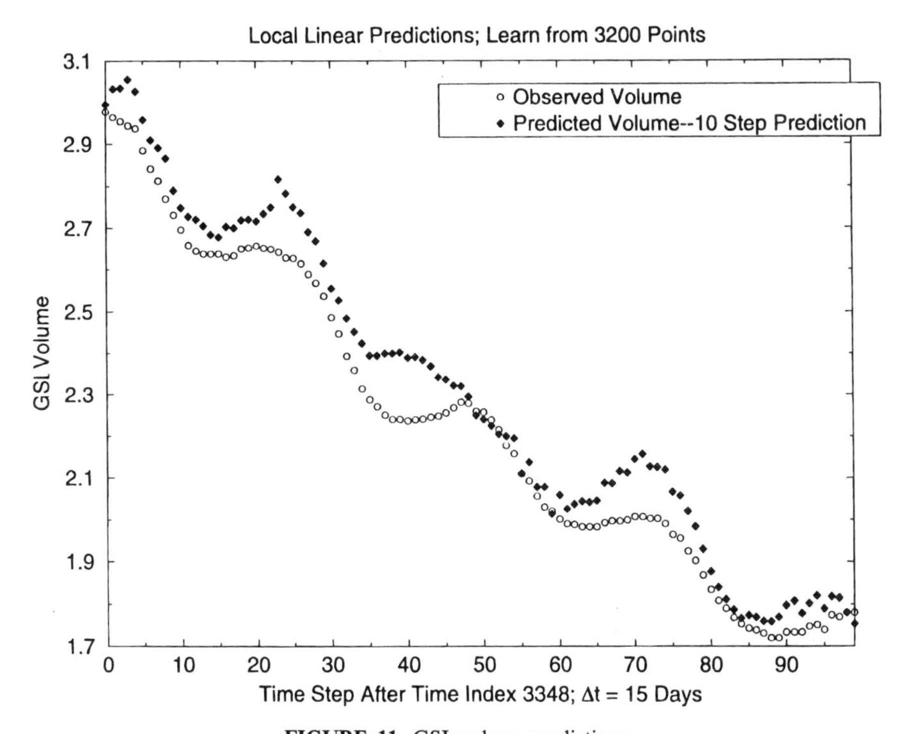

FIGURE 11. GSL volume predictions.

the issues we have thought to be interesting were raised above in the consideration of local false nearest neighbors, and we shall not simply list them again.

CONCLUSIONS

The tools we have developed for the analysis of observed data apply equally well for linear and nonlinear sources of the data. If linear sources are encountered they would reveal themselves in the Fourier spectra and in zero or negative Lyapunov exponents, if not before from a close examination of the periodic structures in the time series. In that case, we can recommend without shame the use of the vast selection of Fourier-based methods for the systematic analysis of such data. If the data come from a nonlinear source operating in a chaotic regime, we can recommend without shame the use of our methods as presented here or improved versions of these methods which we hope the readers will construct and share with us. Fourier methods tuned to linear data sources will prove misleading and without foundation in such a case.

As we have pointed out at various points in this paper, there is an enormous body of analysis to be done to determine the limits on the geometric methods we have discussed. My colleagues and I hope that the quite striking results of the analyses we have presented will generate interest in the statistical community in establishing firm foundations for the statistical reliability of all of these methods as a function of critical parameters such as number of data, dimension of reconstructed space, time delay used in reconstructed state space vectors, etc.

Two key ingredients recur in the analyses we have presented, and I personally hope that there can be some firm foundations and new ideas for them as well:

- *Multidimensional searches* occur again and again. We have used kd-trees for these searches, and we have made no attempt to optimize these searches to reduce the operations count as much as possible. Doing this and delineating the conditions under which the optimized searches can be expected would be a quite useful contribution.
- *Interpolations in multidimensional space* are required over and over again in this work. It enters in local false nearest neighbors, in determining Lyapunov exponents and in prediction on the attractor. The literature, frankly, is somewhat unclear to the consumer (me) about when to use what kind of basis functions (radial, polynomial, etc.) and what kind of accuracy to expect and how this all scales with the size of neighborhoods, the dimension of the space d_E, etc. Knowledge of this sort would be most valuable both academically and for the planning of useful and efficient algorithms for carrying out the kinds of analysis discussed here.

Finally, let me iterate that the algorithms considered here as well as others in our "toolkit" will be available during Winter 1996 as a package running under Motif (UNIX) and Windows 95/NT. Please write or e-mail the author to find out how to purchase a copy, if you wish. One may also get information on the World Wide Web at http://www.zweb.com/apnonlin/.

REFERENCES

1. ABARBANEL, H. D. I. 1995. Analysis of Observed Chaotic Data. Springer-Verlag. New York/Berlin.
2. ABARBANEL, H. D. I., R. BROWN, J. J. ("SID") SIDOROWICH & SH. TSIMRING. 1993. Rev. Mod. Phys. **65:** 1331.
3. KENNEL, M. B., R. BROWN & H. D. I. ABARBANEL. 1992. Phys. Rev. A **45:** 3403.
4. FRASER, A. M. & H. L. SWINNEY. 1989. Phys. Rev. **33A:** 1134; and FRASER, A. M. 1989. IEEE Trans. Info. Theory **35:** 245.
5. ABARBANEL, H. D. I. & M. B. KENNEL. 1993. Phys. Rev. E **47:** 3057.
6. MOON, Y. I. & U. LALL. 1995. J. Hydrol. Eng. Submitted.
7. MANN, M., U. LALL & B. SALTZMAN. 1994. Geophys. Res. Lett. Submitted.
8. SANGOYOMI, T. B. 1993. Ph.D. dissert. Utah State Univ.
9. LALL, U. & M. MANN. 1994. Water Resources Res. Submitted.
10. OSELEDEC, V. I. 1968. Trudy Mosk. Mat. Obsc. **19:** 197.
11. KAPLAN, J. L. & J. A. YORKE. 1979. Lect. Notes Math. **730:** 228.

Nonlinear Inference and Cluster-Weighted Modeling

NEIL GERSHENFELD

Physics and Media Group
MIT Media Lab
Cambridge, Massachusetts 02139

Unlike most scientific disciplines, astronomical model-building must typically be based on observation alone rather than on direct manipulation of a system. Because of this, there has long been a close connection between astronomy and the study of time series. In fact, the modern form of the latter discipline arguably starts with Yule's analysis in 1927 of the sunspot series, introducing a particular case of what we would now call an ARMA model.[1]

If we measure a quantity $x(t)$ such as the light curve of a variable star, and want to use it to forecast a quantity $y(t)$ (which might just be a future value of x), an ARMA (M, N) (Auto-Regressive Moving-Average) model is

$$y(t) = \sum_{m=1}^{M} a_m y(t - m\tau) + \sum_{n=0}^{N} b_n x(t - n\tau) \tag{1}$$

(where τ is the time delay between samples). This linear model has been the workhorse of time series analysis since Yule's day since it is relatively straightforward to fit to data and to analyze, but there are a number of well-known problems with it. Much of the world is nonlinear, and so a linear model may not be appropriate, or it may apply only to local regions. The choice of the number of terms M and N is crucial, because if it is too low the model might not be able to describe the data (model mismatch error), but if it is too high the model can fit the noise in the data that does not generalize (model estimation error).

An obvious, and traditional, extension to an ARMA model is to include higher-order polynomial terms such as

$$y(t) = \sum_{n_1=0}^{N} \cdots \sum_{n_d=0}^{N} a_{n_1 \cdots n_d} y(t - \tau)^{n_1} \cdots y(t - d\tau)^{n_d} \tag{2}$$

(for the case where the future of y is predicted from the past of y). This still has serious problems. Any terms that are left out of this series means that the lag variables cannot interact at that order. If a given order is not adequate to describe the data, all that can be done is add higher-order terms, which requires precariously balancing ever more rapidly diverging terms to describe a function that might not be at all divergent but that simply is not well approximated by polynomial basis functions.

A polynomial expansion is an example of a linear expansion in nonlinear basis functions

$$y(t) = \sum_{n=1}^{N} a_n f_n(y(t-\tau), \ldots). \tag{3}$$

A more general alternative is to have coefficients inside the nonlinearities:

$$y(t) = \sum_{n=1}^{N} f_n(a_n\,;\, y(t-\tau), \ldots). \tag{4}$$

This difference is profound. Under fairly general assumptions, it can be shown[2] that the typical approximation error of a function of the form of (**4**) is $O(N^{-1})$, while of (**3**) it is $O(N^{-2/d})$. This is called the *curse of dimensionality*: in low dimensions a model with linear coefficients works as well as one with nonlinear coefficients, but in high dimensions the linear coefficient model requires exponentially more terms. The cost of this power is that the optimal values of the linear coefficients can be found in a single matrix inversion, but for the nonlinear case an iterative search is needed and even then the best global solution can rarely be found. This is not as great a problem as it might sound, because there is also what can be called a *blessing of dimensionality*: if the nonlinear model is sufficiently large, there are so many acceptable solutions that a local search can with reasonable certainty find one.

To understand the nature of this search problem, it is helpful to start from a general Bayesian framework for describing modeling.[3] Let me use d to refer to the observed data set, and m to refer to all of the parameters needed to specify a model that is intended to describe the data. A simple model just predicts values for the data; a better model can also predict the uncertainty, making it possible to calculate $p(d\,|\,m)$, the probability that an observed data set was generated by a given model. For experimental time series analysis we start with the data and do not know the model, and so we want to find the most likely model given the data. Bayes' rule can be used to write this as a search to find the maximum likelihood

$$\max_m p(m\,|\,d) = \max_m \frac{p(m, d)}{p(d)} = \max_m \frac{p(d\,|\,m)p(m)}{p(d)}. \tag{5}$$

Since we are searching for the maximum, we will get the same answer for the most likely model if we take the logarithm of the probability. Therefore, to fit the model to the data we can find the maximum log-likelihood

$$\max_m \log \frac{p(d\,|\,m)p(m)}{p(d)} = \max_m \left[\log p(d\,|\,m) + \log p(m) - \log p(d)\right]. \tag{6}$$

The first term on the right-hand side measures how well the model describes the data. If a Gaussian error model is assumed, this reduces to measuring the least-squares error between the data and the model predictions. The second one contains prior beliefs on what is a good versus a bad model, and is crucial in constraining a general model that is underdetermined by the data. The last term contains prior beliefs about how likely the data set is, and is important in comparing analyses of models across different data sets.

Common choices for the $p(m)$ model prior term are (1) the integral square second derivative of the functional form predicted by the model (this expresses the assumption applicable to many physical systems that nearby states behave similarly); (2) the entropy of the distribution of the predictions (Maximum Entropy, based on the belief that a good model should assume as little as possible about the data); or (3) the amount of effort that it takes to describe the model and its parameters (Occam's Razor, which can be expressed as the Minimum Description Length (MDL), or in information theoretic language as Algorithmic Information Theory (AIC)). There is no single best choice for a model prior; it depends on the *a priori* beliefs of the modeler, *a posteriori* experience with data, and on other constraints such as computational effort. However, for many physical systems these common priors (also called regularization terms) may not be reasonable at all. Minimizing curvature can reduce sharp features that should be present in the model. Maximizing entropy has no notion of smoothness, and it seeks to minimize features in the prediction surface, neither of which might be appropriate for some systems.

An alternate way to fit a model to data is Expectation-Maximization (E − M).[3] The idea is to assume that part of a model is correct, use it to evaluate another part of the model, then assume that the latter part is correct and re-evaluate the former part, and so forth. For many problems this converges to some kind of best local solution, and it is used routinely in techniques such as clustering and Hidden Markov Models. However, it is hard to express arbitrary priors with E − M.

Now I present a new framework, Cluster-Weighted Modeling (CWM), that generalizes previous cluster-based models (e.g., reference 4) to make it possible to specify model priors. Let y refer to a scalar quantity that is to be predicted based on a measurement of some quantity $\vec{x} = (x_1, \ldots, x_d)$ (it is easy to generalize to vector predictions). For example, y could be a future sample of a signal and x could be a lag vector of past samples. Let $y = f(\vec{x}; \vec{\beta})$ be a prior on the functional form of their *local* relationship with parameters β in the model (such as a locally linear model, where the β are the slopes); no assumption is being made about the *global* behavior. I am going to write the joint probability distribution of these variables $p(y, \vec{x})$ as a sum of M terms with weights w_i

$$p(y, \vec{x}) = \sum_{i=1}^{M} w_i p_i(y, \vec{x}), \tag{7}$$

where the p_i's are normalized and $\sum_{i=1}^{M} w_i = 1$. Each term in this sum is associated with a model "cluster" (to be explained). Unconditional density estimation, while attractive in principle, is notoriously difficult to do well in practice. Further, it is usually wasted since most of the information is lost when reducing it to predictions. A key feature of CWM is that even though I am starting with the unconditional distribution, I am going to write it in a form that matches how it will be used for forecasting. It is best viewed as being in between density estimation and direct prediction.

The cluster probabilities can be split into a conditional and an unconditional factor

$$p_i(y, \vec{x}) = p_i(y \mid \vec{x}) p_i(\vec{x}), \tag{8}$$

and I am going to take as the form of these

$$p_i(y, \vec{x}) = \frac{1}{\sqrt{2\pi\sigma_{i,y}^2}} e^{(y - f(\vec{x}; \vec{\beta}_i))^2 / 2\sigma_{i,y}^2} \prod_{d=1}^{D} \frac{1}{\sqrt{2\pi\sigma_{i,d}^2}} e^{[(\vec{x} - \vec{\mu}_i) \cdot \hat{v}_i]^2 / 2\sigma_{i,d}^2}. \tag{9}$$

This is a multivariate Gaussian in the input space, and a Gaussian in the output space with the model prior determining the mean. The local model parameters β_i will be found by maximizing the likelihood of the cluster.

To understand my choice for this form, look at the conditional forecast:

$$\begin{aligned} \langle y \mid \vec{x} \rangle &= \int y p(y \mid \vec{x}) \, dy \\ &= \int y \frac{p(y, \vec{x})}{p(\vec{x})} \, dy \\ &= \frac{\sum_{i=1}^{M} w_i p_i(\vec{x}) \int y p_i(y \mid \vec{x}) \, dy}{\sum_{i=1}^{M} w_i p_i(\vec{x})} \\ &= \frac{\sum_{i=1}^{M} w_i p_i(\vec{x}) f(\vec{x}; \vec{\beta}_i)}{\sum_{i=1}^{M} w_i p_i(\vec{x})}. \end{aligned} \tag{10}$$

Locally, the behavior is defined by the nearest model(s) $f(\vec{x}; \beta_i)$, and globally the $p_i(\vec{x})$'s do a Gaussian interpolation among these models. The cluster parameters in the input space $p_i(\vec{x})$ determine the influence of the model in the output space. This means that our prior on the local behavior is enforced, but that the only global assumption we have made is the total number M of features to explain. Therefore, it is equally easy to represent continuous and discontinuous functions. M controls the trade-off between under- and over-fitting, and must be determined by cross-validation with out-of-sample data.

If just a single cluster is used this reduces to a global linear model; as more clusters are used these combine to form a nonlinear prediction function. As with finite elements, there is a trade-off between using fewer clusters with more complex models, and more clusters with simple models. It is also possible to start with multiple kinds of cluster models, and use the E – M procedure to let them find the data that they explain best.

Let $(\vec{x}_1, y_1, \ldots, \vec{x}_N, y_N)$ be an observed data set, and define a factor $\alpha_i(y, \vec{x})$ that measures the fraction of a point explain by a given cluster

$$\alpha_i(y, \vec{x}) = \frac{w_i p_i(y, \vec{x})}{\sum_{j=1}^{M} w_j p_j(y, \vec{x})}. \tag{11}$$

Then the expected value of a quantity with respect to a cluster is given by a normalized sum over α. For example,

$$\langle \vec{x} \rangle_i \equiv \frac{\sum_{n=1}^{N} \vec{x}_n \alpha_i^{\gamma}(y_n, \vec{x}_n)}{\sum_{n=1}^{M} \alpha_i^{\gamma}(y_n, \vec{x}_n)}. \tag{12}$$

γ is an optional factor that controls cluster interactions and the convergence rate, with $\gamma = 1$ being the default, $\gamma > 1$ forcing the clusters apart, and $\gamma \to \infty$ reducing to hard clustering.

The cluster model parameters are determined by the cluster-weighted expectation of the maximum log-likelihood

$$\begin{aligned} 0 &= \frac{\partial}{\partial \vec{\beta}_i} \langle \log p_i(y, \vec{x}) \rangle_i \\ &= \left\langle [y - f(\vec{x}; \vec{\beta}_i)] \frac{\partial f(\vec{x}; \vec{\beta}_i)}{\partial \vec{\beta}_i} \right\rangle_i. \end{aligned} \tag{13}$$

Note that this weighting includes both how close the data are to the cluster in the input space, as well as how well the cluster model predicts the data in the output space. For the local linear case, $f = a + \vec{b} \cdot (\vec{x} - \vec{\mu}_i)$; taking the partials shows that fitted model parameters are then

$$a_i = \langle y \rangle_i \tag{14}$$

and

$$\vec{b}_i = \mathbf{C}^{-1} \cdot (\langle y\vec{x} \rangle_i - a_i \vec{\mu}_i), \tag{15}$$

where

$$\mathbf{C} = \begin{bmatrix} \langle (x_1 - \mu_{i,1})(x_1 - \mu_{i,1}) \rangle_i & \cdots & \langle (x_1 - \mu_{i,1})(x_D - \mu_{i,D}) \rangle_i \\ \vdots & \ddots & \\ \langle (x_D - \mu_{D,1})(x_1 - \mu_{i,1}) \rangle_i & \cdots & \langle (x_D - \mu_{D,1})(x_D - \mu_{i,D}) \rangle_i \end{bmatrix} \tag{16}$$

is the cluster-weighted covariance matrix.

The E − M sequence for Cluster-Weighted Modeling starts by randomly spreading clusters around the data set. The iteration is then:

1. Update the p_i's

$$p_i^{\text{new}}(y, \vec{x}) = \frac{1}{\sqrt{2\pi\sigma_{i,y}^2}} e^{(y - f(\vec{x}; \vec{\beta}_i))^2/2\sigma_{i,y}^2} \prod_{d=1}^{D} \frac{1}{\sqrt{2\pi\sigma_{i,d}^2}} e^{[(\vec{x} - \vec{\mu}_i) \cdot \hat{v}_i]^2/2\sigma_{i,d}^2}. \tag{17}$$

2. Update the α_i's

$$\alpha_i^{\text{new}}(y, \vec{x}) = \frac{w_i\, p_i(y, \vec{x})}{\sum_{j=1}^{M} w_j\, p_j(y, \vec{x})}. \tag{18}$$

3. Find the new means

$$\vec{\mu}_i^{\text{new}} = \langle \vec{x} \rangle_i. \tag{19}$$

4. Find the new weights

$$w_i^{\text{new}} = \frac{\sum_{n=1}^{N} \alpha_i(y_n, \vec{x}_n)}{\sum_{j=1}^{M} \sum_{n=1}^{N} \alpha_i(y_n, \vec{x}_n)}. \tag{20}$$

5. Calculate the new covariance matrix

$$\mathbf{C}^{\text{new}} = \begin{bmatrix} \langle (x_1 - \mu_{i,1})(x_1 - \mu_{i,1}) \rangle_i & \cdots & \langle (x_1 - \mu_{i,1})(x_D - \mu_{i,D}) \rangle_i \\ \vdots & \ddots & \\ \langle (x_D - \mu_{D,1})(x_1 - \mu_{i,1}) \rangle_i & \cdots & \langle (x_D - \mu_{D,1})(x_D - \mu_{i,D}) \rangle_i \end{bmatrix}. \tag{21}$$

6. Diagonalize to find the input eigendirections

$$\{\hat{v}_{i,d}\}^{\text{new}} = \text{eigenvectors of } \mathbf{C}_i. \tag{22}$$

7. The variances are found from the eigenvalues

$$\{\sigma_{i,d}^2\}^{\text{new}} = \text{eigenvalues of } \mathbf{C}_i. \tag{23}$$

8. Fit the model, which for the linear case is a constant of

$$a_i^{\text{new}} = \langle y \rangle_i \tag{24}$$

and a slope

$$\vec{b}_i^{\text{new}} = \mathbf{C}^{-1} \cdot (\langle y\vec{x} \rangle_i - a_i\, \vec{\mu}_i). \tag{25}$$

9. Finally, update the output variance

$$\sigma_{i,y}^2 = \langle (y - f(\vec{x};\, \beta_i))^2 \rangle_i. \tag{26}$$

This cycle is repeated until it converges, and the number of clusters M and the interaction term γ are determined by cross-validation with out-of-sample data (and can vary during the iteration).

Cluster-Weighted Modeling combines many of the best features of techniques for nonlinear modeling, while avoiding many of the attendant liabilities. It is easy to implement, straightforward to specify priors on the local behavior, makes weak assumptions other than the local prior, and is easy to interpret. Further, the cluster probabilities can also be used for classification problems, mixture of models can be implemented by mixing cluster types, and global priors can be introduced by adding fixed clusters (such as one with a huge variance and small weight to enforce asymptotic stability).

REFERENCES

1. WEIGEND, A. S. & N. A. GERSHENFELD, Eds. 1993. Time Series Prediction: Forecasting the Future and Understanding the Past. Santa Fe Institute Studies in the Sciences of Complexity. Addison-Wesley. Reading, MA.
2. BARRON, A. R. 1993. IEEE Trans. Info. Theory **39:** 930.
3. GERSHENFELD, N. 1997. The Nature of Mathematical Modeling. Cambridge University Press. To be published.
4. ROSE, K., E. GUREWITZ & G. C. FOX. 1990. Phys. Rev. Lett. **65:** 945.

Reconstructing a Dynamics from a Scalar Time Series

G. GOUESBET,[a] L. LE SCELLER,[a] C. LETELLIER,[a]
R. BROWN,[b] J. R. BUCHLER,[c,d] AND Z. KOLLÁTH[c]

[a] *LESP/URA CNRS 230/INSA de Rouen/BP 08*
Place Emile Blondel
76131 Mont Saint-Aignan Cédex, France

[b] *Institute for Nonlinear Science*
University of California
La Jolla, California 92093-0402

[c] *Department of Physics*
University of Florida
Gainesville, Florida 32611

INTRODUCTION

Chaotic time series data are now commonly observed in a large variety of fields. The analysis of such time series and the extraction of physical information is therefore a topic of growing interest. The techniques have matured to the point where they have become of practical use. It is the purpose of this review to discuss some such applications. For excellent general recent reviews of the available techniques we refer the reader to references 1 and 2.

This type of analysis presumes that the data have been generated by a low dimensional physical dynamics of *a priori* unknown nature, and that the signal may have been contaminated by noise of various sorts. In most applications we have at our disposal only a single measured quantity, i.e., a scalar time series. Fortunately one can exploit a very powerful redundancy principle that is discussed in the pioneering papers by Packard *et al.*[3] and by Takens[4] (cf. also reference 5) which allow one to reconstruct an image of the *whole* dynamics in an embedding space. These embedding theorems allow one then to extract useful information about the unknown dynamics from this single measured variable.

The first step in the analysis then consists of a phase-space reconstruction in an embedding space. In order to achieve it, we have to determine the embedding dimension d_E of the reconstructed phase space (also called state space). Of particular interest is of course the minimum embedding dimension, i.e., the lowest dimension for which the reconstruction gives an unambiguous image of the attractor. The reconstruction may be carried out in a number of ways, but it is safest to use several techniques because none of them alone is foolproof (for excellent reviews, cf. references 1 and 2). All these methods rely on delay coordinates, i.e., constructing the

[d]Supported by the National Science Foundation.

vectors of length (dimension) d_E

$$\mathbf{Y}^n = (s(t_n), s(t_{n-\tau}), s(t_{n-2\tau}), \ldots, s(t_{n-(d_E-1)\tau})) \tag{1}$$

from a scalar time series $\{s_n\} = \{s(t_n)\}$ that is sampled stroboscopically, i.e., at equal time intervals $t_n = t_o + n\delta t$, where the *delay* $\Delta = \tau\delta t$. We note that in some applications, such as astronomical observations, such a regular sampling may not be possible, and it is generally necessary to interpolate the data prior to the analysis.

Once the dimension d_E is known, we have the freedom to build a coordinate set to span the reconstructed state space. There exists an infinity of linear combinations of these delay vectors that are mathematically equivalent (e.g., reference 6), but because of unavoidable noise, in practice a suitable choice is important. A prolific literature is devoted to this issue. Many applications use the bare $\mathbf{Y}_n$ because this way the d_E coordinates all have the same noise properties, but other linear combinations have also been used, such as derivative coordinates, either direct differences, discrete Legendre polynomials[6] or Krawchuck polynomials (Auvergne, priv. comm.), etc. We note also that some care must be exercised because certain types of filtering of the data can augment the apparent dimension of the dynamics (e.g., reference 1). In whatever form delay coordinates are used, a crucial step is to determine an optimal value of the time delay Δ. However, this optimal value depends on the application. One popular choice is to use the value for which the mutual information[7] has its first zero. When polynomial maps are constructed from real data we shall see that the optimal value is a compromise that limits the nonlinearity of the map and minimizes the effect of noise. The method of *false nearest neighbors*[1] provides an efficient general guideline both for the dimensionality of the attractor and for the delay.

Once we have reconstructed the attractor in an embedding phase space, the next step is to obtain a set of equations which models the evolution of the system in the phase space. More particularly, we are interested in *global models* here, i.e., models which describe the dynamics on the whole state space, rather than models which approximate the dynamics in local balls. These methods then allow what we call a *global vector field or flow reconstruction.*

The chapter is organized as follows. First, we briefly review the general methods based on delay coordinates[8–11] and on derivative coordinates[12] for global vector field reconstructions. Secondly, a section is devoted to applications of these methods to real, i.e., experimental and observational data. Thirdly, a recent extension of global vector field reconstruction methods, which provides a set of equations incorporating a control parameter with an explicit physical meaning, is examined. In particular, preliminary results with data arising from an electrodissolution experiment are discussed. Finally, the conclusions are presented.

THE RECONSTRUCTION METHODS

Let us consider an *a priori* unknown nonlinear dynamical system defined by a set of autonomous ODEs:

$$\dot{\mathbf{x}} = \mathbf{g}(\mathbf{x}; \mu) \tag{2}$$

in which $\mathbf{x} \in \mathbb{R}^d$ and $\mathbf{g}$ is the unknown true vector field associated with the underlying dynamics of the physical system, and we call the solution vector $\mathbf{x}(t; \mu)$ the *state vector* which describes a *trajectory* in phase space. The quantity $\boldsymbol{\mu} \in \mathbb{R}^p$ is the parameter vector with p components, which for a given time series is assumed to be constant in this section. We note that in many situations we do not have an *a priori* knowledge of the dimension d of the dynamics. The dynamics can equivalently be described by a *map*

$$\mathbf{X}^{n+1} = \mathbf{G}(\mathbf{X}^n; \boldsymbol{\mu}), \tag{3}$$

where $\mathbf{X}^n \equiv \mathbf{x}(t_n) \in \mathbb{R}^d$, with $t_n = n\delta t$, represents a stroboscopic sampling of the trajectory at the equal time intervals δt. The measured time series s_n is assumed to be a generic function of the trajectory, $s_n = s(\mathbf{x}(t_n))$.

Depending on the applications, our goal is either to reconstruct a corresponding map $\mathbf{F}$ in a d_E dimensional embedding space with the help of the given delay vectors $\mathbf{Y}^n$ (1)

$$\mathbf{Y}^{n+1} = \mathbf{F}(\mathbf{Y}^n; \boldsymbol{\mu}) \tag{4}$$

or to reconstruct a flow $\mathbf{f}$

$$\dot{\mathbf{y}} = \mathbf{f}(\mathbf{y}; \boldsymbol{\mu}) \tag{5}$$

where $\mathbf{y}(t_n) = \mathbf{Y}^n$. The embedding theorems tell us that there is a one-to-one correspondence between the embedded flow $\mathbf{f}$ of (4) and the physical flow $\mathbf{g}$ (2). We refer to the latter as *physical* because it represents the number of variables that are required to fully describe the dynamics. It is however important to distinguish between the dimensions of the two flows. Because the embedding can cause spurious cusps or intersections, in principle d_E is greater or equal to d; in practice it is often found, or assumed, that $d = d_E$ (perhaps because the cusps and intersection points are rare and do not appear to cause havoc when the map is iterated to create synthetic signals).

Maps

The reconstruction can be very conveniently made with polynomial nonlinearities in such a way that the components of $\mathbf{F}(\mathbf{Y})$ take the form of multivariate polynomials. The Weierstrass theorem[13] guarantees the convergence of the approximation of any smooth function by a polynomial expansion. Consequently, using polynomial expansions to build global models of dynamical systems comes into mind in a natural way. Polynomial techniques for example have been extensively used for *local* modeling, in which the vector fields $\mathbf{F}(\mathbf{Y})$ (or $\mathbf{f}(\mathbf{y})$) are locally approximated by polynomials whose coefficients are determined by least-squares (LS) fits. In contrast, when a global model is desired, the situation becomes more difficult due to the large number of required data points and to the need for the use of higher-order polynomials. In order to overcome such difficulties, several groups have independently proposed using orthogonal, or better, orthonormal polynomials generated by the invariant density on the attractor.[8,9,12,14,15] There are two equivalent ways to construct these polynomials, namely, by a Gram–Schmidt orthonormalization or by a LS fitting with monomials, both of which give complementary insight into the data requirements.

Gram–Schmidt Construction of Orthogonal Polynomials

Interestingly the determination of the coefficients of the polynomials and therefore of the coefficients of the reconstructed vector field, $\tilde{\mathbf{F}}(\mathbf{y})$, only requires the computation of moments of the data. (Henceforth we shall drop the tilde from the reconstructed maps and vector fields when there is no confusion possible.) More specifically, a basis of orthonormal multivariate polynomials $\{\phi_k(\mathbf{y}) = \phi_k(y_1, y_2, \ldots, y_{d_E})\}$ on $\mathbb{R}^{d_E}$ can be built with a conventional Gram–Schmidt (GS) procedure. According to this procedure, the polynomials $\phi_k(\mathbf{y})$ read as:

$$\phi_k(\mathbf{y}) = \frac{\phi_k^*(\mathbf{y})}{(\phi_k^*, \phi_k^*)}, \tag{6}$$

in which $\phi_k^*(\mathbf{y})$ are defined as:

$$\begin{cases} \phi_1^* = P_1 \\ \phi_k^* = P_k - \sum\limits_{\alpha=1}^{k-1} (P_k, \phi_\alpha)\phi_\alpha, \; k > 1, \end{cases} \tag{7}$$

in which P_k's designate the leading multivariate monomials and (,) is a scalar product. The orthonormality condition is $(\phi_i, \phi_j) = \delta_{ij}$. The polynomials of the orthonormal basis are then found to be given by expansions on the monomials P_k with expansion coefficients defined as:[16]

$$A_\alpha^k = A_\alpha^{*k} \Bigg/ \left[\sum_{\beta=1}^{k} \sum_{\gamma=1}^{k} A_\beta^{*k} A_\gamma^{*k} (P_\beta, P_\gamma) \right]^{1/2}, \tag{8}$$

where

$$A_\alpha^{*k} = - \sum_{\beta=\alpha}^{k-1} A_\alpha^\beta \left[\sum_{\gamma=1}^{\beta} A_\gamma^\beta (P_k, P_\gamma) \right]. \tag{9}$$

Each component F_i of the vector field $\mathbf{F}(\mathbf{y})$ which describes the dynamics is approximated as:

$$F_i = \sum_{j=1}^{P} c_{ij} \phi_j, \tag{10}$$

in which the c_{ij} are the Fourier coefficients for the ith component F_i of the vector field $\mathbf{F}$, which are:

$$c_{ij} = (F_i, \phi_j). \tag{11}$$

In practice, the expansion is terminated at some maximum polynomial order P.

Least-Squares Construction of Orthogonal Polynomials

A mathematically equivalent alternative to constructing the orthogonal polynomials and the expansion coefficients is to perform a linear LS fit with all possible monomials up to order P, i.e., to minimize the expression

$$S = \sum_{n=1}^{N-1} |\mathbf{Y}^{n+1} - \mathbf{F}(\mathbf{Y}^n)|^2, \tag{12}$$

where N is the number of delay vectors. The vector function $\mathbf{F}$ is given by

$$\mathbf{F}(\mathbf{Y}) = \sum_{\mathbf{k}} \mathbf{A}^{\mathbf{k}}\, Y_1^{k_1} Y_2^{k_2} Y_3^{k_3} \ldots, \tag{13}$$

where the Y_i are the components of $\mathbf{Y}$, and the vector index $\mathbf{k} = (k_1, k_2, k_3, \ldots)$ runs over all powers with the constraint $\sum_j k_j \leq P$, the maximum order of the expansion.

The LS minimization[17] is most straightforwardly carried out with a QR algorithm. It can be shown that the expansion of $\mathbf{F}$ in monomials (**14**) can be regrouped into the form obtained with the GS procedure (**11**). The two approaches, mathematically, are therefore strictly equivalent, although their numerical stability properties are different. In fact, both the GS and the QR methods run into numerical difficulties when either the embedding dimension or the order of the expansion P gets large, because the number of coefficients gets large and the fit becomes ill-conditioned when used on short data sets.

For that reason Serre *et al.*[9] have found it useful to use another standard method for solving the LS problem, namely, a singular value decomposition (SVD).[17] In the LS problem, which determines a vector of parameters a by minimization of (**12**),

$$S = \|\boldsymbol{A} \cdot \mathbf{a} - \mathbf{b}\|^2, \tag{14}$$

the matrix $\boldsymbol{A}$ is decomposed into a product of orthogonal and diagonal matrices as follows:

$$\boldsymbol{A} = \boldsymbol{U} \cdot \mathrm{diag}(w_i) \cdot \boldsymbol{V}^T \tag{15}$$

and the solution appears in the form

$$a_i = \sum_{j,k} V_{ij} \frac{1}{w_j} U_{kj} b_k. \tag{16}$$

When the problem is well conditioned, all eigenvalues w_i are above machine precision and the SVD gives an answer equivalent to the other solutions of the LS problem. The power of the method resides in the fact that it possesses a good behavior even when the LS design matrix $\mathbf{A}$ is poorly conditioned, i.e., when the parameters are close to being linearly dependent, whether because of a poor choice of parameters or because their number exceeds the number of data points. It suffices then to restrict the sum i over the eigenvalues w_j to the ones above some (numerical) noise threshold. With SVD the number of coefficients can be taken arbitrarily large, in fact even larger than the number of available data. The SVD method thus automatically selects the most important linear combinations of the coefficients and discards the rest.

Flows

Standard Formulation

The preceding methods are not only useful for constructing global polynomial *maps*, but they can also be extended to reconstruct *flows* globally. In fact since the data are given at discrete times one needs to replace the standard flow (**4**) by a discrete approximation anyway and the difference between a map and a flow is not very big. As previously, one uses delay coordinates generating a vector $\mathbf{y}^n$ in the phase space, namely,

$$\mathbf{y}^n = (s_n, s_{n-\tau}, \ldots, s_{n-(d_E-1)\tau}). \tag{17}$$

The practical implementation of the vector field obviously also depends on the integration scheme. For instance, an explicit Euler integration may be used:

$$\mathbf{y}^{n+1} = \mathbf{y}^n + \mathbf{f}(\mathbf{y}^n)\delta t, \tag{18}$$

where now $\mathbf{y}_n = \mathbf{y}(n\delta t)$ with the time step δt. Given this equation, the model works well when the time interval δt between measurements is small. In many cases, δt must be a few hundred times smaller than the characteristic time of the oscillations that occur in $\mathbf{y}$.

In many experimental situations, it is not feasible to use a small enough sampling interval. It is therefore worthwhile to seek a method that is capable of accurately modeling $\mathbf{f}$ when δt, the time step between two successive measurements, is much larger than the Euler method would allow. This issue has been addressed by Brown *et al.*[8] who advocate the use of an Adams predictor–corrector method. Under this method, one formally integrates an initial condition $\mathbf{y}^n$ to the next point $\mathbf{y}^{n+1}$ in one step of size δt via

$$\mathbf{y}^{n+1} = \mathbf{y}^n + \delta t \sum_{j=0}^{M} a_j^{(M)} \mathbf{f}(\mathbf{y}^{n+1-j}), \tag{19}$$

where δt is the time step of the integration and $a_j^{(M)}$ are the implicit Adams predictor–corrector coefficients.[8] M indicates the order of the corrector portion of the integration. The numerical values of the $a_j^{(M)}$ are given by the method described in reference 17. In their study, Brown *et al.*[8] use a time delay Δ in terms of δt that is estimated by using the mutual information method.[1]

The model flow $\mathbf{f}$ is optimized with the help of a minimum description length technique in which an error function

$$\chi^2 = \frac{1}{2N\sigma^2} \sum_{n=1}^{N} \left| \mathbf{y}^{n+1} - \mathbf{y}^n - \delta t \sum_{j=0}^{M} a_j^{(M)} \sum_{i=0}^{N_p} \mathbf{K}_i \phi_i(\mathbf{y}^{n+1-j}) \right|^2 \tag{20}$$

is introduced. This error function is quadratic with respect to the fitting coefficients $\mathbf{K}_i$ giving a LS problem. The full error function involves a length criterion based on the minimum description length principle for truncating a model (references 8 and, more recently, 18 and references therein). It is indeed very elaborate and provides an efficient automatic criterion to determine the model coefficients. The quality of the model depends on the time delay Δ, on the order M of the correction in the integration and on the number N_p of retained polynomials. Unfortunately, the interest and efficiency of the error function are reduced by the need for a serious coding effort and large computing time requirements. However, the use of an SVD approach which would alleviate these problems has not yet been tried.

The Derivative Approach

In many experimental situations, a dynamical system may be described by a variable, say s, and its successive derivatives $\dot{s}$, $\ddot{s}$, ... The first derivative can be called a speed and the second derivative an acceleration, as in the case of mechanical oscillators, i.e., derivatives may receive a direct physical interpretation. For this reason (and others), it is of interest to develop models which use an embedding based on derivative coordinates in which the phase space vector is:

$$\mathbf{z}_n = (s_n, s_n^{(1)}, s_n^{(2)}, \ldots, s_n^{(d_E - 1)}), \tag{21}$$

where $s_n^{(i)}$ designates the ith derivative of the time series variable s_n at time $n\delta t$.[12] The successive derivatives may be estimated either by using a linear discrete filter built on a basis of discrete Legendre polynomials[6] or by using a local fit with an mth-order polynomial (where m is greater than the largest order of derivative required in the model). In both cases, a window size Δ has to be determined and plays a role rather similar to the time delay Δ used in the time delay method. Nevertheless, from our experience, it appears that the quality of the derivative embedding is less sensitive to the value of Δ than the time delay Δ in the case of a time delay embedding. For a good evaluation of the derivatives, the time interval δt must be a few hundred times smaller than the typical characteristic time of oscillations that occur in the recorded time series $\{s_n\}$.

We thus search for a model in the canonical form

$$\begin{cases} \dot{z}_1 = z_2 \\ \dot{z}_2 = z_3 \\ \quad \vdots \\ \dot{z}_{d_E} = h_{d_E}(\mathbf{z}). \end{cases} \tag{22}$$

One may then remark that the reconstructed vector field $\mathbf{f}(\mathbf{z})$ is composed of $(d_E - 1)$ explicitly known components and by a single unknown component h_{d_E} which may be given a polynomial form. Consequently, this method only requires the determination of a single function h_{d_E} in contrast with the previous method which requires the determination of the d_E functions. This advantage is however balanced by the need of a higher-order polynomial which is required to express the scalar h_{d_E}.

The function h_{d_E} is taken to have the form

$$h_{d_E}(\mathbf{z}) = \sum_{i=1}^{N_p} c_i \phi_i(\mathbf{z}) = \sum_{i=1}^{N_p} K_i P_i(\mathbf{z}), \tag{23}$$

where $P_i(\mathbf{z})$ are the monomials up to order N_p that can be formed from the d_E components of the vector $\mathbf{z}$ according to a relationship defined in reference 12, namely, $P_i(\mathbf{z}) = (z_1^{i_1} z_2^{i_2} \cdots z_{d_E}^{i_{d_E}})$. The biunivocal relationship used between the subscript i and superscript set $\{i_1, \ldots, i_{d_E}\}$ is extensively described in reference 12 when $d_E = 3$. In their applications[12] $\{\phi_i(\mathbf{z})\}$ is a basis of orthonormal polynomials constructed via a GS procedure starting from $\phi_1(\mathbf{z}) = 1$ and using a scalar product defined on a data set, and the vector field (**22**) **F** is integrated by using a fourth-order Runge–Kutta scheme with an adaptive stepsize control.[17]

The model is optimized by using an error function

$$E_r = \sum_{i=1}^{N_q} \left| y_i^{(d_E)} - \sum_{p=1}^{N_p} K_p P_p \right| \Big/ \sum_{i=1}^{N_q} | y_i^{(d_E)} | . \tag{24}$$

This error function is calculated by using absolute values for computational efficiency. It is found that the quality of the model depends on reconstruction parameters, namely, the number N_q of vectors $\mathbf{y}_n$, the number N_s of $\mathbf{y}_n$'s sampled per pseudo-period and the number N_p of retained multivariate polynomials. The error function E_r is used as a guideline to determine the values of the reconstruction parameters as follows. First, optimal values of N_s and N_p are found by minimizing the error function. However, for a given N_q, the obtained approximated system is not necessarily successful, in particular it cannot necessarily be successfully integrated, although the error function passes through a (possibly local) minimum. Consequently, the search of a successful global modeling needs systematical trials on N_q's which can nevertheless be automatically done with computational help. More generally, finding an optimal modeling of nonlinear data series remains a tricky problem for which Brown *et al.*[8] proposed a more elaborated error function than the one of Gouesbet and Letellier, which however is computationally faster.

It is the ultimate goal of the analysis and the nature of the available data set which determine whether one prefers to use a map or a flow, and in the second alternative which of the alternate forms (**5**) or (**23**). Generally it seems that maps give more stable reconstructions than flows (e.g., references 10 and 11).

As far as the choice of GS approach versus the LS approach with SVD is concerned, both approaches work for large data sets, but the first method can be ill-conditioned for high embedding dimensions and high-order polynomials. Furthermore the SVD LS approach is faster.

Noise

Measured time series are of course always contaminated with noise, and the question arises how these methods hold up under realistic situations. We note that the noise problem can be particularly bad in astronomy. Not only is the signal to noise ratio often small, the data generally are not or cannot be obtained at the equal space time intervals that the methods require. The necessary interpolations introduce additional noise.

References 1, 9–11 and 21 address the problems introduced by noise. It should be noted though that noise can also have a beneficial effect, namely, in moderate amounts it stabilizes the maps without affecting their properties, and in fact it has been found useful to add low-level noise to the data for this reason.[9,11,21]

APPLICATIONS TO PHYSICAL SYSTEMS

The global reconstruction techniques discussed above have been successfully applied to experimental and observational data, as well as to data generated by realistic numerical models of complex physical systems as reviewed below. In each case, a brief discussion concerning the validation of the models will be given. We note that not all variants of the reconstruction method have been applied to the same data. It is therefore not possible to make direct comparisons between them.

Nonlinear Vibrations in a Thin Wire

Brown *et al.*[21] have applied their techniques to experimental data obtained from an apparatus used to study nonlinear vibrations in a thin wire.[19] The apparatus basically consists of a mount holding a tensioned wire fed by an alternating current. The frequency of the current is close to the fundamental frequency of the free oscillations of the wire. This current excites forced vibrations when the wire is placed in a permanent magnetic field. As the amplitude and frequency of the current are varied, the system can undergo a torus-doubling route to chaos.[20] Optical detectors are used to measure the transverse amplitude of the wire. The transverse displacements of the wire generate the time series to be analyzed. Control parameters of the experimental set-up may be found in reference 19.

The amplitude oscillations have a pseudo-period equal to about 0.1 s. The time step δt between two successive measurements is equal to 7.7×10^{-4} s. Thus, 130 points per cycle are available. A three-dimensional embedding (FIG. 1) of a chaotic trajectory is created out of the scalar amplitude measurements made by a single detector by using delay coordinates. The time delay is fixed to $\Delta = 39\delta t$, based on the mutual information criterion. The embedding dimension d_E is found to be equal to 3.

The model (FIG. 2) is validated by comparing Lyapunov exponents. From the experimental data, the positive Lyapunov exponent is $\lambda_1 = 4.97 \times 10^{-2}$ and the negative one is $\lambda_3 = -0.702$, while, from the synthetic data, they are $\lambda_1 = 4.31 \times 10^{-2}$ and $\lambda_3 = -0.576$. These values differ by approximately 15% for the positive exponent and 20% for the negative exponent.[19]

Another method for checking the model is synchronization.[8] The modified Fuyisaka and Yamasada method of synchronization is given by

$$\begin{Bmatrix} \dot{\mathbf{x}} = \mathbf{g}(\mathbf{x}) \\ \dot{\mathbf{y}} = \mathbf{f}(\mathbf{y}) - \mathbf{e}(\mathbf{y} - \mathbf{x}) \end{Bmatrix}, \tag{25}$$

where **e** is the coupling matrix between the two dynamical systems involved in (**26**). The $\mathbf{x}(t) \in \mathbb{R}^{d_E}$ denotes the trajectory of the experimental system in its phase space, and $\mathbf{y}(t) \in \mathbb{R}^{d_E}$ the trajectory generated by the model in the embedding phase space, namely the response system in the jargon of synchronization.[21] The driving

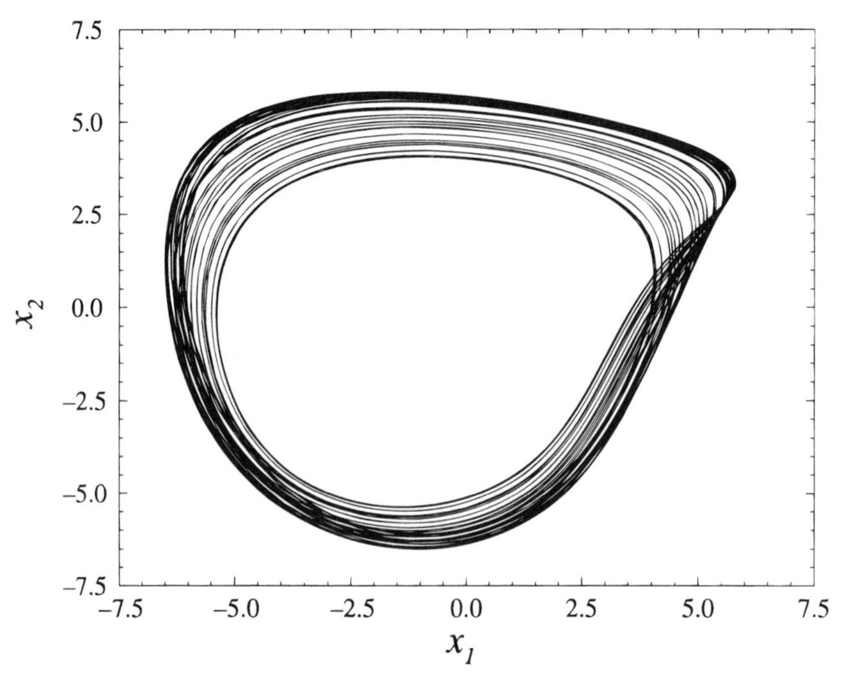

FIGURE 1. Plane projection of the three embedded time series from the string experiment.

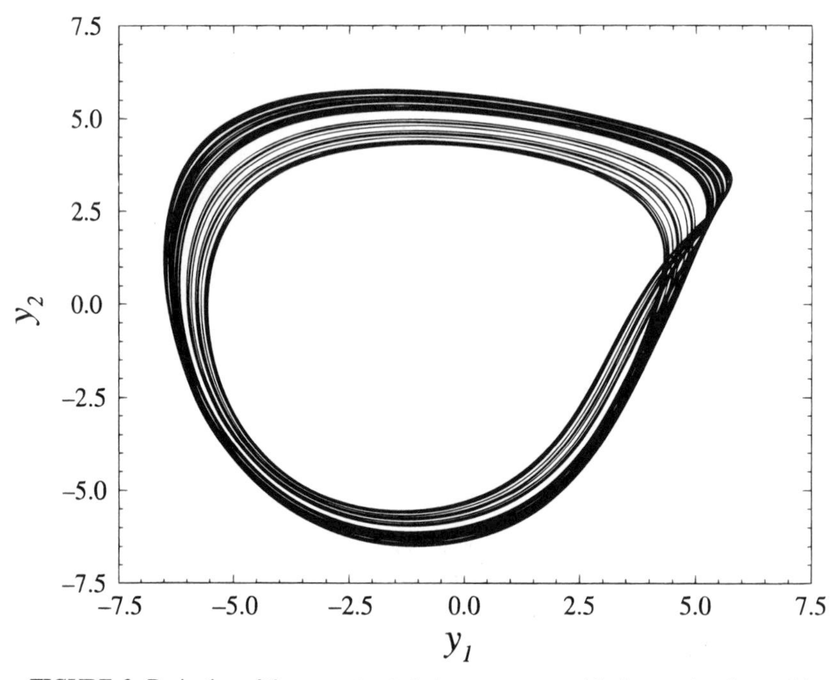

FIGURE 2. Projection of the reconstructed attractor generated by integrating the model.

and the response systems are said to be synchronized if there exists a coupling matrix with a few nonzero components ensuring that the response trajectory is slaved to the driving trajectory. Brown *et al.*[21] used only one nonzero component lying on the leading diagonal which may be written as $e_{\beta\beta} = \varepsilon$. They showed that, depending on the value of β, setting ε to a too small or to a too large value does not produce any synchronous motion. More important, synchronization could only be obtained if the reconstructed vector field **f** was a good enough approximation to the true dynamics **f**.[8] By using the third component of **x** for coupling, they found that the model is synchronized for $\varepsilon \geq \varepsilon_c \approx 0.5$. Tests were given for $\varepsilon = 0.75$ and 3.0.

The topological analysis has been also used as a third test to check the model. A review concerning topological characterization may be found in reference 24. Tufillaro *et al.*[19] showed that the model induces the same template as the experimental data, i.e., the relative organization of the periodic orbits is the same for both the experimental data and the model data. Such a topological analysis guarantees that the original attractor and the reconstructed attractor are topologically equivalent. We think that topological analyses may provide the best criterion for checking a model, particularly if the populations of periodic orbits are also compared.

Brown *et al.*[8] have also obtained models for experimental data arising from an electronic circuit for two values of a control parameter α. The corresponding attractors are displayed in FIGURES 3 and 5. They form a so-called single attractor and a symmetric attractor with two wings, respectively. The models contain $N_p = 4$ and

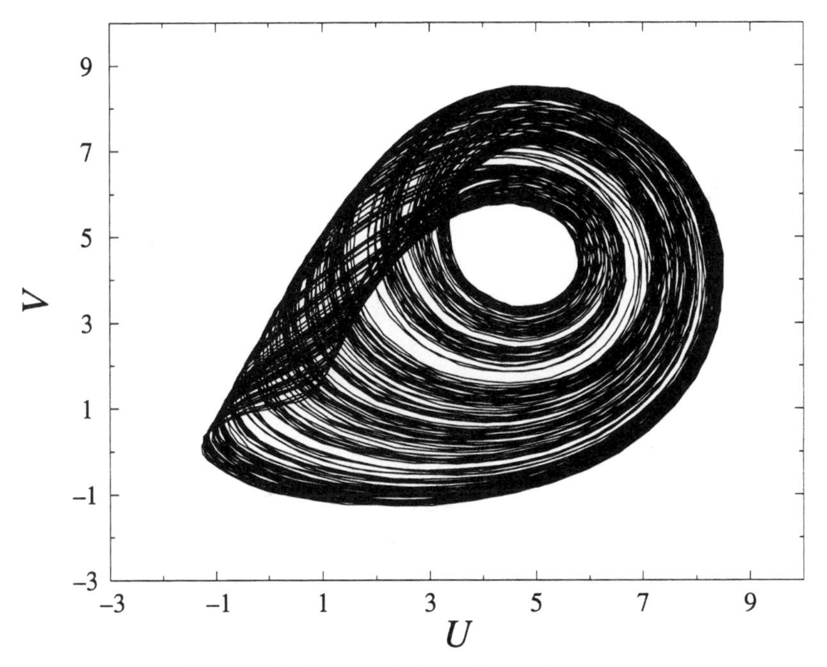

FIGURE 3. Plane projection of the single attractor.

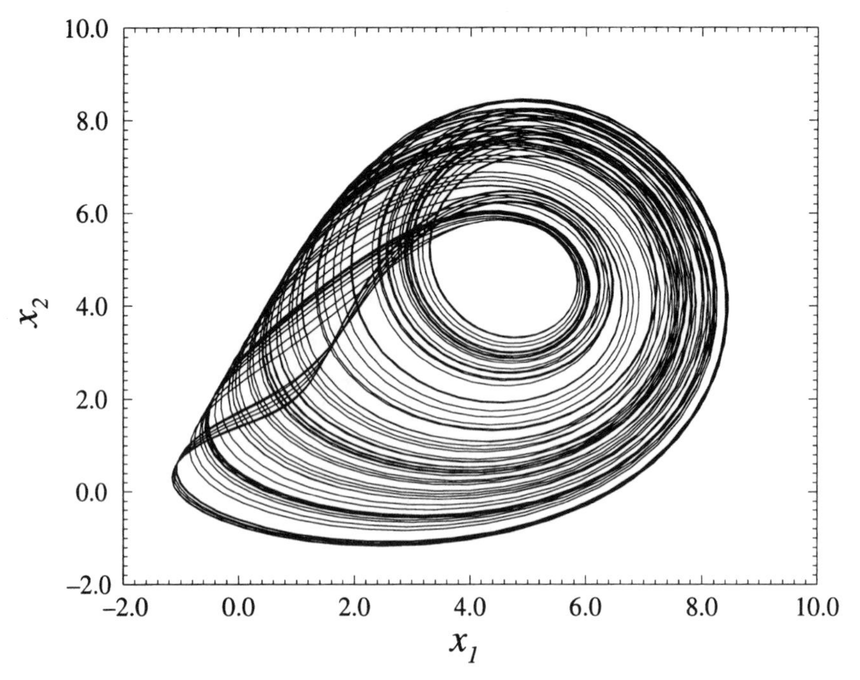

FIGURE 4. Model of the single attractor.

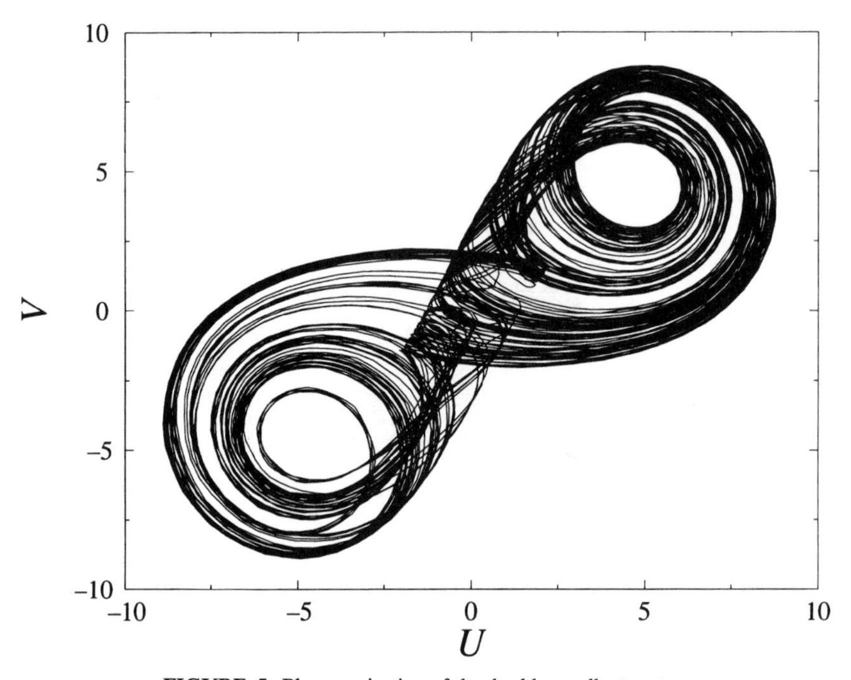

FIGURE 5. Plane projection of the double scroll attractor.

$N_p = 5$ polynomials for $\alpha = 17.4$ and $\alpha = 18.9$, respectively, i.e., chaotic trajectories have been successfully reconstructed with rather simple mathematical expressions (FIGS. 4 & 6).

Finally, we mention the reconstruction of a model of the Belousov–Zhabotinskii reaction[8] whose topological analysis has been given by Mindlin *et al.*[23] A projection of the corresponding attractor (cf. reference 23 for a specific embedding procedure) is displayed in FIGURE 7. Here again, the model is found to be rather simple, containing $N_p = 7$ polynomials with a 3D-embedding (FIG. 8).

Electrodissolution of a Rotating Cu Electrode

We start with a time series which has been obtained from current measurements during the potentiostatic electrodissolution of a rotating Cu electrode in phosphoric acid. The experiments have been described in detail in references 25 and 26. Briefly, the experimental set-up consisted of a rotating disc electrode which had a copper rod, 8.26 mm in diameter, embedded in a 2 cm diameter Teflon cylinder. The rotating speed was maintained at 4400 rpm. The cell was a 500 ml flask with a side neck in which the capillary probe was fixed. The cell contained 250 ml of 85% phosphoric acid and a water bath was used to maintain its temperature at 20°C. A Potentiostat (Princeton Applied Research model 273) was used to regulate the potential of the

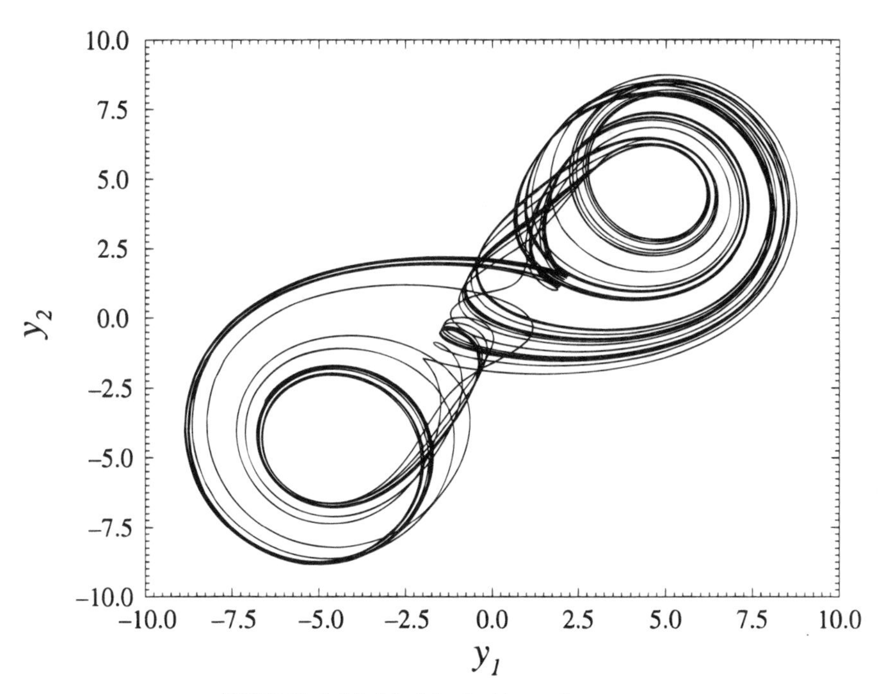

FIGURE 6. Model of the double scroll attractor.

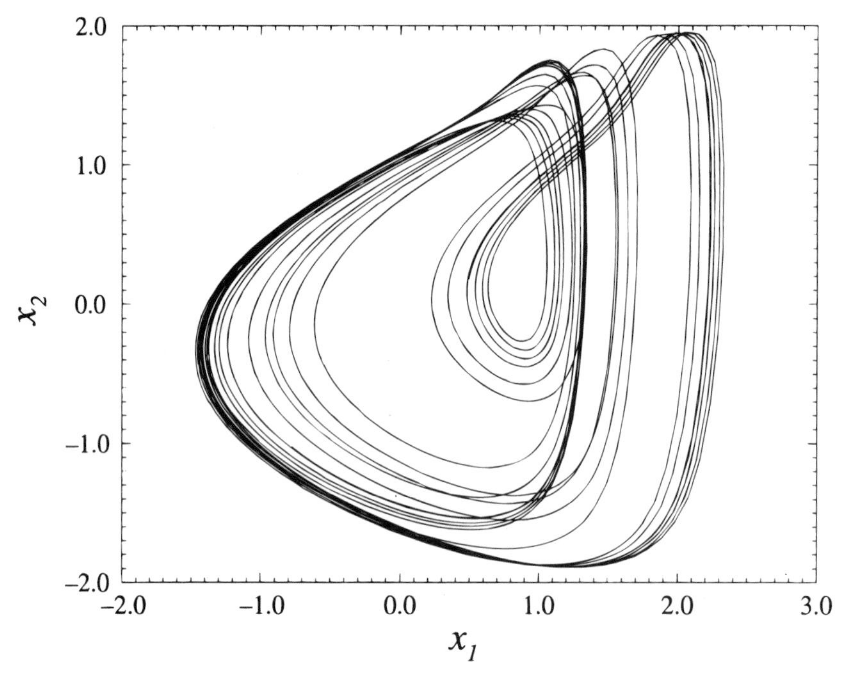

FIGURE 7. A differential-integral phase space embedding of the Belousov–Zhabotinskii data.

working disc electrode with respect to the reference electrode and to monitor the current. The data were recorded at a frequency f_e of 1500 Hz.

The state space of the electrodissolution has been reconstructed from the current time series $I(t)$ by using successive time derivatives. The embedding dimension d_E is found[25] to be equal to 3 from the estimation of the correlation dimension by using the Grassberger–Proccacia algorithm.[32] In the reconstructed phase space, the asymptotic motion settles down onto a chaotic attractor displayed in FIGURE 9. The so-called *copper* attractor is found to be characterized by a horseshoe template (cf. references 25, 26 or 24). The window size Δ used to evaluate the derivatives is taken to be equal to $21\delta t$, where $\delta t = 6.67 \times 10^{-4}$ s. The pseudo-period of the oscillations is about 0.102 s. Thus, we have approximately 150 points per pseudo-period. The number of retained polynomials in the model is found to be equal to 26 which is rather large in comparison with N_p's obtained in the cases discussed in the previous section. As explained earlier, it is inherent to the form of the model, all the nonlinear dynamics being stored in a single standard function.

By numerically integrating the reconstructed model, a reconstructed attractor (FIGURE 10) is obtained which looks rather similar to the original copper attractor. The model is checked by using topological analysis. Both attractors indeed induce the same template. Moreover, a slight difference which has been found between the corresponding two orbit spectra has been explained in terms of contamination of the experimental data by noise.[26]

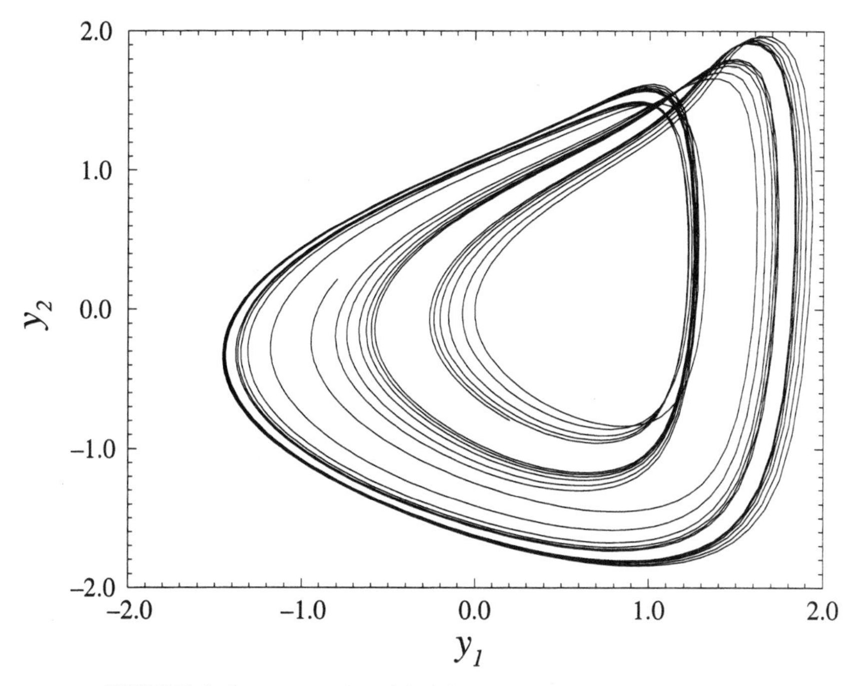

FIGURE 8. Reconstructed model of the Belousov–Zhabotinskii reaction.

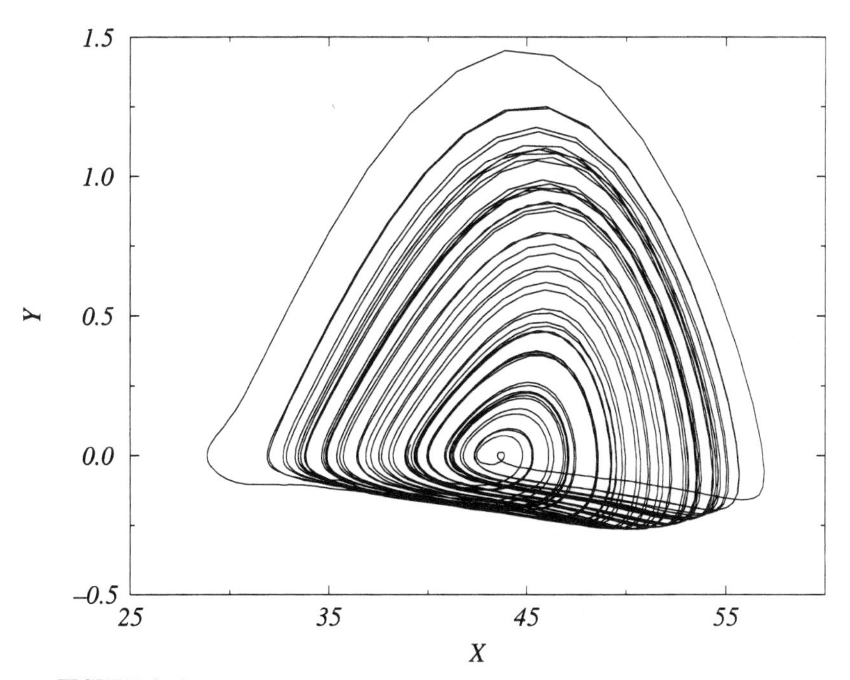

FIGURE 9. Attractor generated by the copper electrodissolution (copper attractor).

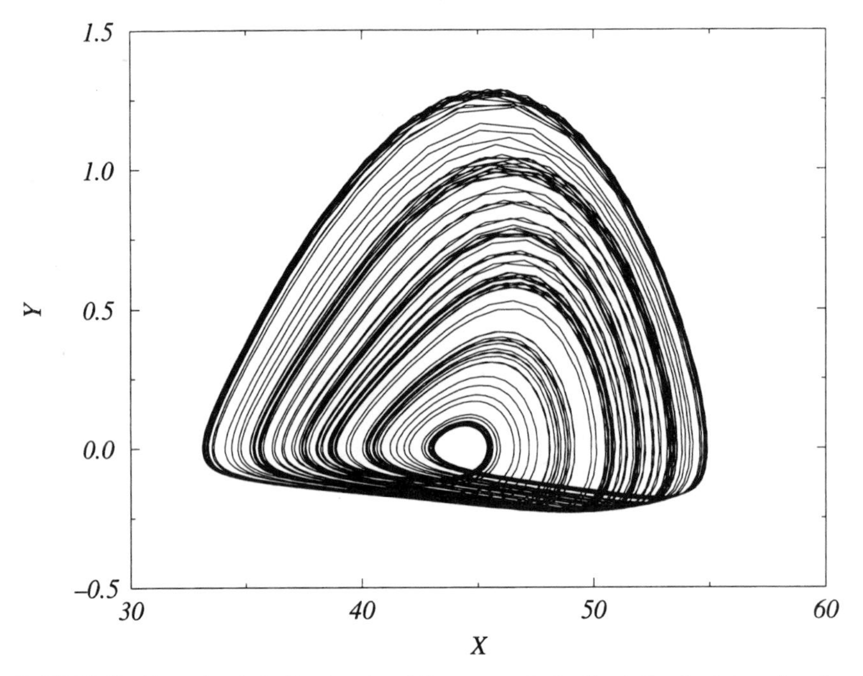

FIGURE 10. Reconstructed attractor A_R of the copper electrodissolution by integrating the reconstructed vector field.

A second electrodissolution has also been studied subsequently. The experimental set-up is very similar to the previously described one except that it concerns an iron electrodissolution in sulfuric acid. The embedding dimension d_E has been found to be equal to 3 by using the mutual information criterion analysis (kindly performed by R. Brown). The window size Δ is equal to $35\delta t$. The attractor generated by the experimental data is displayed in FIGURE 11.

In a model for this dynamics, the number N_p of polynomials has been found to be equal to 78. The attractor obtained by integrating the model (FIGURE 12) looks very similar to, but also slightly different from the original attractor. The model has then been validated, but only partially[27] insofar as the topological analysis could not be safely performed. In particular, the absence of a hole in the middle of the attractor prevents a safe enough definition of a Poincaré section inducing many difficulties in performing the topological characterization. For these reasons, synchronization has also been used. The model has been successfully synchronized with the experimental time series by using $E_{11} = 0.05$ and $E_{33} = 0.05$ as nonzero elements in the coupling matrix of (**25**). We may then state that the model dynamics is close to the experimental dynamics.

Astrophysical Applications

Two applications of the global flow reconstruction method have been made to problems of astrophysical interest; the first to the irregular pulsations of a model of

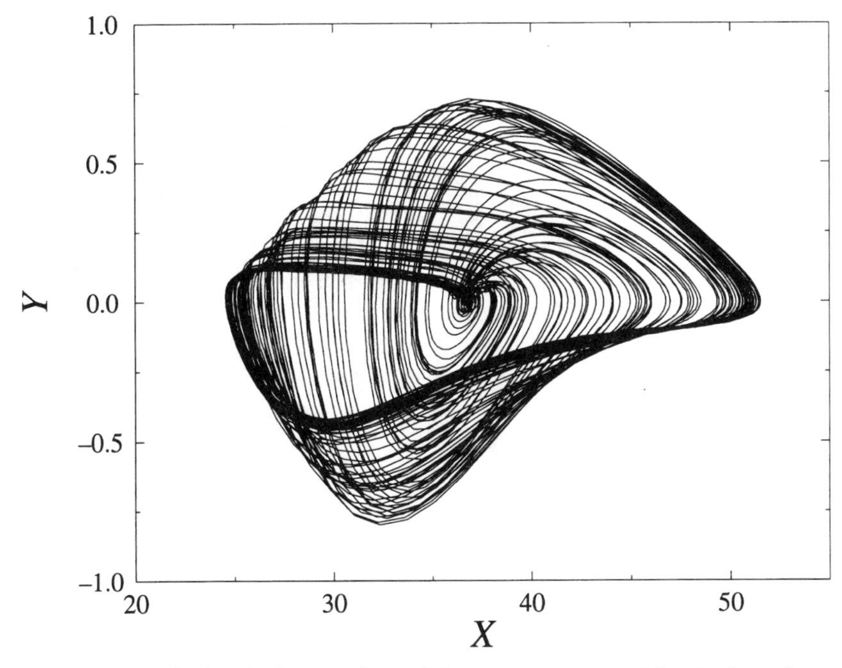

FIGURE 11. Projection in the XY-plane of the attractor generated by the iron electrodissolution.

a star of the W Vir type that were obtained with a state-of-the-art numerical hydrodynamical code, and the second to the observed irregular light-curve data of two stars of the RV Tau type.

W Vir Model Pulsations

The reconstruction of Serre *et al.*[11] confirmed that the irregular pulsations of W Vir models[28] are indeed chaotic, and they furthermore could show that the dynamics could be embedded in a three-dimensional space ($d_E \geq 3$). Lyapunov exponents were also computed and led to a fractal Lyapunov dimension $d_L \approx 2.02$ for the reconstructed attractor, with d_L independent of the embedding dimension. With the aid of the embedding theorem they thus concluded that the physical attractor also has a fractal dimension of ≈ 2.02. From these two bounds $d_L < d \leq d_E$ it follows that the *physical dimension d* is also 3.

These results were robust with respect to changes in the delay and the maximum order P of the polynomials of the map. They also found that as d_E is increased, the highest polynomial degree P can be decreased. A convenient and in some sense optimal representation of the attractor can be made in Broomhead–King coordinates ξ_k (projections onto the eigenvectors of the correlation matrix[34]). The projections onto the planes spanned by the first three principal components are shown in

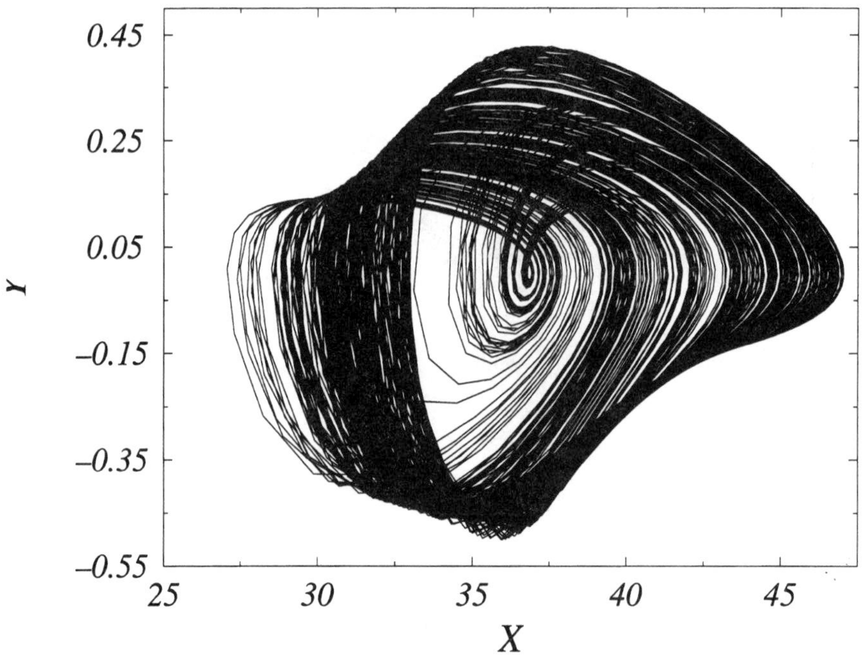

FIGURE 12. Reconstructed attractor of the iron electrodissolution obtained by integrating the reconstructed vector field.

FIGURE 13. Once the map has been determined it can be iterated to produce synthetic signals. FIGURE 13 shows that the synthetic signals are very similar to the original data set, independently of the embedding dimension. For further details we refer to reference 11.

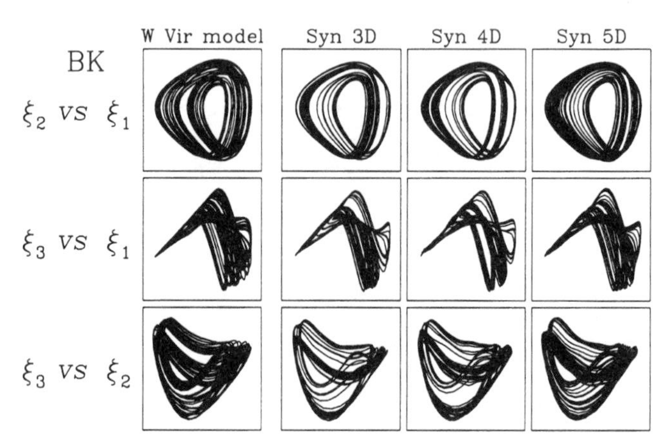

FIGURE 13. W Vir model, lowest three principal component projections. *Column 1*: data set; *2–4*: synthetic signals from noisy data set in 3D, 4D and 5D.

One notes that the numerical hydrodynamical code replaces the PDEs of fluid dynamics and heat flow by a discrete approximation consisting of N mass shells. In the above-mentioned work a set of $3N = 180$ coupled nonlinear ODEs were thus solved. (These ODEs are very stiff in the mathematical sense because of the sharp temperature sensitivity of the equation of state and opacity law.) It is therefore nontrivial that the resultant pulsation can be fully described by a mere three ODEs. This dimensional shrinkage from 180D (or any other number of zones N that sufficiently resolves the spatial structure of the star) to 3D shows that there is an *inertial manifold* in which the dynamics takes place. This implies that the time-dependence of all the physical variables throughout the star (e.g., velocity, density and temperature fields) can be expressed in terms of only three basic variables.

While the *physical* picture is not entirely clear yet, these results taken together with a Floquet analysis of the hydrodynamical models[35] suggest that the linearly unstable fundamental mode of oscillation of the star interacts nonlinearly with a single overtone through a parametric resonance to produce the complex behavior. How the additional dimensional shrinkage from 4D (two complex modes) to 3D comes about is still under investigation.

R Scuti Observational Data

The recent availability of a suitably large and densely sampled observational data set has made it possible to apply the global flow reconstruction method to astronomical data[10,15] and to probe the properties of the irregular pulsation cycles of an actual star, namely star R Scuti. The light-curve is shown in FIGURE 14 on top. The authors demonstrated first that the observational light-curve data of R Scuti, a star of the RV Tau type, is not multi-periodic, and that it cannot have been generated by a linear stochastic (AR) process. The nonlinear reconstruction analysis shows that this star's complicated light-curve is captured by a simple 4D polynomial map or flow (four first-order ODEs) and that the bulk of the signal consists of low dimensional chaos. FIGURE 14, at the bottom, shows a typical synthetic light-curve that has been generated from the reconstructed 4D map.

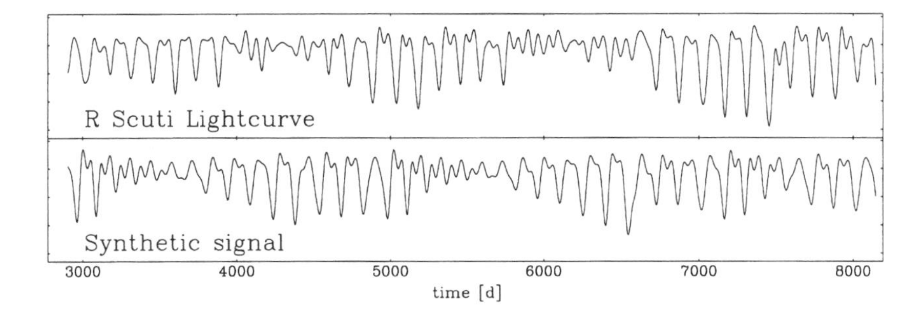

FIGURE 14. Top: R Scuti light-curve; **bottom**: synthetic light-curve.

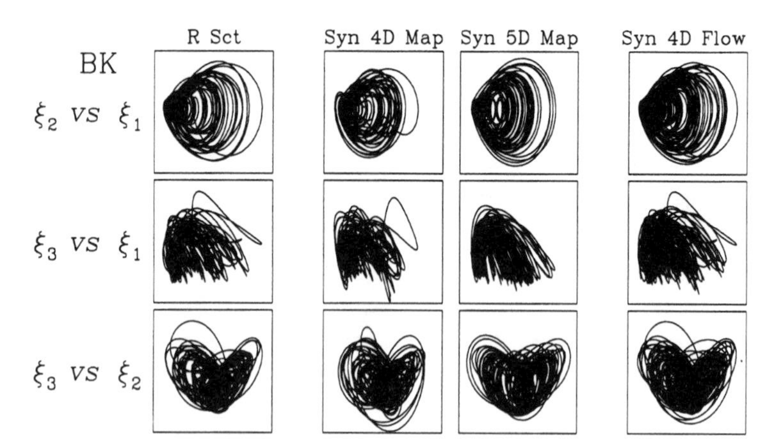

FIGURE 15. Lowest three Broomhead–King coordinate projections. *Column 1*: data set; *2* & *3*: synthetic signals from 4D and 5D maps; *4*: synthetic signal from 4D flow.

FIGURE 15 shows the lowest Broomhead–King projections of the R Scuti light-curve together with synthetic light-curves in 4D and 5D. The rightmost figures show a synthetic signal from a reconstructed flow (using the Adams integration scheme). The synthetic signals are not only very similar in appearance to the R Scuti light-curve as FIGURE 14 shows, but the more discriminating BK projections in FIGURE 15 show that they are really very similar, and robust with respect to dimension. They also show that a flow really underlies the dynamics.

For astrophysical purposes it is important that the method allows us to *quantify an irregular signal*, in terms of Lyapunov exponents and dimension, which has the potential novel benefit for extracting novel stellar constraints from irregular light-curves.

The low dimensionality 4 of the flow again suggests a simple physical picture of the pulsations, namely that the pulsations of R Scuti are the result of the nonlinear interaction of two vibrational "normal" modes of the star[10] (see also Kolláth & Buchler in this volume for additional evidence).

A preliminary study of another star of the same type, but not quite as irregular, AC Her, again indicates a low dimensional dynamics of 4.[33]

Following the earlier discussion, we also note that in a first study[15] the orthonormal polynomials were constructed with the GS procedure. This led to difficulties, in particular concerning the highest polynomial degree as their figure 3 and associated comments indicate. (Numerical cancellations cause negative norms.) A subsequent more thorough study resorted to the LS procedure with the help of an SVD algorithm. This procedure is not only a factor of 10 faster, but it remains numerically stable for all values of d_E and P (cf. reference 10).

RECONSTRUCTION WITH A CONTROL PARAMETER DEPENDENCE

We have discussed some applications of global vector field reconstruction with *given* control parameters to experimental systems. An additional step can be taken in the direction of reconstructing models as a function of a control parameter dependence. Actually such a goal has already been reached with neural networks,[29] but a study of control parameter dependence with a global vector field reconstruction technique such as those discussed in the present paper is still very new.

Principle

We assume that the dynamical system of (**2**) is studied under the variation of a control parameter α taken from the control parameter vector $\boldsymbol{\mu}$. With the method described earlier, the standard system reconstructed from the recorded variable x with α-dependence reads as

$$\begin{cases} \dot{z}_1 = z_2 \\ \dot{z}_2 = z_3 \\ \dot{z}_3 = f_s(z_1, z_2, z_3, \alpha), \end{cases} \quad (26)$$

where the standard function f_s now depends on the control parameter α. The basis polynomials therefore involve monomials of the form $(z_1^i\, z_2^j\, z_3^k\, \alpha^l)$, generalizing the previous discussion. The error function is furthermore modified to account for the control parameter α (for more details, see reference 30).

Numerical Check

A numerical check of the validity of this approach is now given in the case of the Rössler system

$$\begin{cases} \dot{x} = -y - z \\ \dot{y} = x + ay \\ \dot{z} = b + z(x - c). \end{cases} \quad (27)$$

This system has been extensively studied along a line in the control parameter space defined by ($a \in [0.33, 0.557]$, $b = 2$, $c = 4$).[22] Moreover, we have shown that there exists a diffeomorphism between the y-induced attractor and the original Rössler attractor.[31] Consequently, the y-variable for the Rössler system provides a good checkpoint for the present extension. The exact standard function, which may be analytically derived, then has a polynomial form involving 51 monomials.[30]

The learning set of data is composed of four time series $\{y_i,\ i \in [1, N_c = 1000]\}$, generated for four different values of the control parameter a taken to be equal to 0.2, 0.2625, 0.325 and 0.3875. Let us remark that these a-values correspond to three different limit cycles of the period-doubling cascade and to a chaotic behavior just beyond the accumulation point at $a_\infty = 0.386$. The sampling rate of the time series is taken to be equal to $\delta t = 10^{-2}$ s. A good approximation of the standard function is easily found with $N_p = 51$ and $\Delta = 7\delta t$, for which the generalized error function is $E_r = 3.8 \times 10^{-7}$. The estimated coefficients of the reconstructed standard function are reported in Table II of reference 30.

The integration of the reconstructed model generates attractors which may be shown to be topologically equivalent to the original Rössler attractors. In order to check our model with respect to the control parameter dependence, let us now compare the bifurcation diagrams of the reconstructed and of the standard exact systems along the line $a \in [0.33, 0.557]$ (FIGS. 16 and 17). The agreement between both diagrams is very satisfactory indeed.

Preliminary Results from Experimental Data

We now turn to a discussion of preliminary results concerning the use of the previously described technique with control parameter dependence to the case of experimental data. The data arise from copper electrodissolution experiments in which the potential between the reference electrode and the copper rod is varied. These data (from J. Hudson and Z. Fei, University of Virginia) are made of 18 time series which have been recorded for different values of the potential, taken as the control parameter. The pseudo-period of the current oscillations is about 0.1 s and the time step δt is equal to 5×10^{-4} s. Thus, 200 points per pseudo-period are available. Each time series contains 5000 points and is smoothed.

The experimental and reconstructed bifurcation diagrams are displayed in FIGURES 18 and 19, respectively. Although the agreement is not perfect, there is a general resemblance which is certainly encouraging. In particular, the model gener-

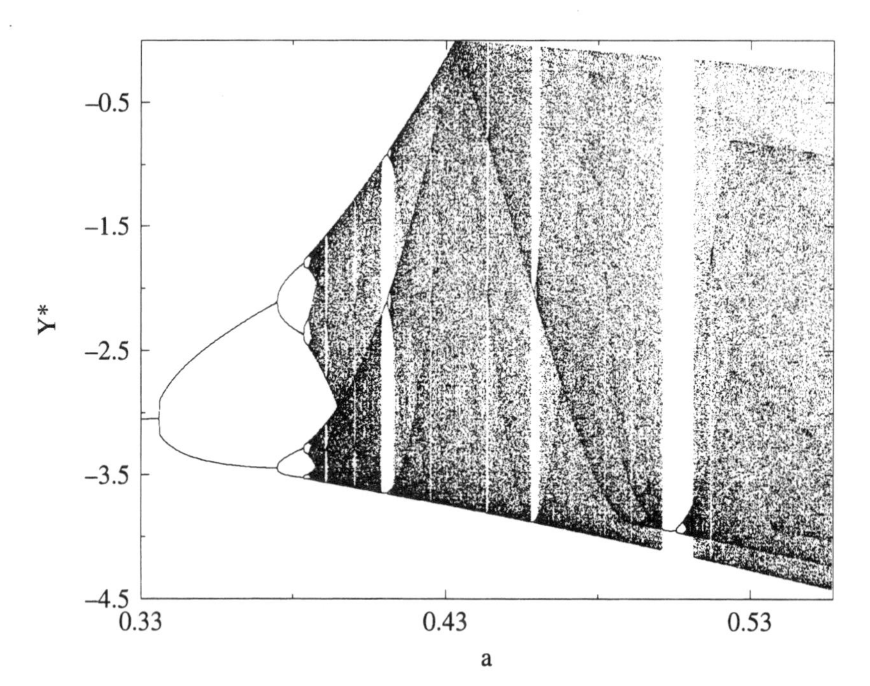

FIGURE 16. Bifurcation diagram for the original Rössler system with y-coordinate.

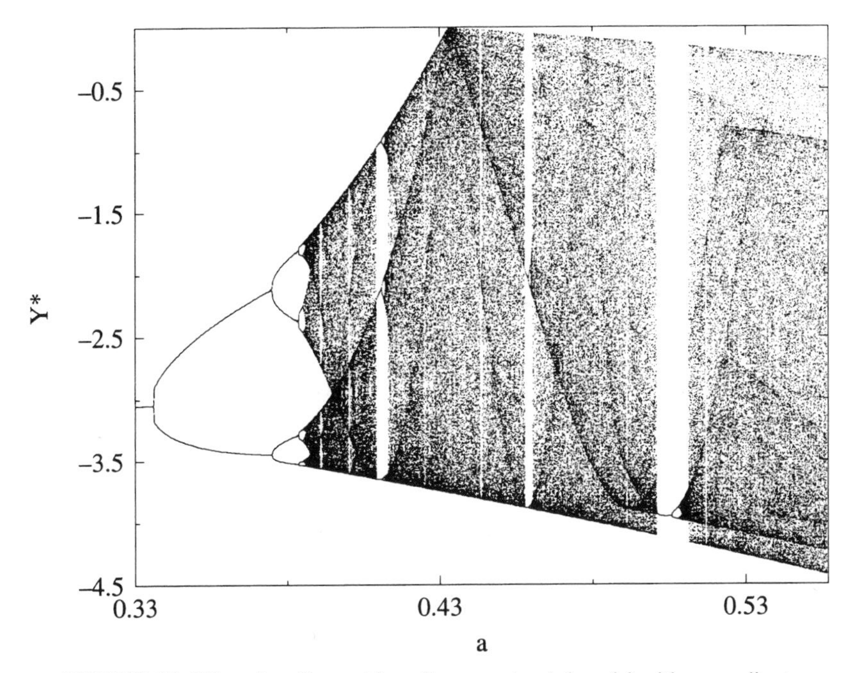

FIGURE 17. Bifurcation diagram from the reconstructed model, with y-coordinate.

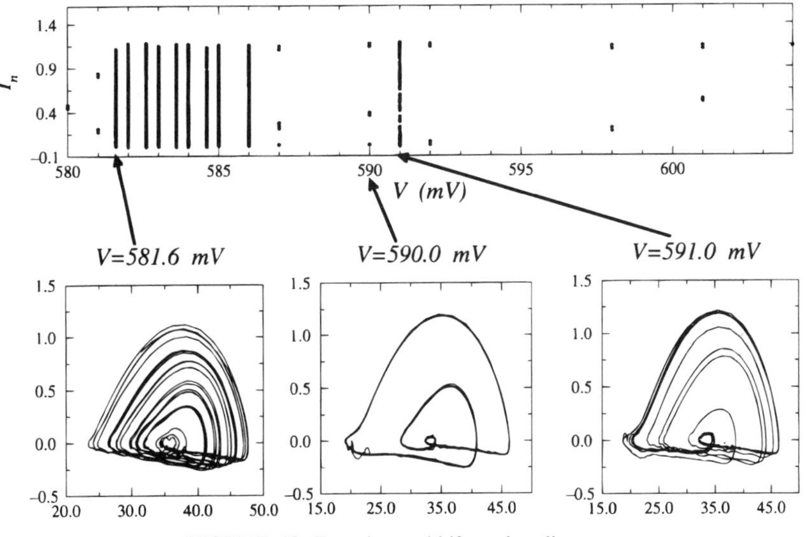

FIGURE 18. Experimental bifurcation diagram.

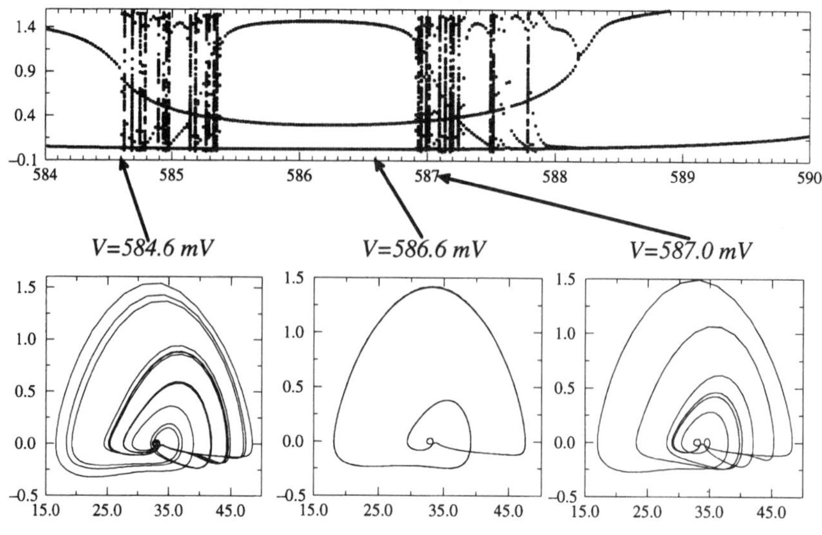

FIGURE 19. Reconstructed bifurcation diagram.

ates a sequence of behavior which is similar to the experimental one. We hope that a further effort will significantly improve these preliminary results.

CONCLUSIONS

Global vector field reconstructions are a topic of growing interest. They allow one to obtain a global model from measurements carried out on a *single* variable scalar time series. In other words, they provide relevant information on variables from which no measurements have been taken. Several variants of this technique have been reviewed and applications have been discussed. It is important to remark that global vector field reconstruction is nowadays developed enough to allow applications to (short, noisy) experimental data. We have also demonstrated that a control parameter dependence can be included in the technique. Then, measurements on one variable for a few values of the control parameter may allow us to get information on variables and for control parameter values which were not experimentally investigated. We may imagine many uses of the phenomenological models so obtained, such as for prediction, encoding of long time series, and control, to only quote a few key words, and it is expected that the interest for this topic will continue to grow.

ACKNOWLEDGMENTS

We wish to thank J. Hudson and Z. Fei for their beautiful data from the copper electrodissolution and N. Rul'kov for the data from the electronic circuit. We would also like to thank these coworkers for their helpful discussions, and Janet Mattei for the AAVSO observational data.

REFERENCES

1. ABARBANEL, H. D. I., R. BROWN, J. J. SIDOROWICH & L. SH. TSIMRING. 1993. Rev. Mod. Phys. **65:** 1331.
2. WEIGEND, A. S. & N. A. GERSHENFELD. 1994. Time Series Prediction. Addison-Wesley. Reading, MA.
3. PACKARD, N. H., J. P. CRUTCHFIELD, J. D. FARMER & R. S. SHAW. 1980. Phys. Rev. Lett. **45:** 712.
4. TAKENS, F. 1981. *In* Dynamical Systems and Turbulence, Warwick 1980, Lecture Notes in Mathematics. D. A. Rand & L. S. Young, Eds. Vol. 898: 366. Springer-Verlag. New York.
5. SAUER, T., J. A. YORKE & M. CASDAGLI. 1991. J. Stat. Phys. **65:** 5.
6. GIBSON, J. F., J. D. FARMER, M. CASDAGLI & S. EUBANK. 1992. Physica D **57:** 1.
7. FRASER, A. M. 1989. IEEE Trans. Info. Theory **35:** 245.
8. BROWN, R., N. F. RUL'KOV & E. R. TRACY. 1994. Phys. Rev. E **49:** 3784.
9. SERRE, T., Z. KOLLÁTH & J. R. BUCHLER. 1995. Astron. Astrophys. In press.
10. BUCHLER, J. R., Z. KOLLÁTH, T. SERRE & J. MATTEI. 1996. Astrophys. J. In press.
11. SERRE, T., Z. KOLLÁTH & J. R. BUCHLER. 1996. Astron. Astrophys. In press.
12. GOUESBET, G. & C. LETELLIER. 1994. Phys. Rev. E **49:** 4955. See also references therein discussing the use of models based on rational functions.
13. RICE, J. R. 1964. The Approximation of Functions, Vol. 1. Addison-Wesley. Reading, MA. Also: 1969. Vol. 2.
14. GIONA, M., F. LENTINI & V. CIMAGALLI. 1991. Phys. Rev. A **44:** 3496.
15. BUCHLER, J. R., T. SERRE, Z. KOLLÁTH & J. MATTEI. 1995. Phys. Rev. Lett. **74:** 842.
16. LETELLIER, C., E. RINGUET, B. MAHEU, J. MAQUET & G. GOUESBET. 1995. 4e Journées Européennes de Thermodynamique Contemporaine, Nancy, to be reprinted in Entropie.
17. PRESS, W. H., B. P. FLANNERY, S. A. TEUKOLSKY & W. T. VETTERLING. 1988. Numerical Recipes. Cambridge University Press. Cambridge, U.K.
18. JUDD, K. & A. MEES. 1995. Physica D **82:** 426.
19. TUFILLARO, N. B., P. WYCKOFF, R. BROWN, T. SCHREIBER & T. MOLTENO. 1995. Phys. Rev. E **51:** 164.
20. TUFILLARO, N. B., T. ABBOTT & J. REILLY. 1992. An Experimental Approach to Nonlinear Dynamics and Chaos. Addison-Wesley. Reading, MA.
21. BROWN, R., N. RUL'KOV & N. B. TUFILLARO. 1994. Phys. Rev. E **50:** 4488.
22. LETELLIER, C., P. DUTERTRE & B. MAHEU. 1995. Chaos **5:** 271.
23. MINDLIN, G. B., H. G. SOLARI, M. A. NATIELLO, R. GILMORE & X. J. HOU. 1991. J. Nonlin. Sci. **1:** 147.
24. LETELLIER, C. & G. GOUESBET. 1997. Topological structure of chaotic systems. This issue.
25. LETELLIER, C., L. LE SCELLER, E. MARÉCHAL, P. DUTERTRE, B. MAHEU, G. GOUESBET, Z. FEI & J. L. HUDSON. 1995. Phys. Rev. E **51:** 4262.
26. LETELLIER, C., L. LE SCELLER, P. DUTERTRE, G. GOUESBET, Z. FEI & J. L. HUDSON. 1995. J. Phys. Chem. **99:** 2016.
27. LETELLIER, C., L. LE SCELLER, G. GOUESBET, Z. FEI & J. L. HUDSON. 1996. Topological analysis and global vector field reconstruction of multi-branched chaos: the case of iron electrodissolution. In preparation.
28. KOVÁCS, G. & J. R. BUCHLER. 1988. Ap. J. **334:** 971.
29. RICO-MARTÍNEZ, R., K. KRISCHER, I. G. KEVREKIDIS, M. C. KUBE & J. L. HUDSON. 1992. Chem. Eng. Comm. **118:** 25.
30. LE SCELLER, L., C. LETELLIER & G. GOUESBET. 1996. Phys. Lett. A **211:** 211.

31. LETELLIER, C. & G. GOUESBET. 1996. J. Phys. II (France) **6:** 1615.
32. GRASSBERGER, P. & I. PROCCACIA. 1983. Physica D **9:** 189.
33. BUCHLER, J. R., Z. KOLLÁTH & T. SERRE. 1995. Ann. N.Y. Acad. Sci. **773:** 1.
34. BROOMHEAD, D. S. & G. P. KING. 1986. Physica D **20:** 217.
35. MOSKALIK, P. & J. R. BUCHLER. 1990. Ap. J. **355:** 590.

Topological Structure of Chaotic Systems

C. LETELLIER AND G. GOUESBET

LESP/URA CNRS 230/INSA de Rouen/BP 08
Place Emile Blondel
76131 Mont Saint-Aignan Cédex, France

INTRODUCTION

It is now well known that chaotic dynamics may be generated by many kinds of experimental systems from various fields of science such as hydrodynamics, astrophysics, chemistry, biology, electronics, etc. It is therefore of particular interest to possess an extended toolbox to characterize the dynamics of the studied physical systems. From the pioneering paper by Lorenz,[1] which is often considered as providing the first example of a deterministic system generating a long-term unpredictable behavior, many analyses have been developed which may be separated into two branches.

First, the geometrical properties of attractors on which the asymptotic motion settles down have been used. Such methods are based on a notion of distance in the reconstructed state space according to the pioneering paper by Packard *et al.*[2] and the very mathematical paper by Takens.[3] Geometrical properties present the attractive feature of being usable in an n-dimensional space. For instance, the correlation dimension[4] and Lyapunov exponents[5] have been commonly used to characterize attractors. The knowledge of Lyapunov exponents gives an indication about the predictability of a system and also on the number of active dynamical degrees of freedom involved in it.[6,7] Nevertheless, the computation of such geometrical quantities is very sensitive to noise perturbations.

A second approach, based on the knowledge of the population of periodic orbits which constitutes the skeleton of the attractor, allows us to quantify how the chaotic behavior is developed. In particular, a partition of the attractor induces an identification of each orbit by a symbolic sequence.[8,9] This encoding, called the symbolic dynamics, allows us to predict a theoretical order of creation of periodic orbits under the change of a control parameter.[10]

In the earlier 1990s, the idea arose that topological properties could be used as a complement to the characterization of geometrical properties. Indeed, it has been demonstrated that the relative organization of periodic orbits provides a fine characterization of an attractor.[11–13] The topological characterization of an attractor is based on topological invariants, namely linking numbers, which are robust under control parameter changes. The topological characterization associated with the use of the symbolic dynamics then constitutes a powerful tool to study attractors. In particular, it is robust with respect to noise perturbations since it essentially involves the low period orbits which are not really affected by noise and also because results are expressed in terms of integers which are not very sensitive to measurement errors. Unfortunately, such a characterization is nowadays restricted to tridimensional spaces.

In this chapter we focus our attention on the applications of topological characterization in the study of dynamical systems. A brief review of the concepts used in characterizing the topology of attractors is given, followed by an introduction to a specific procedure which is required when the attractor possesses symmetry properties. A further section then exemplifies the concepts by discussing a few applications on physical systems, namely an electrodissolution experiment, a model of W Vir Cepheid and an experimental electronic circuit. Finally, the conclusions are presented.

TOPOLOGICAL CHARACTERIZATION OF ATTRACTORS

In the last few years several works discussed the topological description of chaotic attractors. In particular, the idea has arisen that an attractor can be described by the population of periodic orbits, their related symbolic dynamics and their linking numbers.[11] In three-dimensional cases, periodic orbits may be viewed as knots[13] and, consequently, they are robust with respect to smooth parameter changes allowing the definition of topological invariants under isotopy (continuous deformation).

The topological approach is based on the relative organization of periodic orbits in the state space. We now present the basic concepts of topological characterization and symbolic dynamics. For the sake of simplicity, we use the well-known Rössler attractor as an example.

Template

The Rössler system[14] reads as:

$$\begin{cases} \dot{x} = -y - z \\ \dot{y} = x + ay \\ \dot{z} = b + z(x - c), \end{cases} \qquad (1)$$

where a, b, c are the control parameters. When $(a, b, c) = (0.398, 2, 4)$ the asymptotic motion settles down onto a strange chaotic attractor (FIG. 1).

The attractor may be viewed as a simply stretched and folded band. Two different stripes may be exhibited from this attractor (FIG. 2): (1) one which is located in the center of the attractor is a very simple stripe without any π-twist (FIG. 2a) and (2) a second stripe which presents a negative π-twist (FIG. 2b) and is therefore similar to a Moebius band. We may then distinguish two topological regions on the attractor. Following a pioneering paper by Birman and Williams,[15] it has been shown[11,13,16] that a template which encodes the topological properties of an attractor may be built in tridimensional state spaces. Such a template is a convenient view of the attractor to exhibit the different stripes within the attractor and their relative organization. From the Rössler attractor, a template constituted by two stripes is then extracted and displayed in FIGURE 3(a). The band is split into two stripes, one without any π-twist and one with a negative π-twist (FIG. 3a). Due to a standard

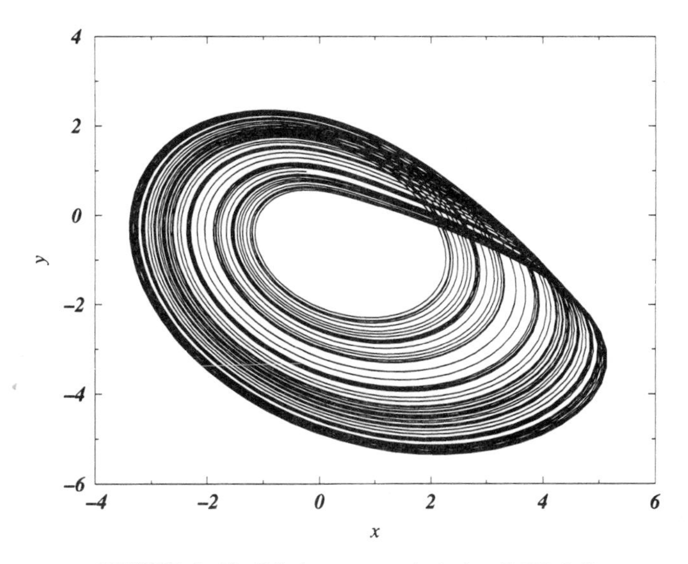

FIGURE 1. The Rössler attractor $(a, b, c) = (0.398, 2, 4)$.

insertion convention,[16] stripes must be reinjected in the bottom band from back to front, and from left to right. Consequently, a permutation between the stripes is required, thereby leading to FIGURE 3(b).

This convention allows an unambiguous description of the template by defining a linking matrix[16] as follows. Diagonal elements $M(i, i)$ are equal to the number of π-twists of the ith stripe and off-diagonal elements $M(i, j)$, $i \neq j$ are given by the algebraic number of intersections between the ith and the jth stripes. One then may

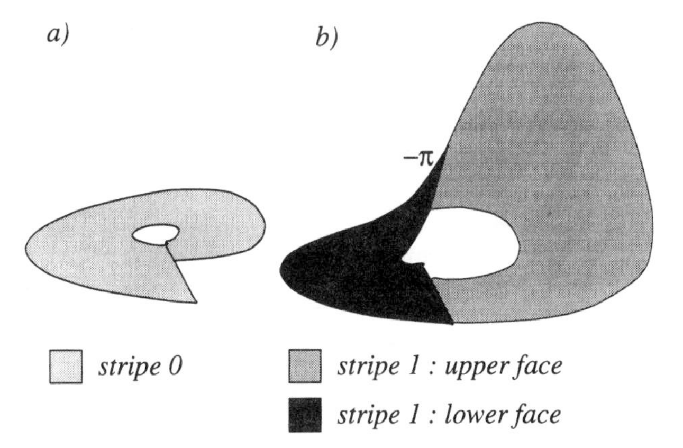

FIGURE 2. The two stripes of the Rössler attractor.

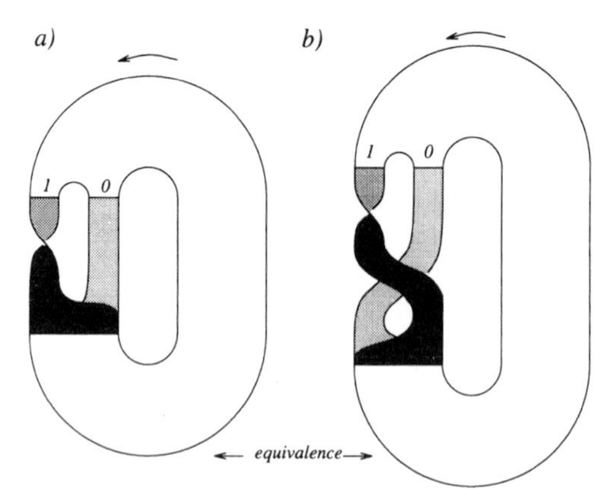

FIGURE 3. Template of the Rössler attractor. A permutation between the stripes is required to meet the standard insertion convention.

check that the Rössler template is defined by the linking matrix:

$$M \equiv \begin{pmatrix} 0 & -1 \\ -1 & -1 \end{pmatrix}. \tag{2}$$

Each stripe may be labelled: symbol 0 designs the simple stripe, while symbol 1 is associated with the stripe which presents a negative π-twist. In this way, trajectories are encoded by a string of 0's and 1's. In particular, periodic orbits may then be encoded in a one-to-one way. We have here defined a symbolic dynamics. This procedure needs a precise partition of the attractor which is given by a first-return map to a Poincaré section.

First-Return Map

A Poincaré section is here defined as the set of intersections of a chaotic trajectory with a plane transverse to the flow. For the Rössler system, such a Poincaré section is suitably defined as

$$P \equiv \{(y, z) \in \mathbb{R}^2 \,|\, x = x_-, \dot{x} > 0\}, \tag{3}$$

where $x_- = c - \sqrt{c^2 - 4ab}/2$ is the x-coordinate of the central fixed point.

The first-return map is then computed with the y-variable and displayed in FIGURE 4. It presents two monotonic branches: an increasing branch associated with stripe 0 and a decreasing branch associated with stripe 1. The critical point y_c which separates the branches precisely defines the partition. In our case, $y_c = -3.04$. Thus, each intersection y_i with the Poincaré plane corresponds to a code $K(y_i)$:

$$K(y_i) = \begin{cases} 0 & \text{if} \quad y_i > y_c \\ 1 & \text{if} \quad y_i < y_c. \end{cases} \tag{4}$$

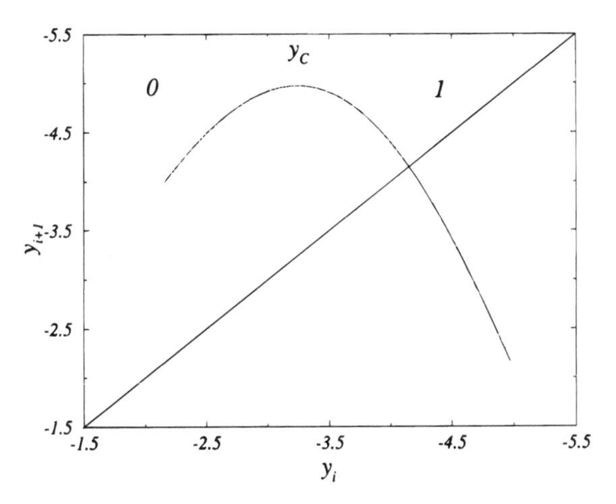

FIGURE 4. First-return map of the Rössler system: $(a, b, c) = (0.398, 2, 4)$.

Once periodic orbits are extracted by detecting the close returns in a Poincaré section by a Newton–Raphson iteration scheme, periodic points in this Poincaré section may be encoded. An orbit of period p has p periodic points and is represented by a string S of p-codes:

$$S = K(y_1)K(y_2) \ldots K(y_p)$$

where y_i's are the y-coordinates of the periodic points.

Unimodal Order

Each period-p point is represented by a symbolic sequence of p symbols. The ith point of a period-p orbit is labelled by the string:

$$S_i = K(y_i)K(y_{i+1}) \ldots K(y_p)K(y_1) \ldots K(y_{i-1}). \tag{5}$$

All periodic points are then ordered by the unimodal order.[9,10]

Definition 1: The unimodal order $\prec_1$ on the symbol set 0, 1 is defined as follows. Let us consider two symbolic sequences:

$$W_1 = \sigma_1\sigma_2 \ldots \sigma_k\sigma_{k+1} \ldots$$

and

$$W_2 = \tau_1\tau_2 \ldots \tau_k\tau_{k+1} \ldots$$

where σ_i's and τ_j's designate the codes. Suppose $\sigma_i = \tau_i$ for all $i < k$ and $\sigma_k \neq \tau_k$. Let $W^* = \sigma_1 \cdots \sigma_{k-1} = \tau_1 \cdots \tau_{k-1}$ be the common part between W_1 and W_2. We set that a string $\sigma_1\sigma_2 \cdots \sigma_{k-1}$ is even (odd) if the sum $\sum_{i=1}^{k-1} \sigma_i$ is even (odd), and that W^*

is even when no common part is found between W_1 and W_2. Then, we have:

$$\begin{cases} W_1 \prec_1 W_2 & \text{if} \quad W^* \quad \text{is even} \quad \text{and} \quad \sigma_k < \tau_k \\ W_1 \prec_1 W_2 & \text{if} \quad W^* \quad \text{is odd} \quad \text{and} \quad \tau_k < \sigma_k \\ W_2 \prec_1 W_1 & \text{if} \quad W^* \quad \text{is odd} \quad \text{and} \quad \sigma_k < \tau_k \\ W_2 \prec_1 W_1 & \text{if} \quad W^* \quad \text{is even} \quad \text{and} \quad \tau_k < \sigma_k . \end{cases}$$

When $W_1 \prec_1 W_2$, we say that W_2 implies W_1.

A period-p orbit will be denoted by the symbolic sequence W_i (without any parentheses) which implies the $(p-1)$ others. This sequence is noted (W_i), included between parentheses, and here called *orbital sequence.* In a similar way, two orbital sequences may be ordered following the unimodal order. When orbital sequence (W_1) implies the orbital sequence (W_2), we say that (W_1) *forces* (W_2) and we note $(W_2) \prec_2 (W_1)$ where $\prec_2$ is the forcing order.

By this way, all periodic orbits are ordered. The orbital sequence which forces all orbital sequences extracted from the attractor is called the *kneading sequence.* Within the Rössler attractor, the kneading sequence (among the orbits of period less than 12) is found to be (10111101010).[17] All orbits forced by the kneading sequence are then found to be present within the attractor up to period 11.

Symbolic Plane

With experimental systems, however, an orbit spectrum may be determined within the limits imposed by the available time series length. Indeed, in such cases, due to the limited amount of data and to the influence of external noise, the orbit spectrum is rarely well-known. As shown by Tufillaro *et al.*[18] the retrievable information on the population of periodic orbits crucially depends on the length of the time series. Consequently, the determination of the kneading sequence is rather unaccurate when using short experimental time series.

Fang[19] has shown that an empirical procedure (also used by Tufillaro *et al.*[18]) may, however, exhibit the pruning front and, consequently, the kneading sequence. To explain this procedure, let us first recall that a chaotic trajectory forms a string

$$\mathbf{s} = \ldots \sigma_{-3} \sigma_{-2} \sigma_{-1} \sigma_0 \sigma_1 \sigma_2 \sigma_3 \ldots$$

where σ_0 is the present, σ_{-i}'s the past and σ_i's the future $(i > 0)$.

Symbolic coordinates which span a symbolic plane are then defined on the future and the past as follows:

$$\begin{cases} x_\sigma(\mathbf{s}) = \sum_{i=1}^{D} \frac{b_i}{2^i} \quad \text{where} \quad b_i = \sum_{j=1}^{i} \sigma_j \pmod 2 \\ y_\sigma(\mathbf{s}) = \sum_{i=0}^{D} \frac{c_i}{2^i} \quad \text{where} \quad c_i = \sum_{j=0}^{i-1} \sigma_{-j} \pmod 2, \end{cases} \tag{6}$$

where

$$\mathbf{s} = \sigma_{-D} \ldots \sigma_{-3} \sigma_{-2} \sigma_{-1} \sigma_0 \sigma_1 \sigma_2 \sigma_3 \ldots \sigma_D .$$

If **s** is an infinite symbol string generated by a chaotic orbit, then D is infinity in the above definition. However, since we are dealing with finite data sets, Tufillaro *et al.*[18] approximate the symbolic plane coordinates of a point by taking $D = 16$. In this way, we can use a finite symbol string from a chaotic trajectory to generate a sequence of points on the symbolic plane. The symbolic plane for the Rössler attractor is given in FIGURE 5. In the present case of an orbit spectrum governed by the unimodal order, the pruning front is suitably estimated by a line.[19]

The symbolic coordinate x_σ of the pruning front allows us to determine the kneading sequence. Indeed, after having computed the orbital sequences of periodic orbits, the kneading sequence is associated with the orbital sequence whose x_σ is closest to the pruning front. For reference, symbolic coordinates x_σ of orbital sequences, for orbits with a period less than 9, are reported in Table I of reference 27. In the Rössler case, the pruning front is located at $x_\sigma = 0.8376$. From the orbit spectrum of the Rössler attractor, the kneading sequence is then found to be (10111101010) whose symbolic coordinate is 0.8375, in good agreement with the pruning front location, when only orbits with periods smaller than 12 are considered.

Template Validation

A template of the Rössler attractor has been given earlier in the chapter and the orbit spectrum is extracted. The template must now be checked by comparing the

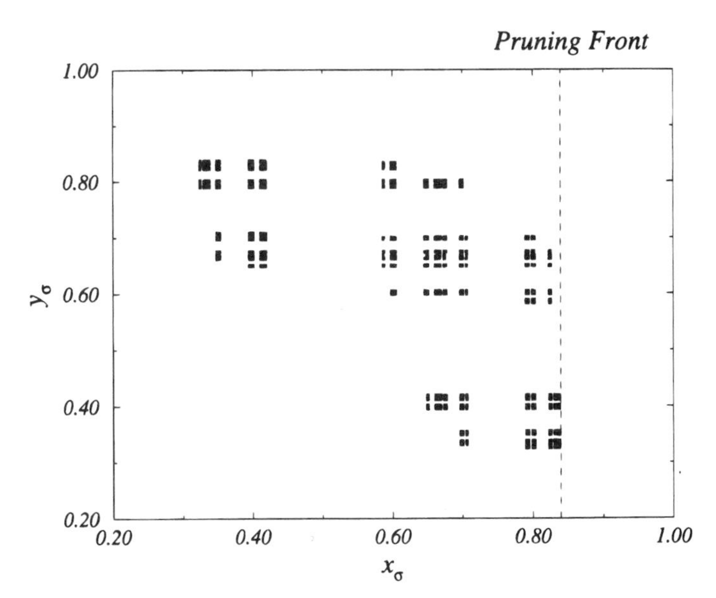

FIGURE 5. Symbolic plane of the Rössler attractor: the orbit spectrum is governed by the unimodal order as shown by the pruning front which is well estimated by a line.

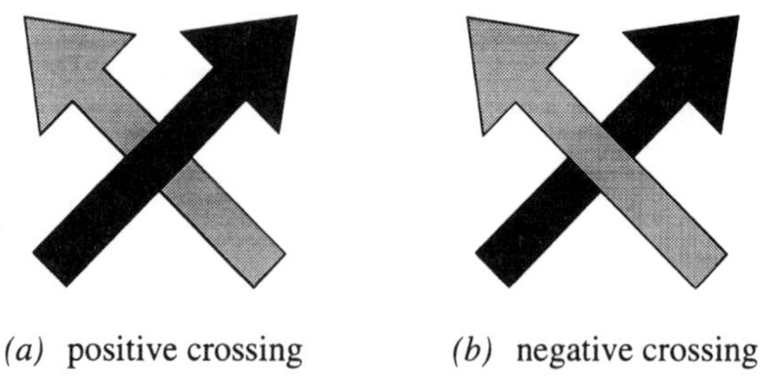

FIGURE 6. Crossing conventions.

linking numbers predicted by the template and the ones counted on the attractor. The linking number $L(N_1, N_2)$ of an orbit pair is given by the half-sum of the oriented crossings (following the convention, given in FIGURE 6, used by Melvin and Tufillaro[16]) on a regular plane projection of orbits N_1 and N_2. For example, linking number $L(1011, 1)$ is equal to -2 (FIGURE 7).

With utmost rigor, a few linking numbers are needed to completely check the template. As the template which carries the periodic orbits is identified, the organization of the orbits within the attractor is known. For a complete discussion about equivalence between periodic orbits embedded within a strange attractor and orbits of the template, see references 12 and 13.

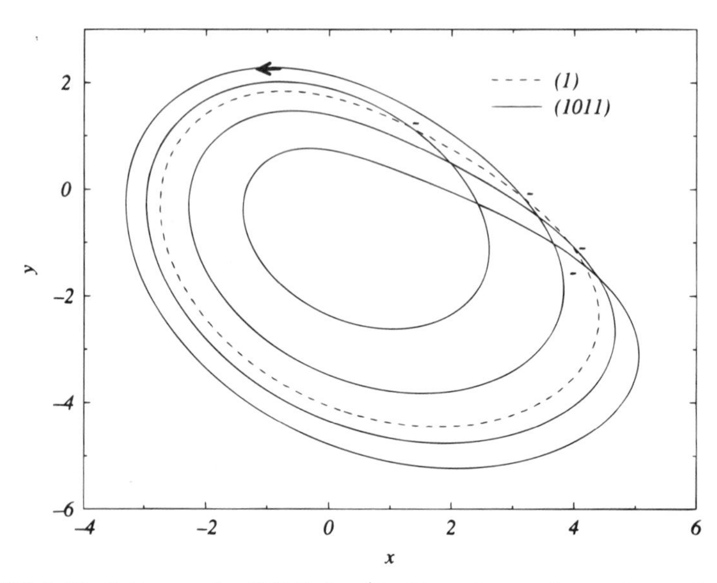

FIGURE 7. The linking number $L(1011, 1) = \frac{1}{2}[-4] = -2$ counted on a plane projection of the orbit couple (1011, 1). Crossings are signed by inspection on the third coordinate.

TOPOLOGY OF EQUIVARIANT SYSTEMS

The procedure to characterize the topology of a chaotic attractor has been discussed in the previous section. This procedure exhibits the topological properties of attractors generated by any vector field. Nevertheless, when the vector field is equivariant, the attractor presents symmetry properties which may be used to obtain a more representative characterization. A specific procedure is then required, as discussed below in the case of the Lorenz system.

Physical Interpretation of the Specific Procedure

The Lorenz equations describe the Rayleigh–Bénard convection which arises when a fluid is placed in a small cell which limits the number of rolls after the onset of instability from the basic motionless state. Relying on a truncature of the governing equations in terms of a few modes, Lorenz proposes a set of three ordinary derivative equations which reads as:[1]

$$\begin{cases} \dot{x} = \sigma(y - x) \\ \dot{y} = Rx - y - xz \\ \dot{z} = -bz + xy, \end{cases} \tag{7}$$

where $\sigma = 10$ is the Prandtl number of the fluid, R is a reduced dimensionless Rayleigh number and $b = 8/3$ is an aspect ratio of the cell. The three variables x, y and z designate physical quantities as follows: First, x is proportional to the circulatory fluid flow velocity. By convention, positive x's (resp. negative x's) correspond to clockwise (resp. anticlockwise) rotation. Next, y characterizes the temperature difference between ascending and descending fluid elements. As for x, a convention concerning the sign must be chosen, consistently with the rotation convention. Then, y is taken to be proportional to the temperature difference between the left part and the right part of the roll, i.e., $y > 0$ when the left part is warmer than the right part. Last, z is proportional to the deviation of the vertical temperature profile from its basic equilibrium value. Depending on the signs of the variables x and y, four configurations are obtained (FIG. 8).

Nevertheless, there are only two kinds of physical processes. One circulatory motion is governed by heat convection (FIG. 8a) and one circulatory motion is driven by hydrodynamical instabilities (FIG. 8b). For each of these motions, we have two configurations which, to some extent, are physically equivalent, one being converted to the other by a change in the sign conventions. This may easily be viewed from the structure of the Lorenz system when R is a bit greater than 1. For such an R-value, the two fixed points $F_{\pm}$ given by

$$\begin{cases} x_{\pm} = \pm\sqrt{b(R-1)} \\ y_{\pm} = \pm\sqrt{b(R-1)} \\ z_{\pm} = R - 1 \end{cases} \tag{8}$$

are stable focus points, i.e., the asymptotic behavior in the state space settles down on one of the two fixed points. Each fixed point physically corresponds to a convection roll whose rotation sense depends on the initial conditions. In both cases,

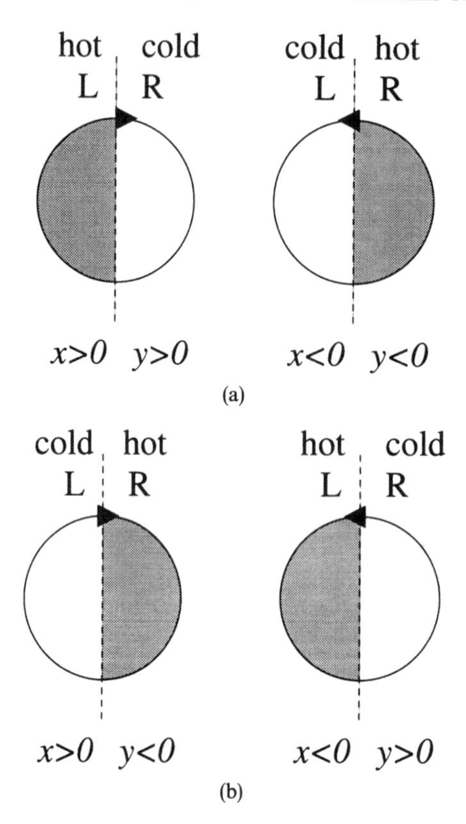

FIGURE 8. Four configurations of rotation may be distinguished, depending on the signs of x and y, but there are only two kinds of physical processes.

rotation is governed by the heat convection process as experimentally observed.[20] Therefore, dynamically speaking, we take the point of view that there is no difference between both asymptotic solutions since they are physically equivalent.

These observations are also in agreement with the equations insofar as the vector field f defined by the Lorenz system is equivariant, i.e., $\mathbf{f}(\gamma\mathbf{X}(t)) = \gamma\mathbf{f}(\mathbf{X}(t))$ where $\mathbf{X} = (x, y, z)$ and γ is a matrix defining the equivariance. In the Lorenz case, γ reads as:

$$\gamma \equiv \begin{bmatrix} -1 & 0 & 0 \\ 0 & -1 & 0 \\ 0 & 0 & 1 \end{bmatrix} \tag{9}$$

and defines a rotation by $\pm\pi$ (the rotation sign is irrelevant). The order of symmetry is two, since $\gamma^2 = I$ where I designates the identity matrix. When $R > 24.74$, the asymptotic motion settles down onto a strange attractor which presents a symmetry property (FIG. 9a). We are now concerned by the dynamical analysis of this attractor, taking into account the equivariance of the vector field.

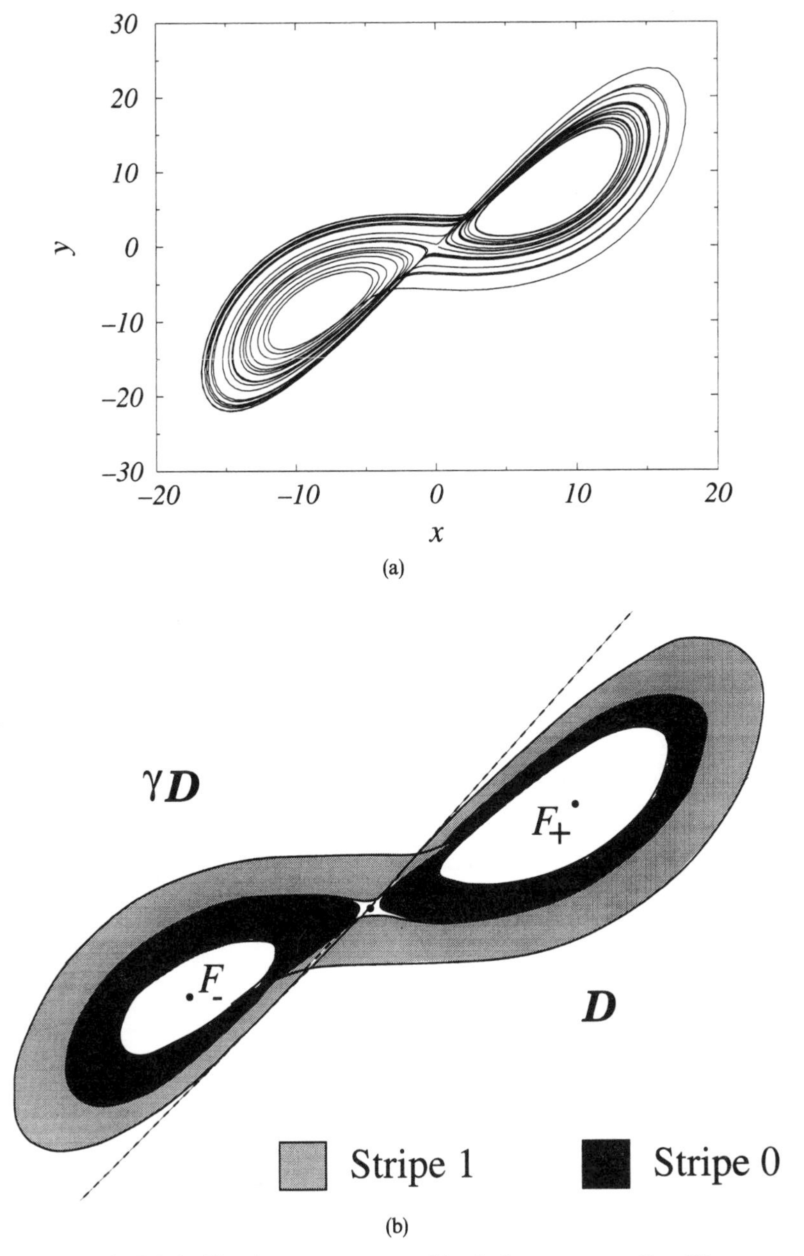

FIGURE 9. Chaotic attractor governed by the Lorenz system ($R = 28$).

On the Whole State Space

Starting from the plane projection of the attractor (FIG. 9a), a mask is built and displayed in FIGURE 9(b). Such a mask exhibits two local torsions equal to $+\pi$. It may be synthesized by a double template displayed in FIGURE 10. Each sub-template
may be associated with a wing.

By inspection on the xy-plane projection, one may remark that projection of the trajectory in the same wing is essentially associated with a heat convection, while the transition from one wing to the other is characterized by an excursion in the quadrants where x and y have opposite signs, i.e., by a motion driven by hydrodynamical instabilities. Consequently, to these two kinds of behaviors may be associated two symbols in order to build a symbolic dynamics. Assigning symbol 0 to heat convection and symbol 1 to hydrodynamical instabilities, stripes of each sub-template are labelled 0 and 1, respectively. By using this symbolic dynamics, periodic orbits present within the attractor may be labelled. We showed that all symbolic sequences on $\{0, 1\}$ are associated with one orbit within the attractor up to period 8.[21]

Two kinds of periodic orbits may then be distinguished: (1) symmetric orbits (FIG. 11a), which are globally invariant under the action of γ, and (2) asymmetric orbits, which may be mapped to a symmetric configuration by the action of γ and therefore appear in a pair (FIG. 11b). Consequently, the two symmetric configurations of a pair of asymmetric orbits are both encoded by the same symbolic sequence. In other words, the behavior associated with a configuration cannot dynamically be distinguished from its symmetric configuration.

If we intend to characterize the dynamical behavior by inspection of the topological properties of the attractor, linking numbers must therefore not distinguish between the two configurations of an asymmetric orbit. Unfortunately this is not observed when one counts linking numbers between two asymmetric orbits. For instance, we displayed the plane projection of a configuration of orbit (1001) with the two configurations of the asymmetric orbit encoded by (101) (FIG. 12). It is

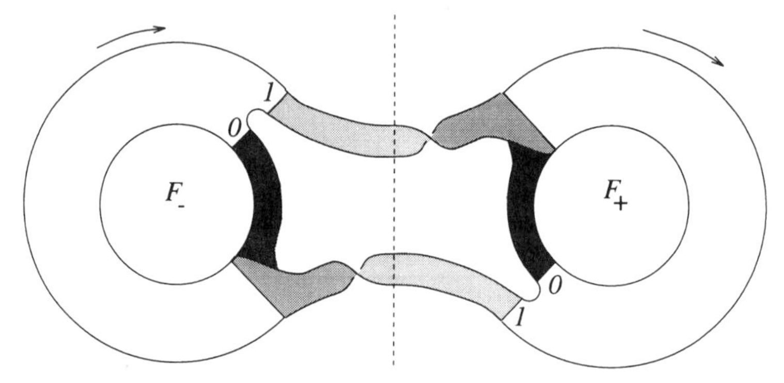

FIGURE 10. Double template of the Lorenz attractor. Reinjection of the trajectory in the same wing is encoded by 0 and transition from one wing to the other is encoded by 1 and associated with a positive π-twist.

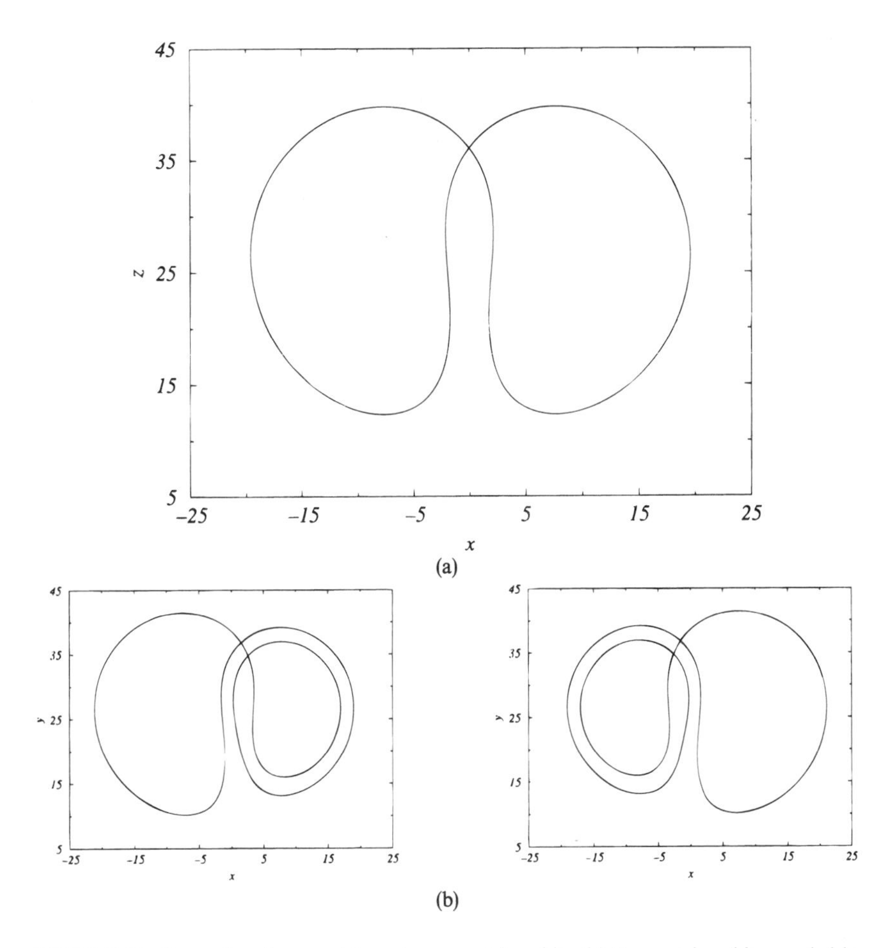

FIGURE 11. Examples of symmetric and asymmetric orbits. (**a**) Symmetric orbit encoded by (1). (**b**) Asymmetric orbit encoded by (101).

found that the linking number $L(1001, 101)$ depends on the used configuration of orbits. Consequently, such a topological characterization on the whole state space does not provide a coherent tool, with respect to the symbolic dynamics, to understand the dynamical behavior of the Lorenz system.

On a Fundamental Domain

In the previous section we showed that the topological characterization of the whole attractor does not take into account the equivariance of the Lorenz model. However, as pointed out by Cvitanović and Eckhardt,[22] symmetric attractors may

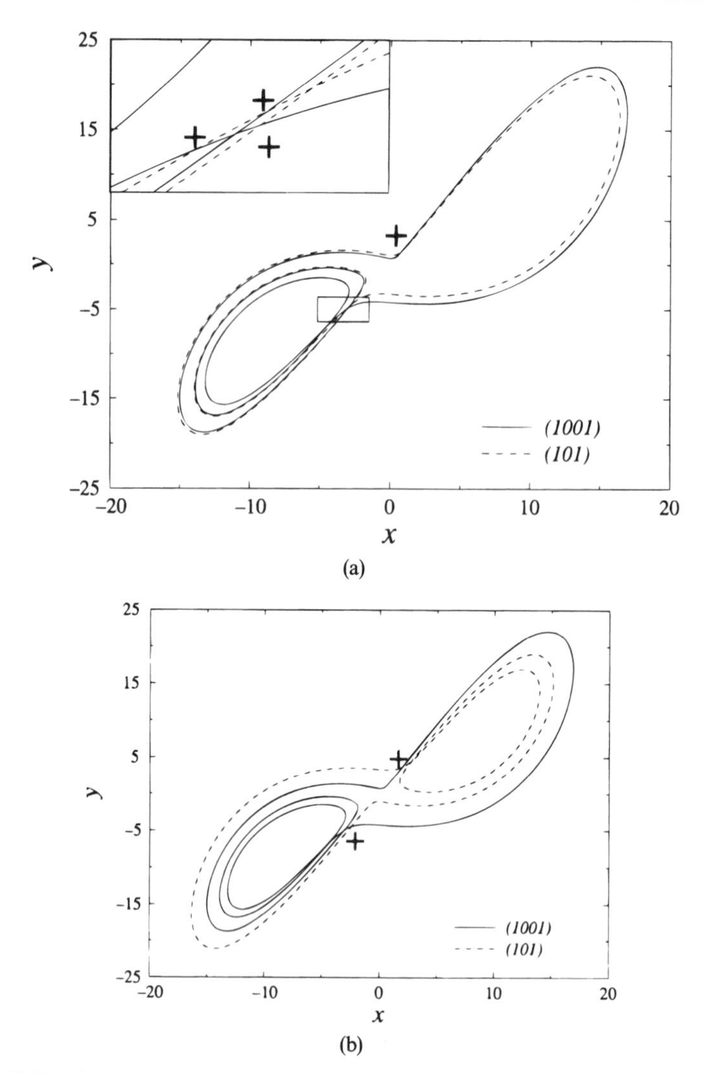

FIGURE 12. The linking number between asymmetric orbits encoded by (1001) and (101) depends on the chosen configuration. (**a**) $L(1001, 101) = +2$. (**b**) $L(1001, 101) = +1$.

be tasselated by a fundamental domain and many of its copies. In the Lorenz case, the symmetry being of order 2, the fundamental domain $\mathcal{D}$ and one copy are sufficient. Roughly speaking, a wing may be viewed as a good representation of the fundamental domain $\mathcal{D}$.

In order to build a template which synthesizes the topology of the fundamental domain, we must keep only one sub-template from the double template displayed in FIGURE 10. Cutting along the dashed line in FIGURE 13 and reinjecting the outcoming stripe 1 in the same wing by a rotation of π, a fundamental template is obtained.

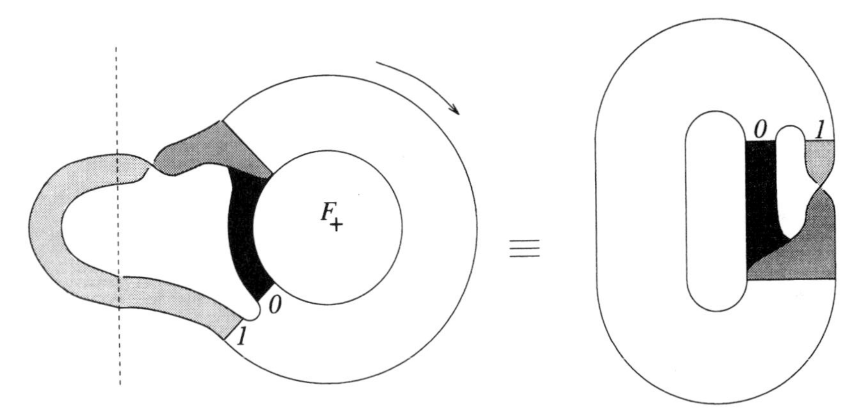

FIGURE 13. Fundamental template obtained by isolating a sub-template and gluing the outcoming stripe to the same wing by a rotation of π.

One may check that all relevant dynamical information is preserved on the fundamental template since the two stripes are still present. Stripe 0 is associated with a heat convection (reinjection of the trajectory in the same wing) and stripe 1 with hydrodynamical instabilities (transition from one wing to the other) characterized by a positive π-twist.

As topological properties are studied on the fundamental domain, a fundamental linking number $\mathscr{L}(N_1, N_2)$ between orbits N_1 and N_2 must be introduced. It is

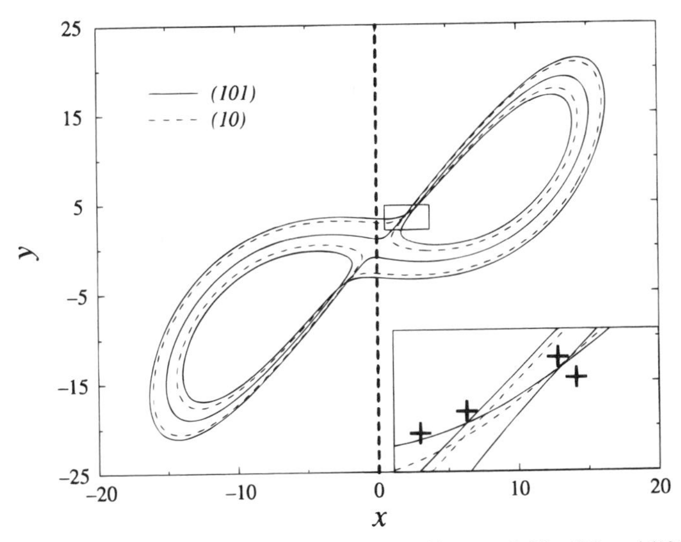

FIGURE 14. Fundamental linking number between orbits encoded by (10) and (101): $\mathscr{L}(101, 10) = +2$.

defined as the half-sum of the oriented crossings between N_1 and N_2 on the fundamental domain. As configurations of asymmetric orbits must not be distinguished, segments of both configurations are taken into account in order to evaluate the fundamental linking number. One may check that fundamental linking numbers $\mathscr{L}(N_1, N_2)$ predicted by the fundamental template are in agreement with fundamental linking numbers counted on the fundamental domain (i.e., a wing) on a regular plane projection.[21] We here give an example for orbits encoded by (10) and (101) (FIG. 14).

The fundamental linking number $\mathscr{L}(101, 10)$ is found to be equal to $+2$ which is indeed found to be in agreement with the template prediction. When a fundamental template is checked by counting oriented crossings on a fundamental domain, we say that the fundamental template is topologically equivalent to the whole attractor in a restricted sense.[23,24]

APPLICATIONS

Validation of Models

One of the main problems of interest in studying an experimental dynamical system is to develop automatic algorithms to extract a set of equations which models its dynamical behavior. This model is directly reconstructed from a time series. Many works have been devoted to such developments (see references 25 or 26 for recent papers and references therein). Thus, starting from an experimental time series, we now assume that a set of equations has indeed been extracted. A relevant task is hereafter to check the model with experimental data. We will discuss an example of validation by using the topological characterization.

The Experiment

The time series was obtained from dissolution current measurement during the potentiostatic electrodissolution of a rotating Cu electrode in phosphoric acid. The experimental set-up consisted of a rotating disc electrode which had a copper rod, 8.26 mm in diameter, embedded in a 2 cm diameter Teflon cylinder. The rotating speed was maintained at 4400 rpm. A cylindrical platinum net band (much larger than the disc) was put around the disc as a counter-electrode to get uniform potential and current distributions. The cell was a 500 ml flask with a side neck in which the capillary probe was fixed. The cell contained 250 ml of 85% phosphoric acid and a water bath was used to maintain its temperature at 20°C. A Potentiostat (Princeton Applied Research model 273) was used to regulate the potential for the working disc electrode with respect to the SCE and to monitor the current. The data were recorded at a frequency f_e of 1500 Hz using a 486 PC and a data acquisi-

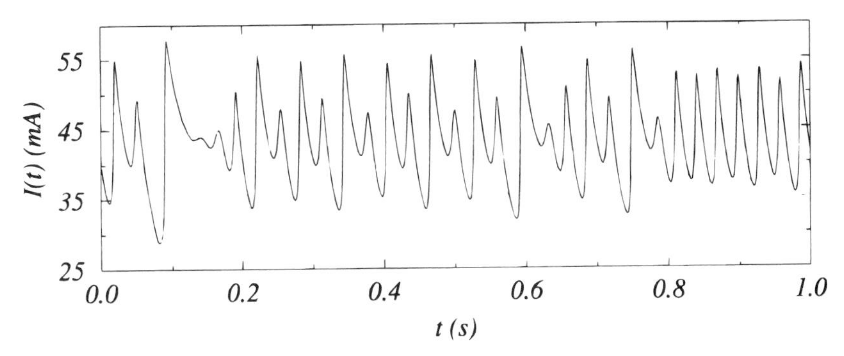

FIGURE 15. Current time series $I(t)$.

tion board (Model DAS-16, Keithley MetraByte's). The current time series $I(t)$ is displayed in FIGURE 15.

The state space of the electrodissolution may be reconstructed from the current time series $X = I(t)$ by using the successive time derivatives of X. The reconstructed state space is found to be tridimensional[27] and may then be spanned by

$$\begin{cases} X = I(t) \\ Y = \dot{I}(t) \\ Z = \ddot{I}(t). \end{cases} \quad \textbf{(10)}$$

In this space, the asymptotic motion settles down onto a chaotic attractor displayed in FIGURE 16. The so-called *copper* attractor is found to be characterized by the template displayed in FIGURE 17.[27,28]

The Reconstructed Set of Equations

Starting from the current time series, we are also able to obtain a system reading as:

$$\begin{cases} \dot{X} = Y \\ \dot{Y} = Z \\ \dot{Z} = F(X, Y, Z). \end{cases} \quad \textbf{(11)}$$

This model is built on the successive time derivatives of the recorded time series which is here taken as X. In this model, all the information we want to retrieve is contained in the so-called standard function $F(X, Y, Z)$ which is estimated by using a multivariate polynomial approximation on nets (see references 25 and 29 for a complete description for the method here used). An approximation $\tilde{F}$ to the function F may be written under the form:

$$\tilde{F}_s = \sum_{l=1}^{N_l} K_l P^l, \quad \textbf{(12)}$$

where $P^l = X^i Y^j Z^k$ and $\{K_l\}$ is the coefficient spectrum. A successful reconstruction has been obtained from the current time series with 26 terms in the function F (the

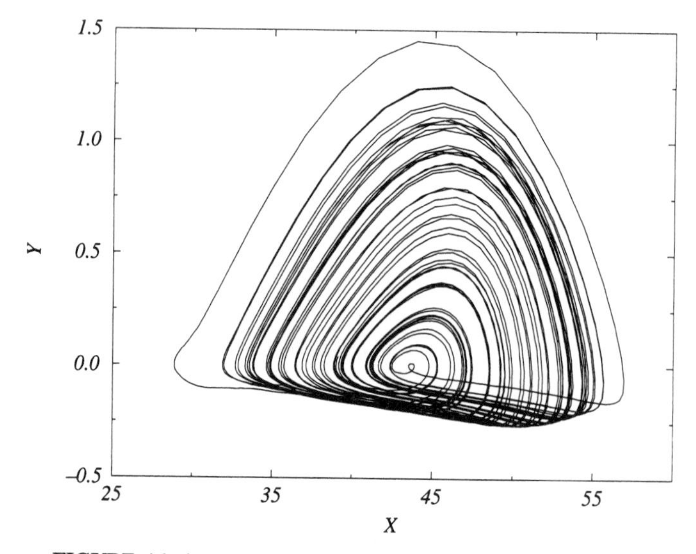

FIGURE 16. Attractor generated by the copper electrodissolution.

whole coefficient spectrum is given in reference 27). By numerically integrating this reconstructed model, a reconstructed attractor (FIG. 18) is obtained which looks rather similar to the copper attractor.

The model will be checked (from a topological point of view) if the topological properties of this attractor A_R are found to be the same as the ones of the original copper attractor. This is easily achieved by counting linking numbers of orbit pairs and checking that they are in agreement with the linking numbers predicted by the

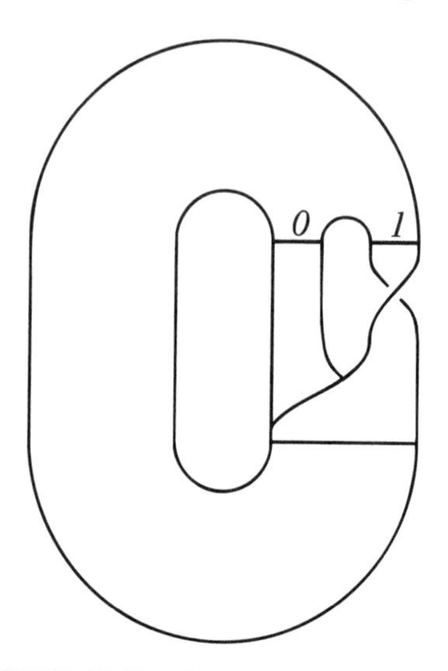

FIGURE 17. Template of the copper attractor.

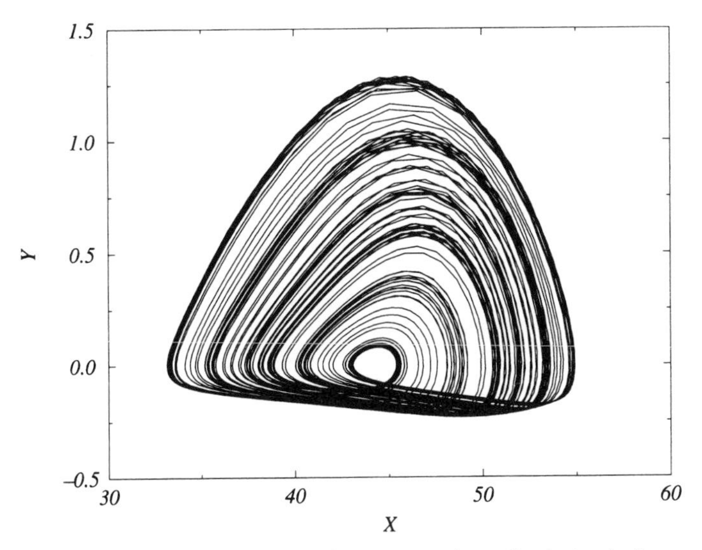

FIGURE 18. Reconstructed attractor A_R of the copper electrodissolution by integrating the reconstructed vector field: projection in the XY-plane.

template induced by the copper attractor. For the copper electrodissolution, such a validation is exemplified in FIGURE 19.

Dynamical Information from Different Variables

In the case where the knowledge of many variables is available, it is of interest to determine whether the different variables contain the same information. Indeed, it may happen that a system is described by variables which provide different kinds of dynamical information. This situation often arises in the presence of symmetry.[23]

A good example of such a problem is constituted by the study of a pulsating star, namely a W Vir Cepheid. In order to study such a star, astrophysicists experimentally record the time evolution of the luminosity L. Nevertheless, when they have numerical models of such a star, they use the time evolution of the radius R of the star since it provides time series which are easier to characterize. Consequently, it is important to check whether the dynamical information given by the R-time series is equivalent to the information contained within the L-time series.

The time series are generated by a hydrodynamical code which constitutes a W Vir model, namely the model D 5200 of reference 30. This model is a realistic and state-of-the-art model insofar as it contains all the complications that arise both in the equations of state and in the opacity due to partial H and He ionizations. The star has a mass of 0.6 $M_\odot$, a luminosity of 500 $L_\odot$, an abundance of hydrogen $X = 0.7$ and an abundance of all the heavier elements $Z = 0.005$. This particular model is chosen because it generates quite irregular pulsations, exhibiting a developed chaos.

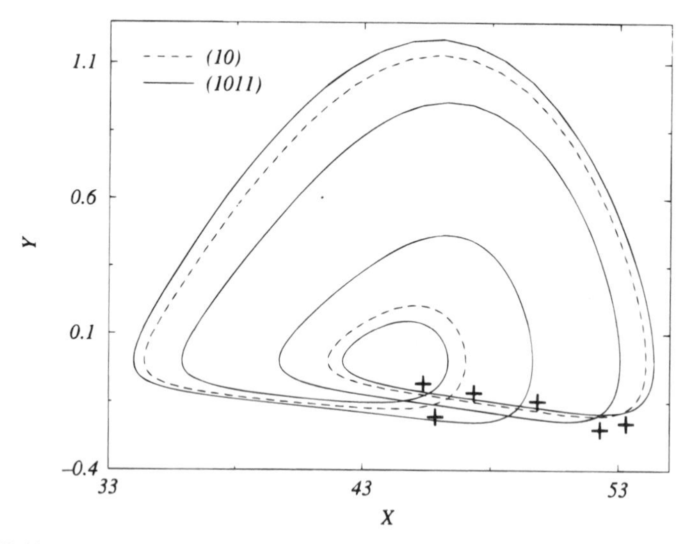

FIGURE 19. *XY*-plane projection of the couples of periodic orbits encoded by (1011) and (10). The linking number $L(1011, 10)$ is found to be equal to $\frac{1}{2}(+6) = +3$ as predicted by the template induced by the copper attractor.

Three time series are provided by integrating the hydrodynamical code, namely the stellar radius $R(t)$, the radial velocity $v_R(t)$ and the luminosity $L(t)$. These variables are displayed in FIGURE 20, with a sampling time $\delta t = 10^3$ s.

An important issue for astrophysicists is then to know if the luminosity is typical, i.e., whether it carries the same dynamical information as the temporal variations of the surface radius or of the radial velocity. A topological answer to this problem may be given by using the topological analysis. Actually, if the topological properties of the three attractors induced by the R-, v_R- and L-time series are the same, the three time series contain the same information on the dynamical behavior, from a topological point of view.

Starting from the R-time series, a chaotic attractor is found to be embedded in a tridimensional space (FIG. 21).[31] This attractor induces a first-return map with a small layered structure. Due to this layered structure, the topological analysis will only give an approximated characterization since critical points cannot accurately be determined.[32] Nevertheless, we showed that the first-return map can conveniently be viewed as an unidimensional map which then confirms the period-doubling cascade observed on this model as a route to chaos.[33] The topology of the R-attractor is found to be characterized by the template displayed in FIGURE 22.

This template has been validated by comparing linking numbers predicted from the template with linking numbers counted on plane projections. The population of periodic orbits has been found to be governed by the unimodal order and the kneading sequence is (10110).

According to the embedding theorem, it should be possible to use any physical variable for the reconstruction. However, we exemplified that a variable of a given system may not contain any pertinent information about the dynamical behavior

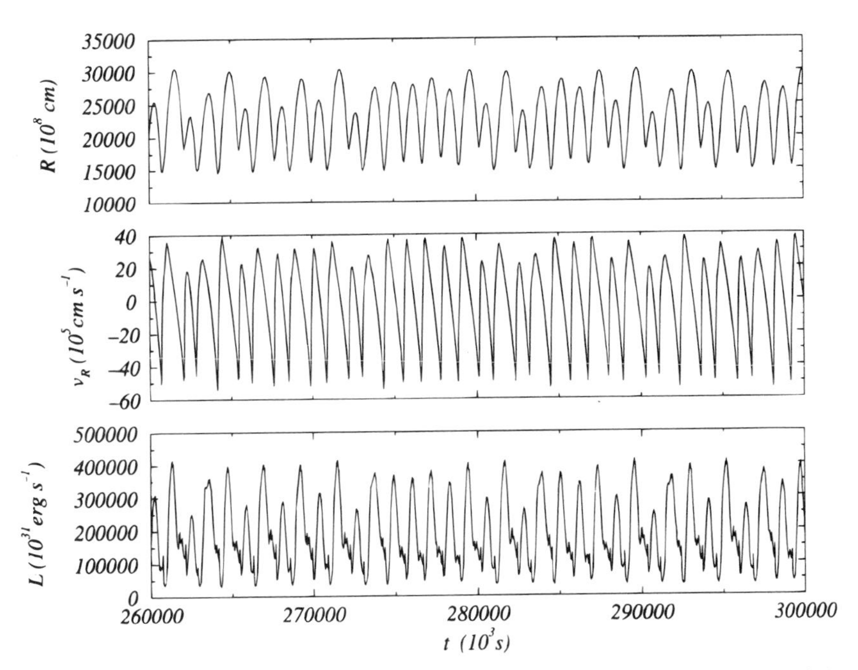

FIGURE 20. Time series of the radius R, the radial velocity v_R and the luminosity L of the star.

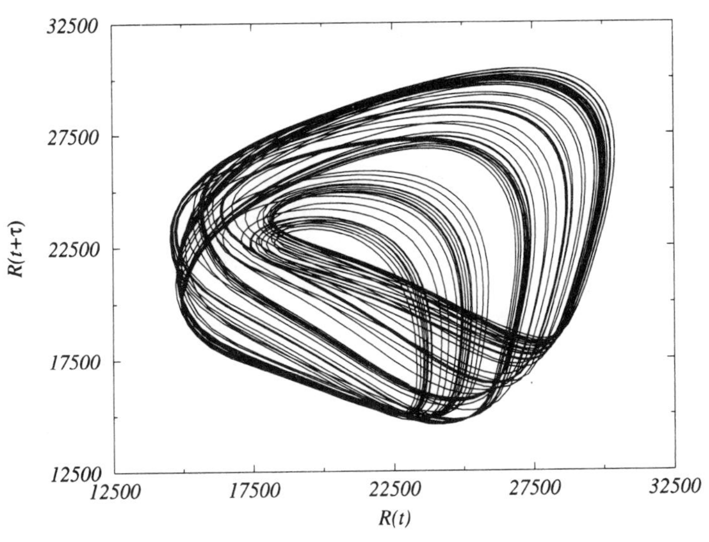

FIGURE 21. Projection of the attractor induced by the R-time series in the plane defined by $R(t)$, $R(t + \tau)$.

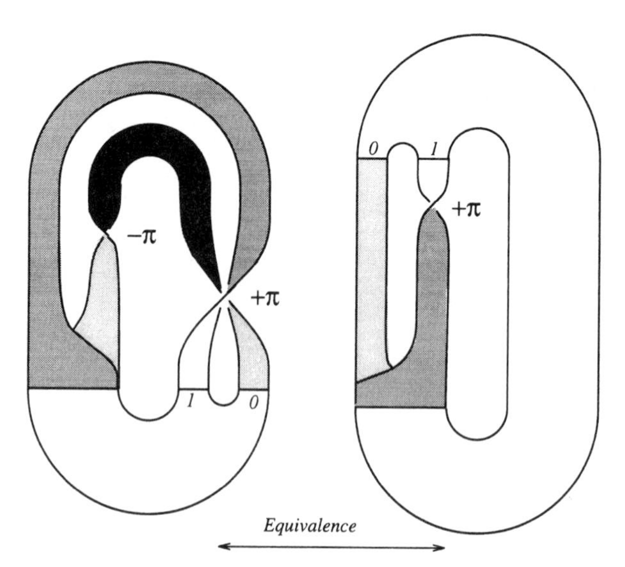

FIGURE 22. Template of the R-induced attractor. Two stripes undergoing through a positive global π-twist are exhibited. One of them has an additional negative π-twist. The global torsion may be reduced within the ribbon graph. The proposed template is represented with respect to the standard insertion convention.

when one is forced to use low dimensional phase spaces.[23] Such is the case when one deals with topological characterization which is limited to 3D-phase spaces. Thus, the cases of the radial velocity v_R and of the luminosity L must be investigated, at least as a checking of our previous results.

The case of the radial velocity v_R of the stellar surface is easy to solve since it is shown[34] that derivative coordinates are equivalent to delay coordinates. v_R being the time derivative of the stellar radius R, the v_R-time series obviously contains the same information about the dynamical behavior as the R-time series.

If the R- or v_R-time series are smooth enough, the temporal behavior of the luminosity, however, is very jittery (FIG. 23) because Lagrangian hydrocodes have great difficulty in resolving the motion of the sharp H partial ionization regions. As seen from FIGURE 23, one then expects more difficulties in studying the L-attractor than the R-attractor. It is therefore of particular interest to check whether the L-induced attractor A_L reconstructed in a 3D space by a time delay method can be found to be topologically equivalent to the R-induced attractor A_R. As for the R-induced attractor, the first-return map of the L-induced attractor exhibits a layered structure[35] and the shape looks rather similar to that of a unimodal map. By computing the symbolic plane, the pruning front is found to be located at 0.8499 inducing the kneading sequence (10110) as for the attractor A_R. Moreover, the orbits extracted from the L-induced attractor are the same as the orbits within A_R. Nevertheless, due to the jittery character of the luminosity time series, the topology of the L-induced attractor cannot be safely characterized. However, relying on the obtained results, the topologies of the R- and of the L-induced attractors are found

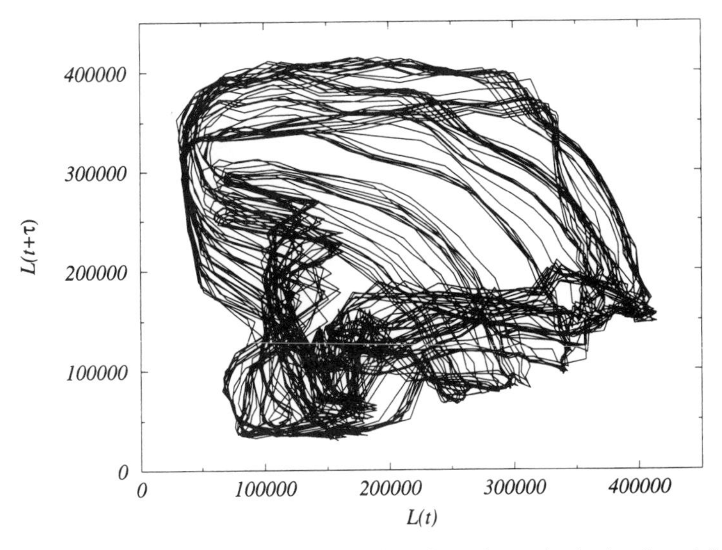

FIGURE 23. Projection of the attractor induced by the L-time series in the plane defined by $L(t)$, $L(t + \tau)$.

to be compatible. It is likely that both attractors are actually equivalent. An astrophysicist may then analyze the star luminosity which is an easy experimental observable, knowing that it is indeed representative of all dynamical variables of the star, including the temporal variations of the surface radius or of the surface radial velocity. The star then appears as a strongly connected object in which all variable behaviors are accessible from the knowledge of a single one.

Departure from Symmetry of an Electronic Circuit

The system is an electronic circuit whose block diagram is shown in FIGURE 24. It has been shown that this circuit can exhibit a transition from periodic oscillations to chaos via period-doubling cascades, intermittency and crises of chaotic attractors.[36] Under certain conditions corresponding to $\alpha = 18.9$ as defined in reference 26, the asymptotic motion settles down onto a double-scroll attractor (FIG. 25) which presents an inversion symmetry. The topological analysis is therefore conveniently achieved by working in a fundamental domain as stated in the previous section. The time series $x(t)$ under study is measured from the capacitor C as indicated in FIGURE 24.

Consequently, a Poincaré set P_{EC} is defined as the union of two Poincaré sections P_{X_+} and P_{X_-} which are defined by

$$P_{X_+} = \{(V, W) \in \mathbb{R}^2 \mid U = 4.0, \dot{U} < 0\}$$

and

$$P_{X_-} = \{(V, W) \in \mathbb{R}^2 \mid U = -4.0, \dot{U} > 0\}, \qquad (13)$$

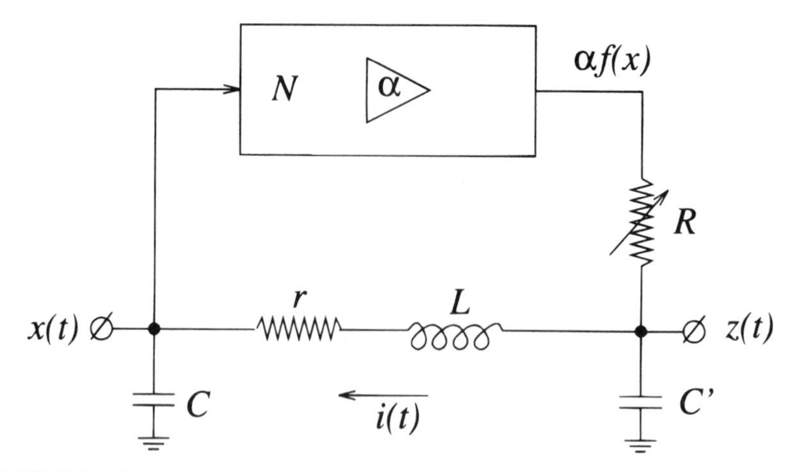

FIGURE 24. A schematic diagram of the electronic circuit here studied. For this circuit, data are collected at $R = 3.38$ kΩ, $L = 145$ mH, $C = 343$ nF, $C' = 225$ nF, $r = 347$ Ω, with a sampling period of 20 μs and $\alpha = 17.4$ and 18.9.

where $U = x(t)$, $V = x(t + \tau)$ and $W = x(t + 2\tau)$ with $\tau = 200$ μs. The first-return map is then computed with an invariant variable taken as

$$\tilde{V} = \begin{cases} V & \text{if} \quad U = 4.0 \\ -V & \text{if} \quad U = -4.0. \end{cases} \tag{14}$$

The first-return map is displayed in FIGURE 26.

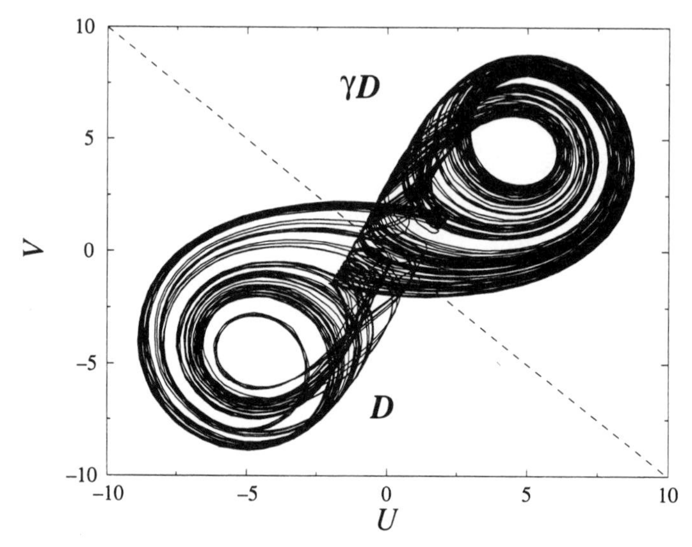

FIGURE 25. Plane projection of the symmetric attractor generated by the experimental electronic circuit.

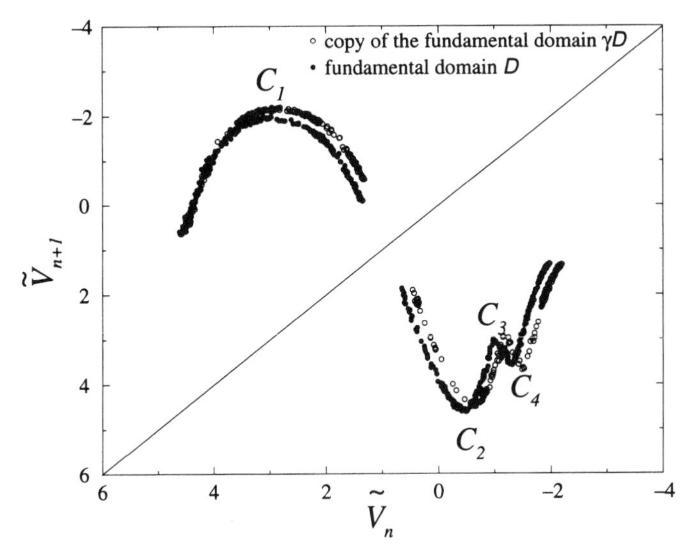

FIGURE 26. First-return map to the Poincaré set P_{EC} computed with the invariant variable $\tilde{V}$.

One may then remark that a small departure from the symmetry is exhibited by the first-return map since the points associated with the fundamental domain $\mathscr{D}$ define a curve which is slightly different from the one constituted by the points associated with the copy $\gamma\mathscr{D}$ of the fundamental domain. This slight difference may

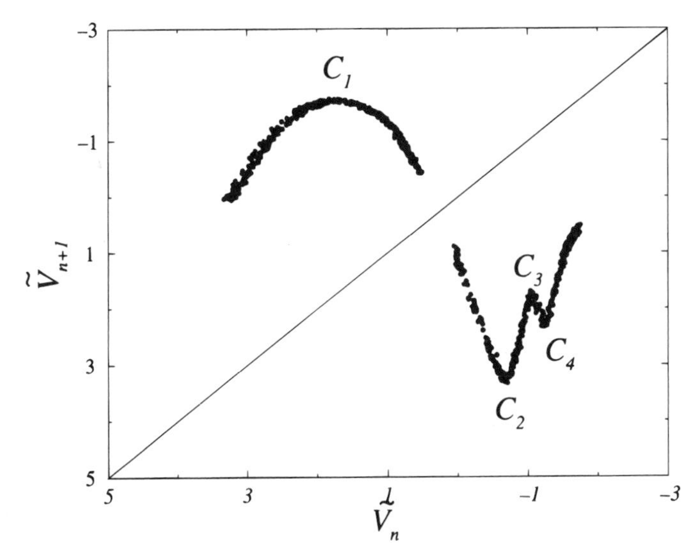

FIGURE 27. First-return map to the Poincaré set computed with the new definition of the invariant variable $\tilde{V}$.

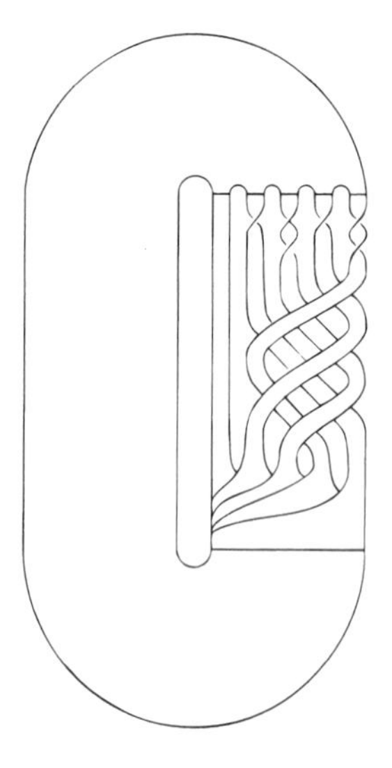

FIGURE 28. Template of the fundamental domain of the symmetric attractor.

come from a small residual defect in the experimental set-up. Nevertheless, in order to perform a pertinent topological analysis we require to eliminate this departure from symmetry. This may be achieved rather well by redefining the invariant variable $\tilde{V}$ as

$$\tilde{V} = \begin{cases} V & \text{if} \quad U = 4.0 \\ -V + 0.18 & \text{if} \quad U = -4.0. \end{cases} \tag{15}$$

The first-return map computed with this new definition of the invariant variable is displayed in FIGURE 27. The first-return map is then found to be constituted by five monotonic branches. Accordingly, the topology of the symmetric attractor is characterized by the five-stripe template displayed in FIGURE 28.

This electronic circuit is also found to be topologically compatible with the well-known Chua's circuit and with the model derived by A. Volkovski and N. Rul'kov.[36]

CONCLUSIONS

In this chapter, it has been shown that the topological characterization could be useful to analyze the dynamical behavior of systems coming from different areas of research. When the studied system is very dissipative and tridimensional, a fine characterization may be obtained by using such an analysis. In particular, the topological characterization is quite interesting because it provides tools which are weakly sensitive to noise perturbations since the relative organization of low period orbits is not affected by noise. Moreover, as the template is built on topological invariants, namely the linking numbers, the topological characterization is invariant under control parameter changes, at least when new stripes are not implied. Nevertheless, linking numbers are not defined in spaces whose dimensions are greater than 3. This prevents the direct extension of the topological characterization for higher-dimensional systems, although G. Mindlin has recently proposed a characterization in 4D-spaces by working with surfaces rather than with links (private communication). A generalization in higher dimensions remains the most relevant objective in this approach. Nevertheless, when one is interested in only defining whether two attractors are topologically equivalent or not, the nature of the map between these two different attractors may be used, as recently proposed by Pecora *et al.*[37] This method presents the advantage of running in any dimension, but cannot be used to accurately characterize the topology of an attractor.

It seems now that the topological characterization is an important tool in studying dynamical systems which may be conveniently combined with other tools as symbolic dynamics or bifurcation theory to have a better understanding of the dynamical processes involved in experimental systems.

ACKNOWLEDGMENTS

We wish to thank J. Hudson and Z. Fei for the beautiful data from the copper electrodissolution, and J. Buchler and Z. Kolláth for the time series from the hydrodynamical model of the W Vir Cepheid. We also wish to thank N. Rul'kov and R. Brown for the data from the electronic circuit. We would like to offer special thanks to L. Le Sceller for the reconstructed model from the electrodissolution data. We would also like to thank these coworkers for their helpful discussions.

REFERENCES

1. Lorenz, E. N. 1963. J. Atm. Sci. **20:** 130.
2. Packard, N. H., J. P. Crutchfield, J. D. Farmer & R. S. Shaw. 1980. Phys. Rev. Lett. **45:** 712.
3. Takens, F. 1981. *In* Dynamical Systems and Turbulence, Warwick 1980, Lecture Notes in Mathematics. D. A. Rand & L. S. Young, Eds. Vol. 898: 366. Springer-Verlag. New York.
4. Grassberger, P. & I. Proccacia. 1983. Physica D **9:** 189.
5. Eckmann, J. P., S. Oliffson, D. Ruelle & S. Ciliberto. 1986. Phys. Rev. A **34:** 4971.
6. Abarbanel, H. D. I., R. Brown, J. J. Sidorowich & L. Sh. Tsimring. 1993. Rev. Mod. Phys. **65:** 1331.
7. Abarbanel, H. D. I. 1995. *In* Nonlinearity and Chaos Engineering Dynamics. J. M. T. Thompson & S. R. Bishop, Eds.: 379. Wiley. New York.
8. Metropolis, N., M. L. Stein & P. R. Stein. 1973. J. Comb. Theor. A **15:** 25.
9. Collet, P. & J. P. Eckmann. 1980. *In* Progress in Physics. A. Jaffe & D. Ruelle, Eds. Birkhäuser. Basel.
10. Hall, T. 1994. Nonlinearity **7:** 861.
11. Mindlin, G. B., X. J. Hou, H. G. Solari, R. Gilmore & N. B. Tufillaro. 1990. Phys. Rev. Lett. **64:** 2350.
12. Mindlin, G. B., H. G. Solari, M. A. Natiello, R. Gilmore & X. J. Hou. 1991. J. Nonlin. Sci. **1:** 147.
13. Tufillaro, N. B., T. Abbott & J. Reilly. 1992. An Experimental Approach to Nonlinear Dynamics and Chaos. Addison-Wesley. Reading, MA.
14. Rössler, O. E. 1976. Phys. Lett. A **57:** 397.

15. BIRMAN, J. S. & R. F. WILLIAMS. 1983. Topology **22:** 47.
16. MELVIN, P. & N. B. TUFILLARO. 1991. Phys. Rev. A **44:** 3419.
17. LETELLIER, C., P. DUTERTRE & B. MAHEU. 1995. Chaos **5:** 271.
18. TUFILLARO, N. B., P. WYCKOFF, R. BROWN, T. SCHREIBER & T. MOLTENO. 1995. Phys. Rev. E **51:** 164.
19. FANG, H. P. 1994. Phys. Rev. E **49:** 5025.
20. LIBCHABER, A. & J. MAURER. 1982. *In* Nonlinear Phenomena at Phase Transitions. T. Riste, Ed.: 259.
21. LETELLIER, C., P. DUTERTRE & G. GOUESBET. 1994. Phys. Rev. E **49:** 3492.
22. CVITANOVIĆ, P. & B. ECKHARDT. 1993. Nonlinearity **6:** 277.
23. LETELLIER, C. & G. GOUESBET. 1996. J. Phys. II **6:** 1615.
24. LETELLIER, C. & G. GOUESBET. 1995. Phys. Rev. E **52**(5): 4754.
25. GOUESBET, G. & C. LETELLIER. 1994. Phys. Rev. E **49:** 4955.
26. BROWN, R., N. F. RUL'KOV & E. R. TRACY. 1994. Phys. Rev. E **49:** 3784.
27. LETELLIER, C., L. LE SCELLER, E. MARÉCHAL, P. DUTERTRE, B. MAHEU, G. GOUESBET, Z. FEI & J. L. HUDSON. 1995. Phys. Rev. E **51:** 4262.
28. LETELLIER, C., L. LE SCELLER, P. DUTERTRE, G. GOUESBET, Z. FEI & J. L. HUDSON. 1995. J. Phys. Chem. **99:** 2016.
29. LETELLIER, C., E. RINGUET, B. MAHEU, J. MAQUET & G. GOUESBET. 1995. 4e Journées Européennes de Thermodynamique Contemporaine, Nancy, France. To be reprinted in Entropie.
30. KOVÁCS, G. & J. R. BUCHLER. 1988. Ast. J. **334:** 971.
31. SERRE, T., Z. KOLLÁTH & J. R. BUCHLER. 1996. Astron. Astrophys. **311:** 845.
32. GRASSBERGER, P., H. KANTZ & U. MOENING. 1989. J. Phys. A: Math. Gen. **22:** 5217.
33. BUCHLER, J. R. 1990. Ann. N.Y. Acad. Sci. **617:** 17.
34. GIBSON, J. F., J. D. FARMER, M. CASDAGLI & S. EUBANK. 1992. Physica D **57:** 1.
35. LETELLIER, C., G. GOUESBET, F. SOUFI, J. R. BUCHLER & Z. KOLLÁTH. 1996. Chaos **6**(3): 466.
36. VOLKOVSKI, A. R. & N. F. RUL'KOV. 1988. Sov. Teck. Phys. Lett. **14:** 656.
37. PECORA, L. M., T. L. CARROLL & J. F. HEAGLY. 1995. Phys. Rev. E **52:** 3420.

Through a Glass Darkly: Distinguishing Chaotic from Stochastic Resonance[a]

J. GRAF VON HARDENBERG,[b] F. PAPARELLA,[b]
A. PROVENZALE,[b] AND E. A. SPIEGEL[b,c]

[b] *Istituto di Cosmogeofisica*
Corso Fiume 4
Torino, Italy

[c] *Department of Astronomy*
Columbia University
New York, New York 10027

Come quando una grossa nebbia spira,
o quando l'emisperio nostro annotta,
par di lungi un molin che 'l vento gira,
veder mi parve un tal dificio allotta ...
DANTE ALIGHIERI [*Inferno*, Canto 34]

INTRODUCTION

In astrophysics, we deal with inverse problems in which observations of some object have to be understood even when we may not know what the object may be (bursters) or what the mechanism driving it is (quasars) or where in the object the mechanism operates (solar cycle). The problem then is to discover the nature of the underlying chaotic or intermittent mechanism and how it produces the observed effects.

There are standard means of analyzing data to uncover the underlying dynamics and these provide clues about the operation of the basic system,[1–3] at least when they work well. For example, if we can get an idea of the dimensions of the space needed to let the dynamics evolve, we have some clue about how complicated a model is needed. These methods are now well documented and are generally considered reasonably successful, but we have encountered cases where they fail completely when applied according to the directions on the package.[4–6]

The time series generated by stochastic processes with long-time correlations (red noise) may lead to a convergent estimate of the correlation dimension, even in the absence of any low-dimensional chaotic dynamics.[4,5] The mechanism called on/off intermittency,[7] which is useful for modeling bursting in general and the Maunder intermission in particular (C. Pasquero, Laurea thesis, 1996), turns out to make dimensional determination difficult unless special means are devised to analyze the observations.[6] Here we discuss a similar failure of the procedure for

[a]This work was partially supported by CNR, Progetto Coordinato "Climate Variability and Predictability".

dimension determination—detecting the difference between stochastic and chaotic resonance—and then describe how the situation may be saved. But first we discuss the general nature of a class of chaotic models that brings out the origin of these difficulties, before turning to the specific example that points the moral.

SYSTEMS AND THEIR DRIVERS

A Bare System

To illustrate our point simply, we consider a deterministic oscillator with state variable x, governed by the standard equation

$$\ddot{x} = -\partial_x U(x; q_j) - \mu \dot{x}, \tag{1}$$

where the potential U depends on the parameters q_j, $j = 1, \ldots, K$ and the friction μ (which might more generally depend on x) is also a parameter. All the solutions of this system, after some possibly long transients, are either fixed points or limit cycles, for $\mu > 0$. Our interest here is in what happens when the parameters are themselves endowed with time dependence and in whether observations of $x(t)$ can be used to distinguish among the possible forms of $q_j(t)$.

The Driver

In specifying the time dependence of the parameters of the oscillator we are, in effect, defining a second system that we call the driver. There are several ways to speak of this time dependence. We may for example refer to the dimension of the driver. When this dimension is quite high, we will say that the driver is stochastic, but, when the dimension is just a few, the driver may be chaotic or periodic. The meaning of these terms is precise enough for our purposes, so that we need not go into great detail about this here except to say that we shall specify the time dependence of q_j either by using a random number generator or by solving a differential equation for it.

SOME EXAMPLES

The combination of the oscillator and the driver make up the full model that we may use to produce complex behavior. We next provide some illustrations of this for different forms of the time dependence of q_j. For this purpose, we consider the double-well potential

$$U(x; q_j) = \tfrac{1}{4}x^4 - \tfrac{1}{2}px^2 + qx, \tag{2}$$

where $K = 2$ and we have used the notation $p = q_1$ and $q = q_2$. If q alone depends on time we have the so-called additive forcing, whereas if we let p depend on time the forcing would be called parametric.

In the simplest examples, we take the time dependence of q and p to be specified independently of x. However, it is sometimes convenient to break more general dynamical systems into two parts, one of which is the driver. In such cases we may not be able to avoid the feedback of x on what we take as parameter, q (or p). We stick mainly to the former case, with no feedback, which is said to have a skew-product structure.

Chaos

Let $q = A \cos(\Omega t)$ and p be constant. We are then dealing with the so-called Duffing equation. It is known that for this system we can have periodic or chaotic $x(t)$. The production of chaos in this way is seen intuitively, if quite loosely, by thinking of small Ω. The periodic solution of the unforced oscillator is then changed quasi-adiabatically so that the system has a number of osculating periodic orbits. The union of such orbits is a schematic picture of the true trajectories in the chaotic state.

In more complicated cases, we may specify the behavior of q or p in terms of an added equation of the general form[8]

$$\dot{q} = G_1(q, x), \tag{3}$$

or

$$\dot{p} = G_2(p, x), \tag{4}$$

for specified G_i, and then we have a standard third-order dynamical system that may display chaotic behavior. For example, the Lorenz equations (7) can be transformed into the form (1) with $q = 0$ and

$$\dot{p} = \varepsilon[p + a(x^2 - 1)], \tag{5}$$

where a and ε are constants.[9]

Intermittency

Now let q vary aperiodically. The variation can be chaotic or stochastic. In either case, if the variation is of suitable type, the system's behavior will vary between vigorous oscillations and near stasis.[10] When such on/off intermittency is observed, it is difficult to tell whether the driver is stochastic or merely chaotic, though some tests that work have recently been devised.[6] A second example of this trouble is the situation of parametric forcing, in which p is dependent on time. As the situation with on/off intermittency has been closely examined already,[6] we concentrate on another example of current interest, with additive forcing.

Resonance

Besides the various cases we have mentioned, there are situations where a combination of two of the signal types may arise. In that case, it is sometimes useful to

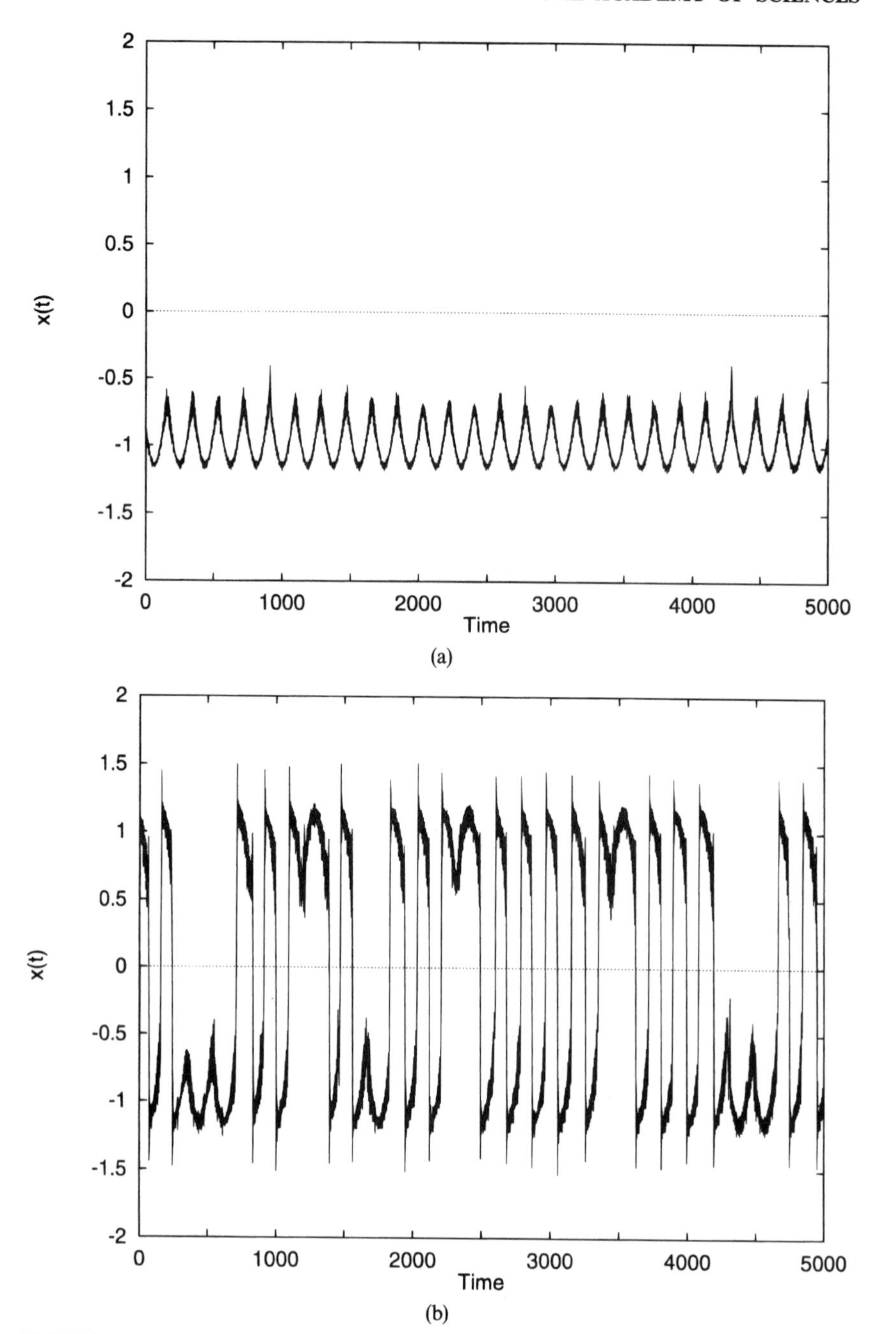

FIGURE 1. Output time series $x(t)$ obtained from the (deterministic) system (**1**), (**2**), (**5**) driven by the output $F_D(t)$ from the Lorenz model (**6**). The amplitude and period of the sinusoidal driver are $A = 0.35$ and $2\pi/\Omega = 187.5$. The noise amplitude is (**a**) $\varepsilon = 0.05$, (**b**) $\varepsilon = 0.15$, (**c**) $\varepsilon = 0.30$ and (**d**) $\varepsilon = 1.0$.

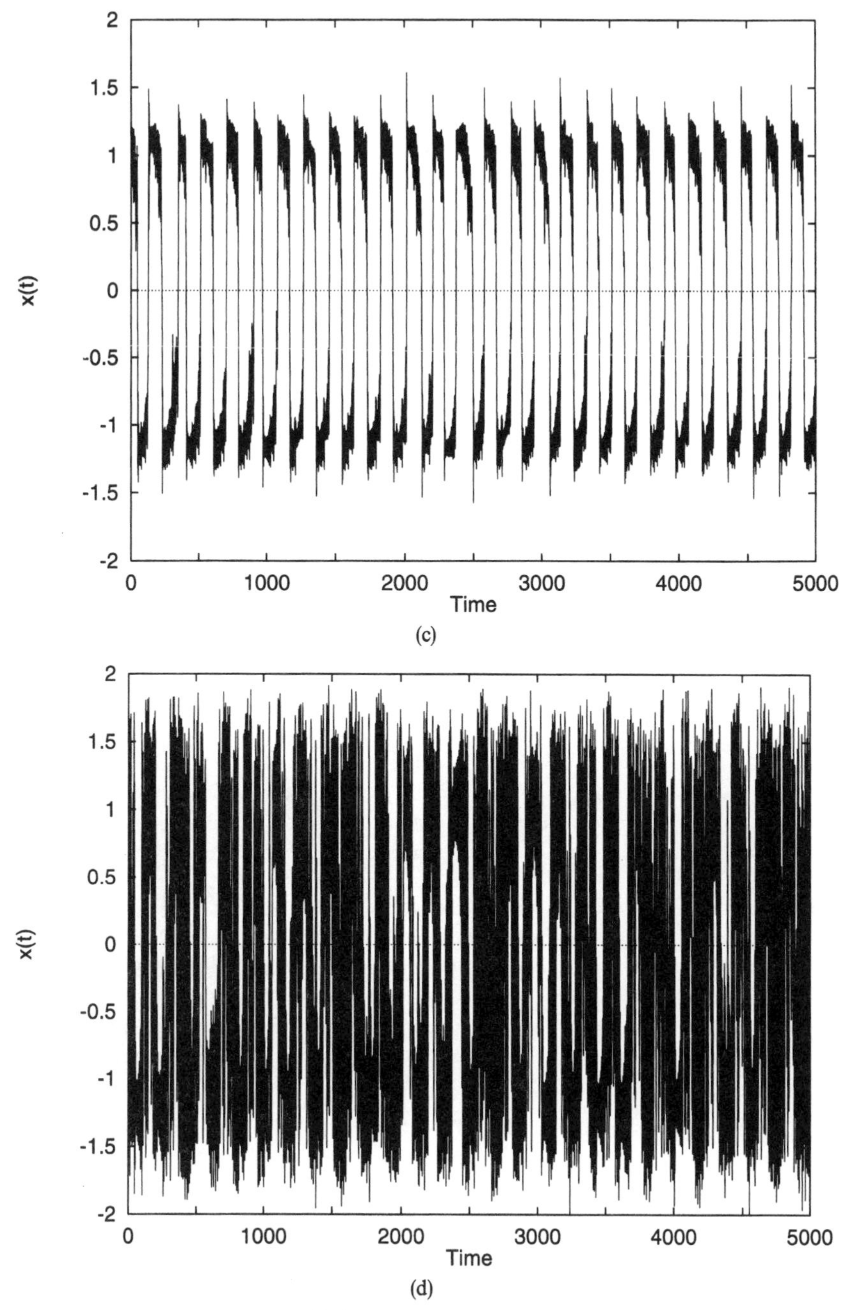

FIGURE 1. Continued.

think of the driver output as a signal that we try to detect with the oscillator. Suppose for example that a periodic signal is fed to the oscillator and that its amplitude is so large as to cause the oscillator to jump periodically from well to well. Then this is a clear detection. At the other limit, where the periodic signal is so feeble as to be unable to induce transitions from one well to another, it may be possible to detect the signal by adding noise to it. The noise will combine with the periodic signal in such a way that the resultant may succeed in sending the oscillator into the other well at certain times, which are multiples of the periodic component. When the periodic signal and the noise are each too weak to do the job alone, this cooperation is often a very successful one, since the barrier crossing occurs with a characteristic phase relation to the maxima of the periodic signal. The signature of the process is clear and we have then a possibility of detecting a very feeble signal. This process is known as stochastic resonance.[11–14]

It is of course quite natural to introduce a chaotic signal as the booster instead of the now conventional stochastic one. Similar effects are produced by this alternate mechanism that we shall call *chaotic resonance.* Examples of this kind of resonance are presented in FIGURE 1, which shows the time series $x(t)$ obtained from the oscillator described by (**1**) when $p = 1$ and

$$q(t) = A\cos(\Omega t) + \varepsilon F(t), \tag{6}$$

where $F(t)$ is determined by the output $Z(t)$ of the Lorenz model

$$\begin{aligned} c\dot{X} &= \sigma(Y - X) \\ c\dot{Y} &= -XZ + rX - Y \\ c\dot{Z} &= XY - bZ \end{aligned} \tag{7}$$

with the standard values $r = 28$, $\sigma = 8/3$ and $b = 10$. We take $F(t) = (Z(t) - Z_m)/Z_0$, where Z_m is the time average of the signal $Z(t)$ and Z_0 is a normalization factor, herein chosen to be $Z_0 = 15$ (so that the driver $F(t)$ oscillates between approximately -1 and 1). The constant c is introduced to rescale the time scale of the Lorenz model with respect to that of the oscillator (**1**); here we have chosen $c = 10$. The different panels in FIGURE 1 correspond to different choices of the "noise" amplitude ε. For low values of ε, the system oscillates around a single well of the potential. For large values of ε, the driver dominates the dynamics and one can clearly see the typical appearance of the Lorenz system. For intermediate values of ε, however, the system alternates in a more or less regular fashion between the two potential wells, quite similar to what is observed in stochastic resonance where a random driver is employed.

THE PROBLEM

We wish to examine here the following problem in data analysis: given a resonant detection of a periodic signal, can we tell whether the enhancer that makes the detection possible is chaotic or stochastic? Thus, if we have a system parameter whose variation is described by (**6**), can we from a measurement on $x(t)$ learn

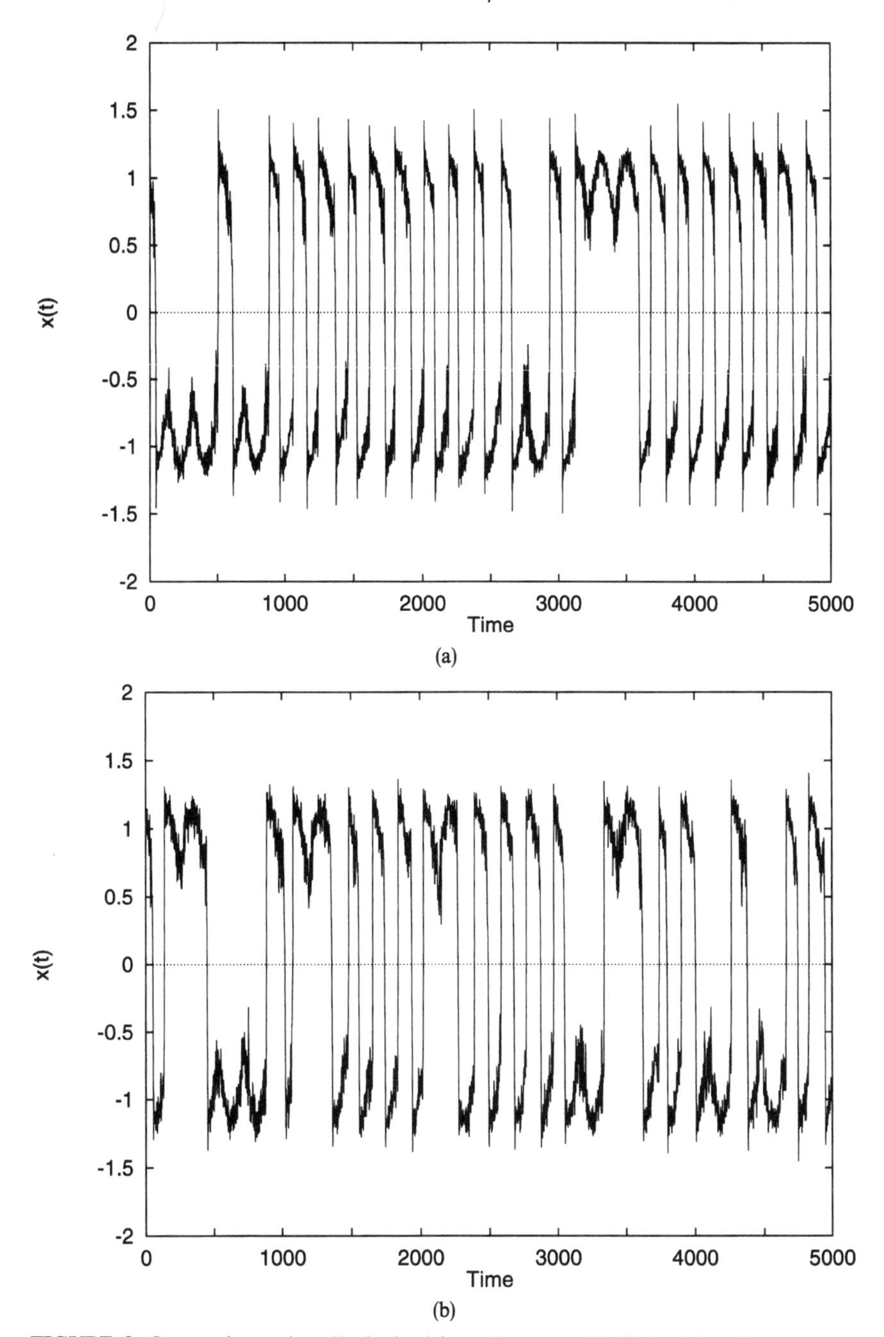

FIGURE 2. Output time series $x(t)$ obtained from the system (**1**), (**2**), (**5**) driven by the phase-randomized driver $F_R(t)$ (**panel a**) and by the stochastic driver $F_L(t)$ generated by the Langevin equation (**8**) (**panel b**). The amplitude and period of the sinusoidal driver are $A = 0.35$ and $2\pi/\Omega = 187.5$; the noise amplitude is $\varepsilon = 0.15$.

whether F is a stochastic or chaotic process, in the sense that we can get a good estimate of the dimension of the system producing it? We shall proceed in the case where the amplitude, A, has a value for which the potential, U, has always two minima and F will be either deterministic or stochastic.

In the adiabatic limit, when the damping time μ^{-1} is much less than the period Ω^{-1}, by choosing the noise amplitude suitably, we can arrange the system switches between two osculating periodic orbits in synchronism with the periodic signal. As we saw in the previous section, the input of the Lorenz signal does the job as well. Moreover, we can compare these results to a stochastic process with the same spectrum as the Lorenz system by taking the same input forcing and randomizing its Fourier phases.[5] That leads also to this kind of resonant process. The question thus posed is, can we from a study of $x(t)$ (in as long a run as we wish) tell what the dimension of the original driver is?

The Examples

To compare stochastic and chaotic resonance and to determine whether we can distinguish between them, we continue to use the two-welled oscillator (**1**)–(**2**) with $p = 1$. With only a sinusoidal input of the form $q = A \cos(\Omega t)$, the critical amplitude above which the system is not confined to a single well is $A_c = 2/\sqrt{27} = 0.3849$. Since we want the forcing to be always subcritical, we chose $A = 0.35$ for this discussion. The value $\mu = 0.5$ is chosen so as to have a damping time μ^{-1} that is much shorter than the period of the forcing, chosen to be $(2\pi)/\Omega = 187.5$.

To obtain the resonance, we consider three different types of irregular forcings in (**5**), namely:

1. A deterministic (presumably chaotic) forcing provided by the output of the Lorenz system, $F_D(t) = (Z(t) - Z_m)/Z_0$.
2. A random forcing $F_R(t)$ obtained by randomizing the Fourier phases of the signal $F_D(t)$. This is a Gaussian signal with the same second-order moment (power spectrum) as the original signal $F_D(t)$.
3. A random forcing F_L obtained from the output of a Langevin equation,

$$\dot{\zeta} = \alpha\zeta + \sigma w, \qquad (8)$$

where $\alpha = -0.9$, $\sigma = 1$ and w is a unit variance, zero mean, Gaussian white noise. By analogy with the Lorenz driver, we define $F_L(t) = (\zeta(t) - \zeta_m)/\zeta_0$, where ζ_m is the time average of the signal $\zeta(t)$ and $\zeta_0 = 0.005$ is a normalization factor.

In all three cases, we have generated the output signal $x(t)$ by integrating system (**1**) with the forcing (**6**), where $F(t)$ is either F_D, F_R or F_L. For this, we used a standard fourth-order Runge–Kutta algorithm, with time step $\delta t = 0.05$ and sampling time $\Delta t = 0.5$. For each run, we generated time series of 10,000 points. FIGURE 2 shows two sample output signals obtained by using the random forcings $F_R(t)$ and $F_L(t)$. Note the close similarity between the output signals obtained with the random forcing (stochastic resonance) and those obtained in the case of a deterministic driver (chaotic resonance), shown in FIGURE 1. This is confirmed by the extreme similarity of the power spectra of the output signals, which are shown in FIGURE 3.

ANALYSIS OF THE SIGNALS

Correlation Dimension

The first step in estimating the dimension of a dynamical system from a measured signal is the procedure of phase-space reconstruction.[1–3] Suppose we have computed or measured $x(t)$ at regular intervals to get a discrete time sequence x_i. We consider a phase space of trial dimension M and construct a trajectory along the set of points given by the M-tuples of delay coordinates denoted by

$$\mathbf{x}_i = (x_i, x_{i-\tau}, \ldots, x_{i-(M-1)\tau}), \tag{9}$$

where τ is an appropriately chosen time delay. The rule of thumb for choosing τ is to assign some characteristic time of the system. A standard choice is about one-fourth of the characteristic oscillation time of the time series, if this shows some form of approximate cycles. Another common choice is to select τ close to the first zero of the signal autocorrelation. We need also to make M so large that the orbit through these points does not cross itself.

One way to quantify the dynamics underlying the time series is to measure the dimension of the set of points. To do this, we shall use the algorithm of Grassberger and Procaccia.[15] First, we compute the correlation integral

$$C(r) = \frac{2}{N(N-w)} \sum_{i=1}^{N} \sum_{j=i+w}^{N} \Theta(r - |\mathbf{X}_i - \mathbf{X}_j|), \tag{10}$$

where Θ is the Heaviside step function and N is the number of points in the reconstructed (vector) time series. The parameter w is introduced to allow for the effect of crowded points produced by a slow motion along the trajectory.[16] Here we have chosen $w = 500\Delta t$.

As r tends to zero, $C(r)$ must do so likewise and the most natural way for this to happen is as a power law: $C(r) \propto r^{\nu}$. The value found for the exponent ν may in general depend on M, the embedding dimension. As M is increased, if the value of ν approaches a limit, this value is called the correlation dimension. If there is no convergence, it is generally assumed that the dimension of the system is so large that the system may be called stochastic.

The point of the present study is that this method, though tried, is not always so true. For example, it has recently been observed that signals produced by on–off intermittency produce a singular clustering in the phase space that leads to false estimates of the dimension.[6] In that case, an ameliorated form of the G-P algorithm was able to do the job. One solution was to choose the right metric in the embedding space, so as to optimally unfold the dynamics of the system. As we shall see here, the problem in distinguishing chaotic from stochastic resonance arises in producing a proper phase space reconstruction. Failure to do that will affect all the subsequent analysis.

Role of the Delay Time

A direct application of the procedure described above leads to the results shown in FIGURE 4, obtained with a value of τ close to the first zero of the signal

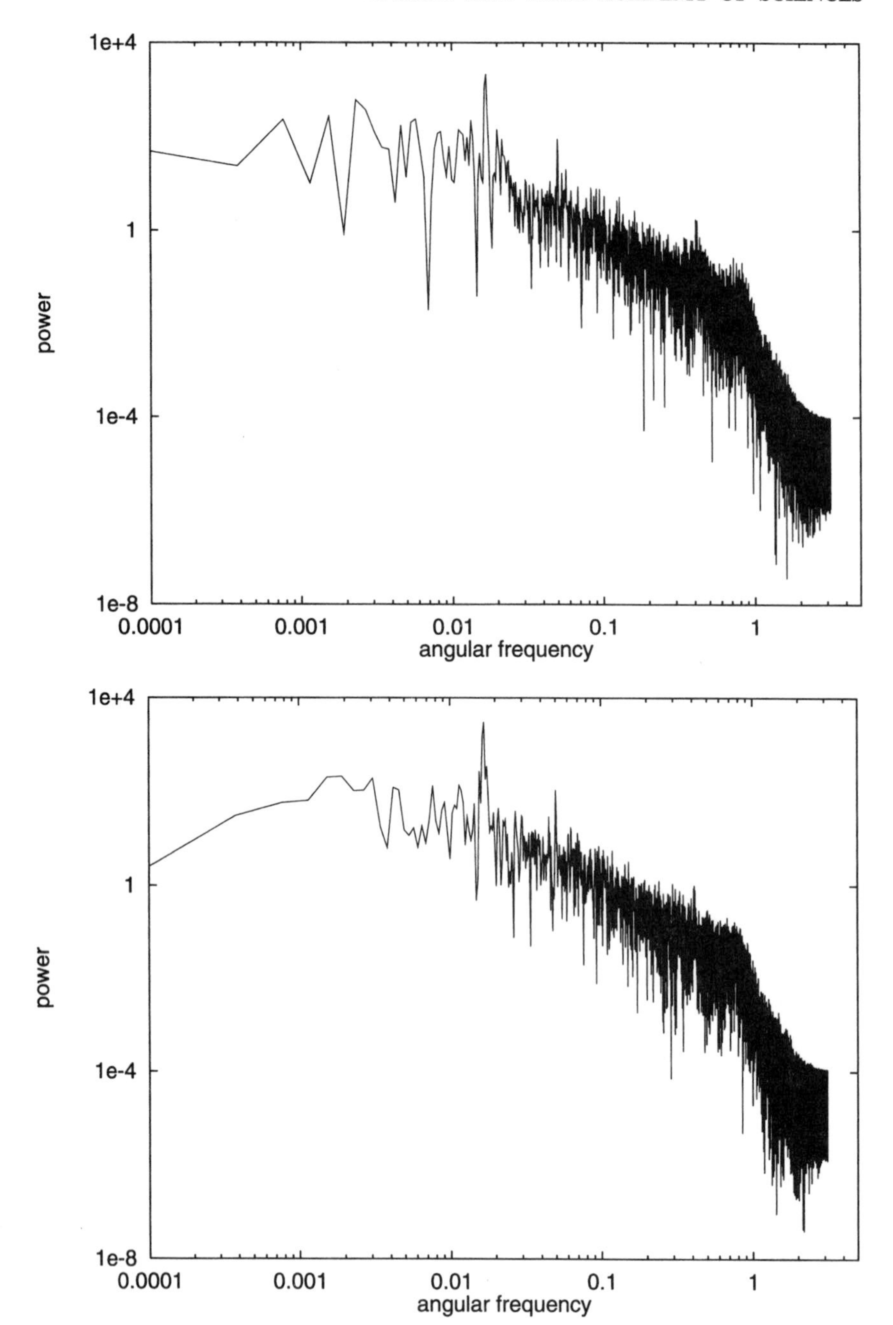

FIGURE 3. The power spectra of the output time series shown respectively in FIGURES 1(b), 2(a) and 2(b). The frequencies are given in units of the sampling time $\Delta t = 0.5$.

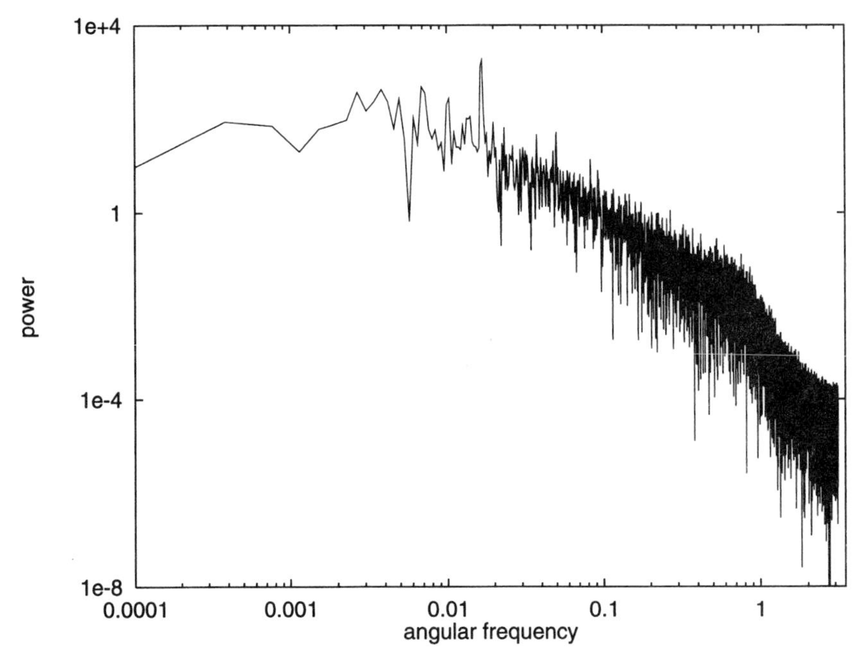

FIGURE 3. Continued.

autocorrelation, $\tau = 120\Delta t = 60$. Both the chaotic and the stochastic resonance signals show a feeble convergence for high embedding dimensions ($M > 10$). Of course, the Lorenz system has a dimension only slightly in excess of two, so this is a sign of something amiss. In this case, the appearance of convergence is in fact meaningless, since with a time series limited to 10^4 points convergence is inevitable sooner or later.

We note that the results are quite sensitive to the choice of the delay time, τ. As we mentioned, there are several possible choices for this quantity that are in common use. In addition to the rough and ready ones mentioned earlier, there is now a growing preference among practitioners for taking τ to be the time at which the mutual information function of the signal[17,18] has its first minimum. The mutual information of a signal $x(t_j), j = 1, \ldots, N$ is defined as[1]

$$I(\tau) = \sum_{j=1}^{N} P(x(t_j), x(t_j + \tau)) \log_2\left[\frac{P(x(t_j), x(t_j + \tau))}{P(x(t_j))P(x(t_j + \tau))}\right], \tag{11}$$

where $P(x)$ is the probability of measuring a value x and $P(x, y)$ is the joint probability of observing the values x and y at times separated by a delay τ.

The mutual information of the resonant signals, which is always positive, is shown in FIGURE 5. In both cases fluctuations are superposed on a gently decreasing trend. Choosing the first minimum of the smoothed version (or running mean) of this curve leads to a value of τ that is comparable to that given by the autocorrelation criterion. Using the small value $\tau = 4\Delta t = 2$, suggested by the first minimum in

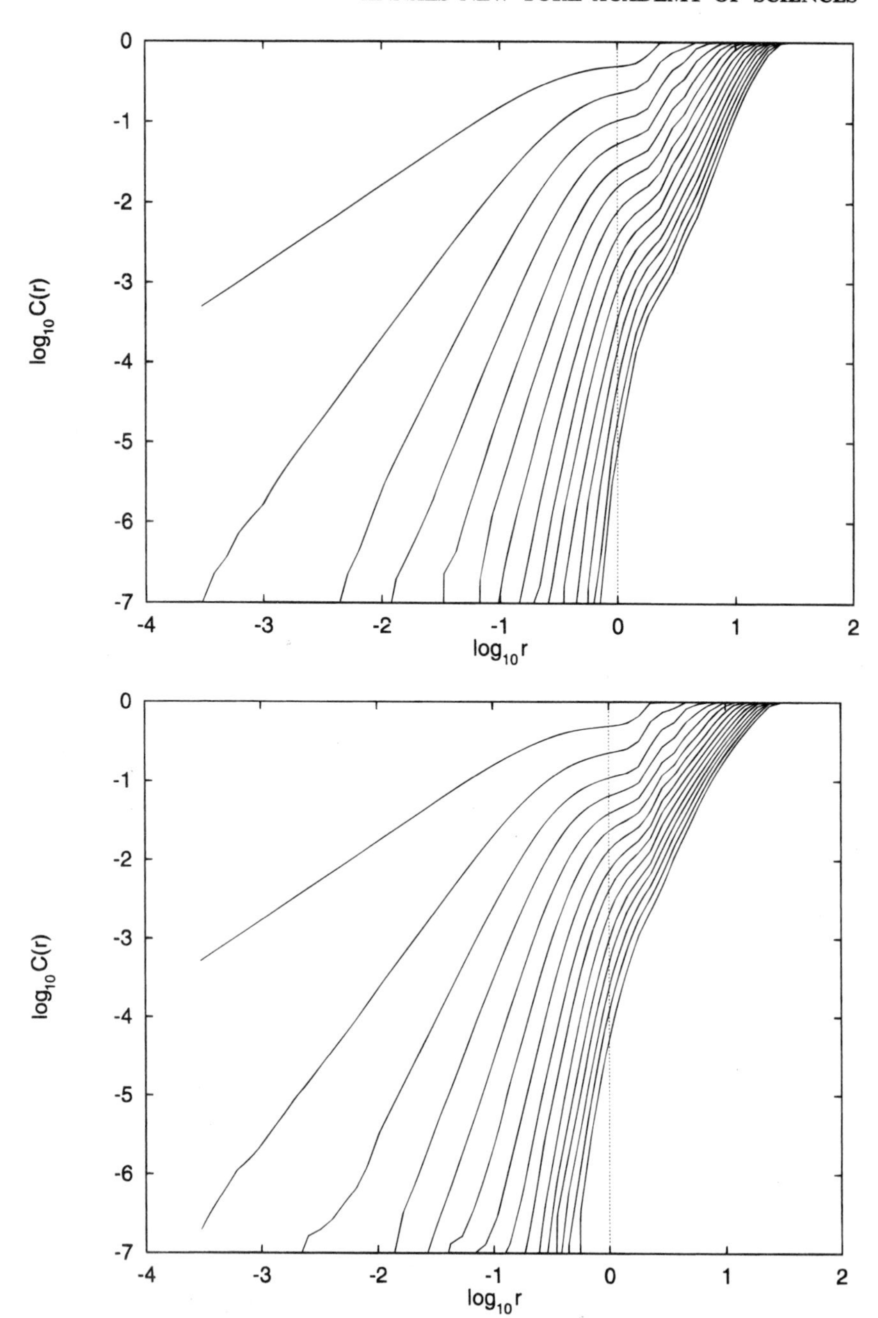

FIGURE 4. The correlation integrals for the time series shown respectively in FIGURES 1(b), 2(a) and 2(b). The embedding dimension M runs from 1 to 15 and the time delay used in the embedding procedure is $\tau = 120\Delta t = 60$.

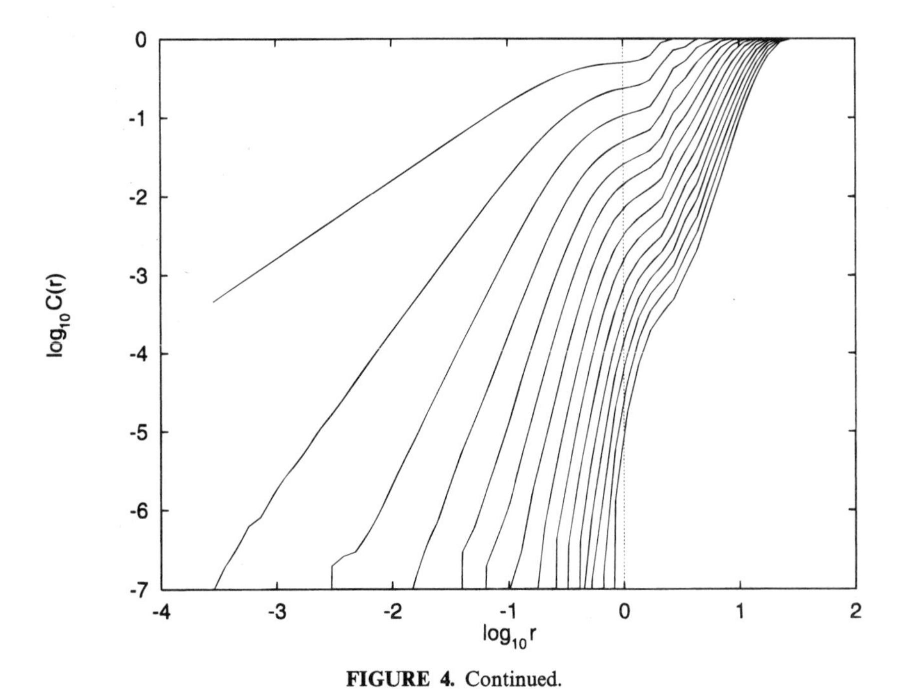

FIGURE 4. Continued.

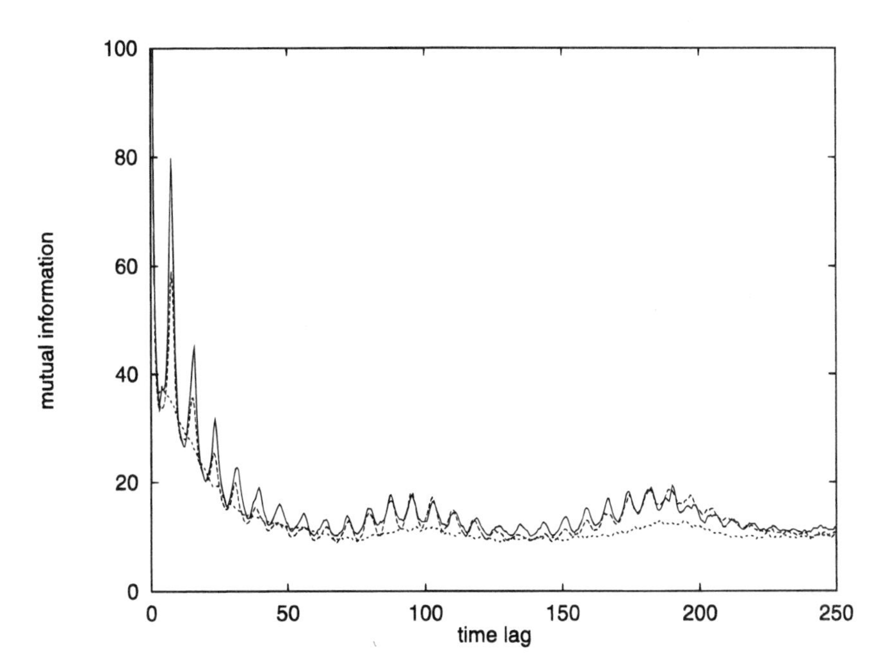

FIGURE 5. Mutual information for the time series shown in FIGURE 1(b) (solid line), FIGURE 2(a) (dashed line) and FIGURE 2(b) (short-dashed line).

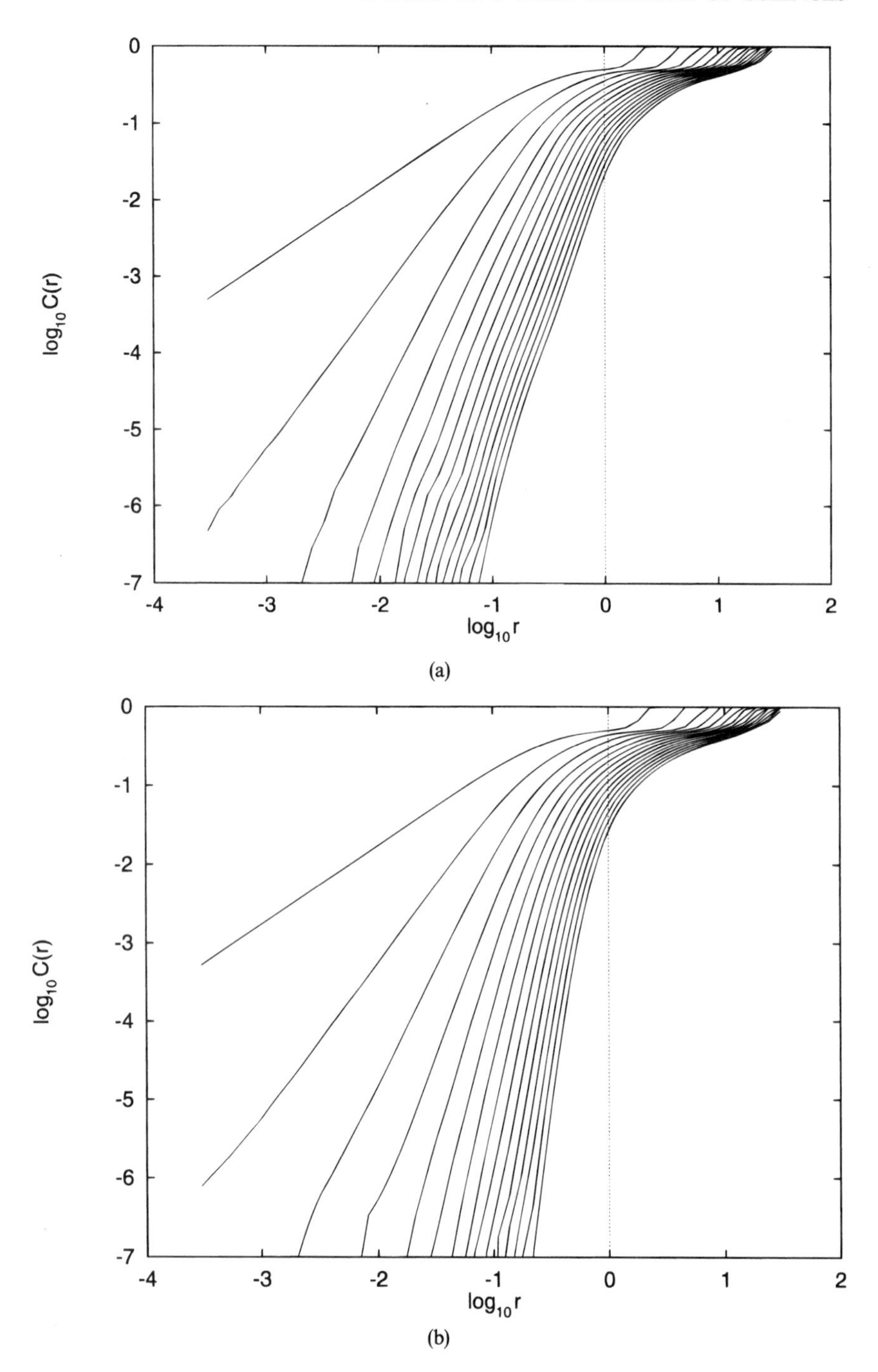

FIGURE 6. Panels (a)–(c) show the correlation integrals for the time series shown respectively in FIGURES 1(b), 2(a) and 2(b). Here the maximum embedding dimension is $M = 15$ and the time delay used in the embedding procedure is $\tau = 4\Delta t = 2$. **Panel d** shows the logarithmic slope ν of the correlation integral as a function of the embedding dimension M.

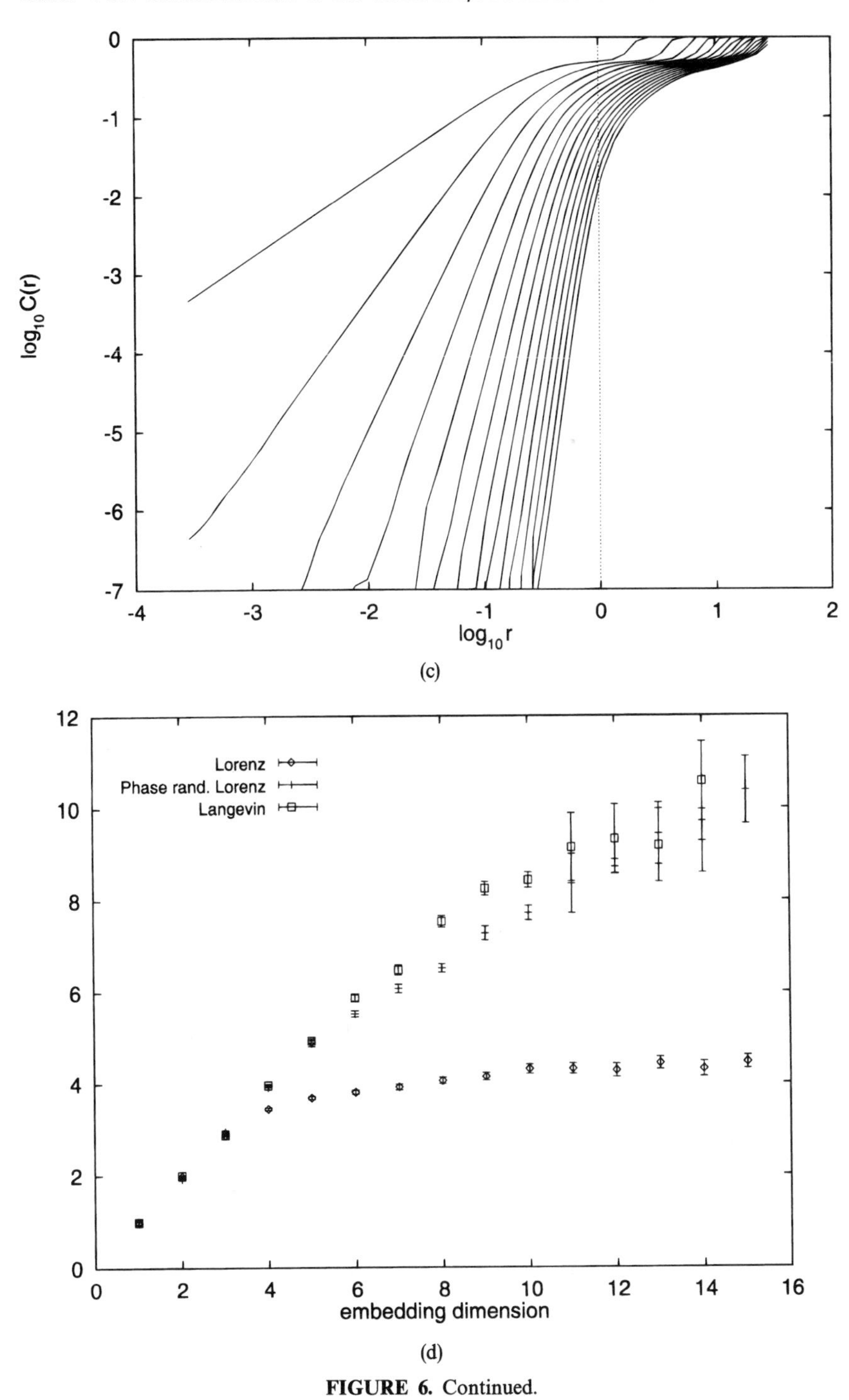

(c)

(d)

FIGURE 6. Continued.

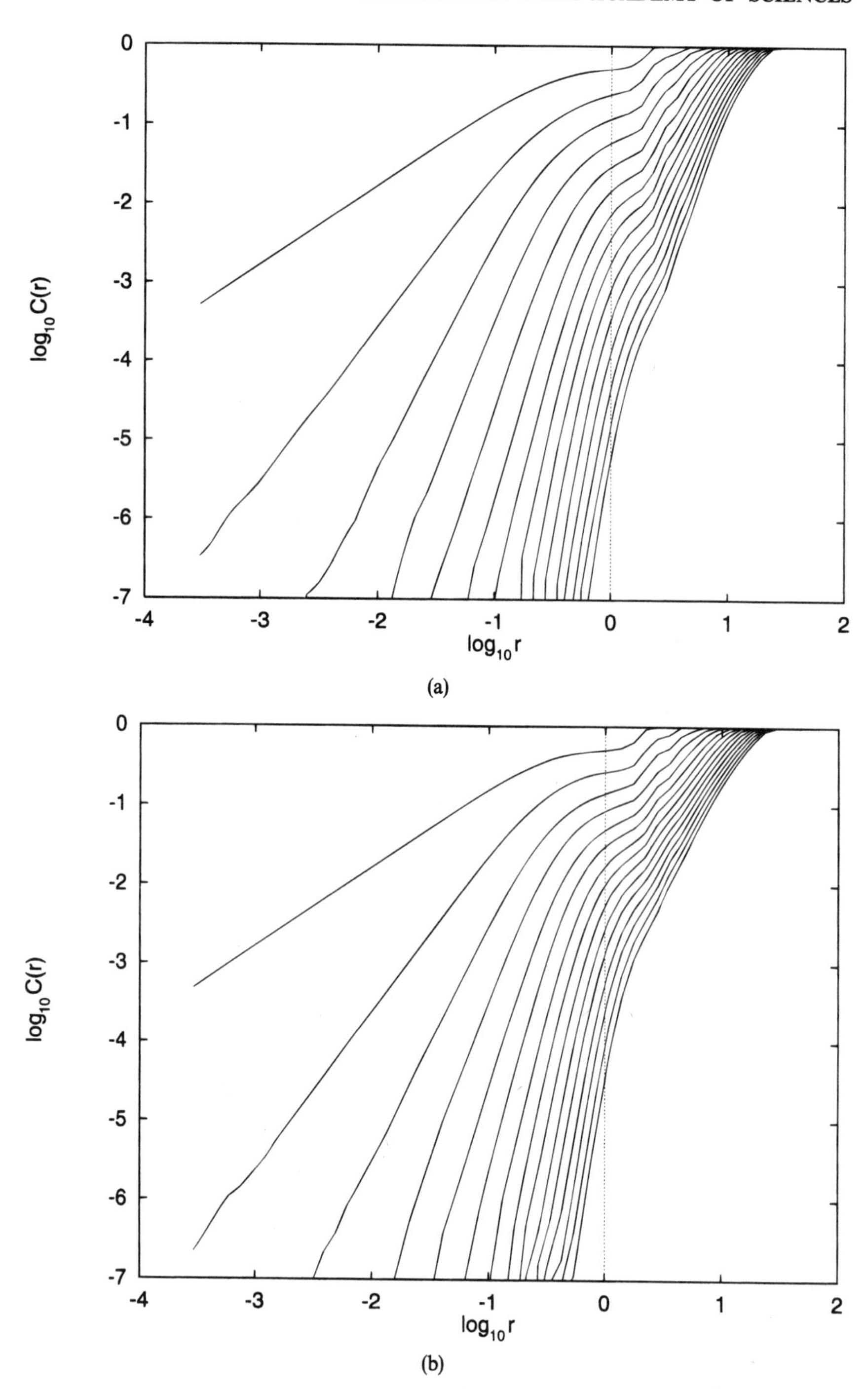

FIGURE 7. (a & b) The correlation integrals for the time series obtained by using a driver generated by respectively the Lorenz model and the Langevin equation. Here the sampling time is $\Delta t = 40$. The maximum embedding dimension is $M = 15$ and the time delay used in the embedding procedure is $\tau = \Delta t'$. **Panel c** shows the logarithmic slope ν of the correlation integrals as a function of the embedding dimension M.

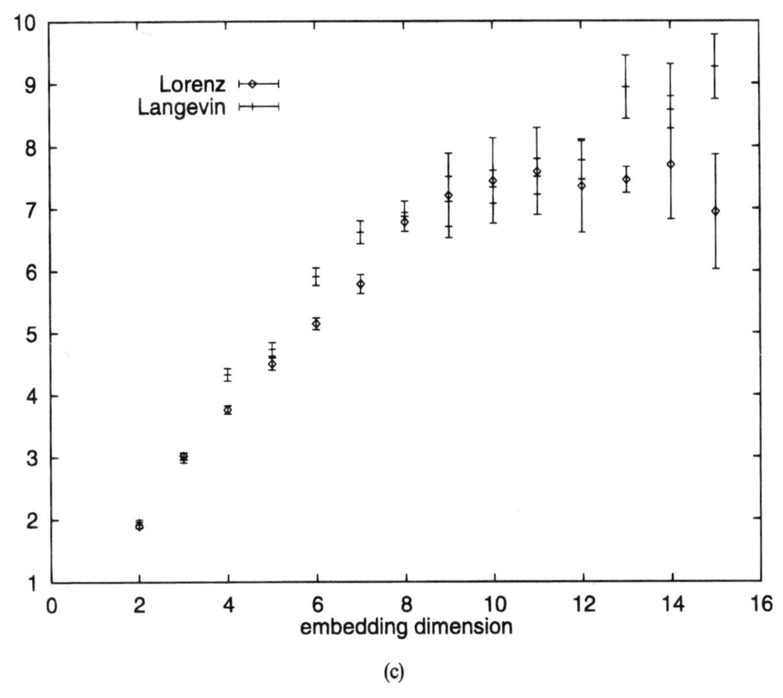

(c)

FIGURE 7. Continued.

the unsmoothed form of the mutual information, one gets the results shown in FIGURES 6(a–c). In this case, two well-separated scaling regions are present. At large scales the stochastic and chaotic signals are still indistinguishable, whereas at small scales the chaotic time series leads to a convergent estimate of the dimension, while the stochastic one does not. FIGURE 6(d) shows the values of the logarithmic slope ν of the correlation integrals versus the embedding dimension M for the small-scale regime.

The Importance of Small-Scale Details

The key to interpreting the behavior illustrated above is the recognition of the existence of two well-separated time scales. The first is the slow modulation time of the potential U, and the other is the time scale of the (chaotic or stochastic) forcing. (Possibly, there is a third significant time scale, $2\pi/\sqrt{p}$, the period of the small oscillations of the system around the minima of U.)

Provided one has access to the small time scales, it is clearly possible to see the nature of the forcing by eliminating the contribution of the slow motion and taking the (numerical) time difference of the signal. The new time series is $y_n = x_{n+1} - x_n$. For chaotic signals it is known that this procedure does not change the correlation dimension.[5] The correlation integrals obtained in this case are very similar to those shown in FIGURE 6. It is then easy to distinguish the stochastic from the chaotic dynamics by analyzing the time series of y_n.

On the other hand, if one cannot resolve the small time scales, it is not possible to distinguish readily between the two types of resonance. In support of this claim, we show in FIGURES 7(a,b) the correlation integrals for two long time series with the (deterministic) Lorenz and the (random) Langevin drivers, but sampled infrequently, with $\Delta t = 40$, much larger than the typical time scales of both the chaotic and the stochastic drivers. No distinction is now visible between the different drivers, nor do other standard methods of time series analysis (such as Fourier phase randomization) ameliorate the situation. FIGURE 7(c) shows the values of the logarithmic slope v of the correlation integrals versus the embedding dimension M.

CONCLUSIONS

In the resonance process studied here, the long time behavior is forced by the short time dynamics, which may be either chaotic or stochastic. When we estimate the correlation dimension of the whole system, employing a large value of τ makes particularly evident the large amplitude variations induced by the slow alternation between the two potential wells. Such a value of τ is much greater than the Lyapunov time of the chaotic system that we use, so the low-dimensional forcing produces as stochastic an output as the random one. A much smaller value of τ, on the other hand, can disentangle the small-scale structures where the information about the forcing resides, allowing us to distinguish between the two types of forcing.

Clearly, in order to properly identify the type of forcing it is necessary to have access to the very small time scales of the signal. Whenever this is not possible, the analysis of only the long time behavior (namely, the slow alternation between the two wells of the potential) does not apparently allow the complete determination of the nature of the forcing system.

REFERENCES

1. ABARBANEL, H. D. I., R. BROWN, J. J. SIDOROWICH & L. TSIMRING. 1993. Rev. Mod. Phys. **65:** 133.
2. WOLF, A. & T. BESSOIR. 1991. Physica **50:** 239.
3. OTT, E., T. SAUER & J. A. YORKE, Eds. 1994. Coping with Chaos. Wiley. New York.
4. OSBORNE, A. R. & A. PROVENZALE. 1989. Physica **35:** 357.
5. PROVENZALE, A., L. A. SMITH, R. VIO & G. MURANTE. 1992. Physica **58:** 31.
6. GRAF VON HARDENBERG, J., F. PAPARELLA, N. PLATT, A. PROVENZALE, E. A. SPIEGEL & C. TRESSER. 1996. Phys. Rev. E. In press.
7. PLATT, N., E. A. SPIEGEL & C. TRESSER. 1994. Phys. Rev. Lett. **70:** 279.
8. SPIEGEL, E. A. 1981. *In* Nonlinear Dynamics. R. H. G. Helleman, Ed.
9. MARZEC, C. J. & E. A. SPIEGEL. 1980. SIAM J. Appl. Math. **38:** 387.
10. POMEAU, Y. & P. MANNEVILLE. 1980. Comm. Math. Phys. **74:** 189.
11. BENZI, R., A. SUTERA & A. VULPIANI. 1981. J. Phys. A **14:** L453.
12. WIESENFELD, K. & F. MOSS. 1995. Nature **373:** 33.
13. MOSS, F., A. BULSARA & M. SHLESINGER, Eds. 1993. Proc. NATO Adv. Research Wksp. on Stochastic Resonance in Physics and Biology. J. Stat. Phys. **70.**
14. BULSARA, A. R. & L. GAMMAITONI. 1996. Physics Today **49:** 39.
15. GRASSBERGER, P. & I. PROCACCIA. 1983. Physica D **9:** 189.
16. THEILER, J. 1986. Phys. Rev. A **34:** 2427.
17. GALLAGER, R. G. 1968. Information Theory and Reliable Communication. Wiley. New York.
18. FRASER, A. M. & H. L. SWINNEY. 1986. Phys. Rev. A **33:** 1134.

Time-Frequency Spatial-Spatial Frequency Representations[a]

LEON COHEN

Hunter College and Graduate Center of CUNY
New York, New York 10021

INTRODUCTION

For many signals in nature the frequency content, the spectrum, changes in time. The basic reason is that the physical processes producing the signal are time dependent. The attempt to understand how the spectrum of signals change in time is called time-frequency analysis. In recent years there has been significant progress made in developing the mathematical and physical ideas for describing a time-varying spectrum.[1] Analogous to time-frequency analysis is spatial-spatial frequency analysis. It generally deals with images and how their spatial frequency content changes with location.[2–4]

It appears, though, that a combined time-frequency spatial-spatial frequency analysis has not been developed and it is the purpose of this chapter to do so. Possible applications are situations where there is considerable variation in both space and time, such as on the surface of the sun. Also, for propagation of pulses we expect there to be a strong relation between spatial frequencies and ordinary frequencies and hence we need a tool that would establish these correlations or lack of them.

Before developing these representations we give a brief outline for the time-frequency case so that we can see how to generalize to this situation.

CHARACTERISTIC FUNCTION OPERATOR METHOD

For a density of two variables $P(a, b)$ the characteristic function is defined by[b]

$$M(\theta, \tau) = \iint P(t, \omega) e^{i\theta t + i\tau\omega}\, dt\, d\omega = \langle e^{i\theta t + i\tau\omega} \rangle \qquad (1)$$

and, inversely, the density can be obtained from the characteristic function,

$$C(t, \omega) = \frac{1}{4\pi^2} \iint M(\theta, \tau) e^{-i\theta t - i\tau\omega}\, d\theta\, d\tau. \qquad (2)$$

[a]Work supported in part by the NSA HBCU/MI Program and the PSC-CUNY Research Award Program. Also, this work was done as part of a NATO Collaborative Research Grants Program.

[b]All integrals go from $-\infty$ to ∞.

In the case where the variables are represented by operators one generalizes the concept of characteristic function by the following idea. First, replace the variables by operators. Second, calculate the characteristic function directly from the signal by the usual quantum mechanical procedure of calculating averages. The logic of this procedure is that the characteristic function is an average, the average of $e^{i\theta t + i\tau\omega}$, where θ and τ are parameters. Therefore we write

$$M(\theta, \tau) = \langle \mathcal{M}(\theta, \tau; \mathcal{T}, \mathcal{W}) \rangle = \int s^*(t) \mathcal{M}(\theta, \tau; \mathcal{T}, \mathcal{W}) s(t)\, dt, \tag{3}$$

where $\mathcal{M}$ is called the characteristic function operator that corresponds to the classical quantity $e^{i\theta t + i\tau\omega}$, and where $\mathcal{T}$, $\mathcal{W}$ are the time and frequency operators. However, since the time and frequency operators do not commute there are an infinite number of correspondences. Some examples are $e^{i\theta\mathcal{T} + i\tau\mathcal{W}}$ or $e^{i\theta\mathcal{T}}\, e^{i\tau\mathcal{W}}$. Clearly, there are an infinite number of choices, hence an infinite number of characteristic functions and hence an infinite number of joint representations. All the possibilities can be characterized by the following form:

$$\mathcal{M}(\theta, \tau; \mathcal{T}, \mathcal{W}) = \phi(\theta, \tau) e^{i\theta\mathcal{T} + i\tau\mathcal{W}}, \tag{4}$$

where $\phi(\theta, \tau)$ is a two-dimensional function, called the kernel.[5,6]

Now, the time and frequency operators are

$$\mathcal{T} = t, \qquad \mathcal{W} = \frac{1}{i}\frac{d}{dt} \qquad \text{in the time domain} \tag{5}$$

$$\mathcal{T} = i\frac{d}{d\omega}, \qquad \mathcal{W} = \omega \qquad \text{in the frequency domain} \tag{6}$$

and therefore

$$M(\theta, \tau) = \int s^*(t) \mathcal{M} s(t)\, dt \tag{7}$$

$$= \phi(\theta, \tau) \int s^*(t) e^{i\tau\mathcal{W} + i\theta\mathcal{T}} s(t)\, dt \tag{8}$$

$$= \phi(\theta, \tau) \int s^*(u - \tfrac{1}{2}\tau) e^{i\theta u} s(u + \tfrac{1}{2}\tau)\, du. \tag{9}$$

Using (2), the distribution (or representation) is[5,6]

$$C(t, \omega) = \frac{1}{4\pi^2} \iint M(\theta, \tau)\, e^{-i\theta t - i\tau\omega}\, d\theta\, d\tau \tag{10}$$

$$= \frac{1}{4\pi^2} \iiint s^*(u - \tfrac{1}{2}\tau) s(u + \tfrac{1}{2}\tau) \phi(\theta, \tau) e^{-i\theta t - i\tau\omega + i\theta u}\, du\, d\tau\, d\theta. \tag{11}$$

A particular choice of kernel determines the particular representation. As we span all possible $\phi(\theta, \tau)$'s we span all possible representations. The Wigner distribution is obtained by taking $\phi = 1$. We would like the distribution to satisfy marginals of time and frequency,

$$\int C(t, \omega)\, d\omega = |s(t)|^2 \tag{12}$$

$$\int C(t, \omega)\, dt = |S(\omega)|^2, \tag{13}$$

where S is the spectrum

$$S(\omega) = \frac{1}{\sqrt{2\pi}} \int s(t) e^{-i\omega t}\, dt. \tag{14}$$

This will be the case if the kernel satisfies $\phi(0, \tau) = \phi(\theta, 0) = 1$.

Space–Time Signals

Although we have used time and frequency, the above can be repeated for spatial-spatial frequency and the same functional form would be obtained. Specifically if we have a signal in space and time, $s(x, t)$, we can use the above procedure on one variable at a time while keeping the other fixed. We define the various Fourier transforms by[c]

$$S(\omega, x) = \frac{1}{\sqrt{2\pi}} \int s(x, t) e^{-i\omega t}\, dt \tag{15}$$

$$S(k, t) = \frac{1}{\sqrt{2\pi}} \int s(x, t) e^{-ikx}\, dx \tag{16}$$

$$S(k, \omega) = \frac{1}{2\pi} \int\int s(x, t) e^{-i\omega t - ikx}\, dt\, dx, \tag{17}$$

where k denotes the spatial frequency. The two individual distributions are

$$C(t, \omega; x) = \frac{1}{4\pi^2} \int\int\int s^*(x, u - \tfrac{1}{2}\tau) s(x, u + \tfrac{1}{2}\tau) \phi(\theta, \tau) e^{-i\theta t - i\tau\omega + i\theta u}\, du\, d\tau\, d\theta \tag{18}$$

and

$$C(x, k; t) = \frac{1}{4\pi^2} \int\int\int s^*(u_x - \tfrac{1}{2}\tau_x, t) s(u_x + \tfrac{1}{2}\tau_x, t) \times \phi(\theta_x, \tau_x) e^{-i\theta_x x - i\tau_x k + i\theta_x u_x}\, du_x\, d\tau_x\, d\theta_x. \tag{19}$$

[c]We use the same symbol $S(,)$ to denote the various Fourier transforms, for example $S(\omega, x)$ and $S(k, t)$. The individuality of the function is determined by the variables. Also, we have chosen a symmetrical definition in (17) for space and time, although for more conventional purposes we would take $e^{-i\omega t + ikx}$ as we do in the final section.

These distributions satisfy the following marginals:

$$\int C(t, \omega; x)\, d\omega = |s(x, t)|^2 \tag{20}$$

$$\int C(t, \omega; x)\, dt = |S(x, \omega)|^2 \tag{21}$$

$$\int C(x, k; t)\, dk = |s(x, t)|^2 \tag{22}$$

$$\int C(x, k; t)\, dx = |S(k, t)|^2. \tag{23}$$

JOINT REPRESENTATIONS

Our aim is to obtain the *four*-dimensional joint representation of x, t, k, ω. We define the operators by

$$\mathcal{T} = t, \qquad \mathcal{W} = \frac{1}{i}\frac{d}{dt} \tag{24}$$

$$\mathcal{X} = x, \qquad \mathcal{W}_x = \frac{1}{i}\frac{d}{dx}. \tag{25}$$

For the characteristic function operator we take

$$\mathcal{M}(\theta_x, \tau_x, \theta, \tau; \mathcal{X}, \mathcal{W}_x, \mathcal{T}, \mathcal{W}) = \phi(\theta_x, \tau_x, \theta, \tau) e^{i\theta_x \mathcal{X} + i\tau_x \mathcal{W}_x + i\theta\mathcal{T} + i\tau\mathcal{W}} \tag{26}$$

and therefore the characteristic function is

$$M(\theta_x, \tau_x, \theta, \tau) = \iint s^*(x, t) \mathcal{M} s(x, t)\, dt\, dx \tag{27}$$

$$= \phi(\theta_x, \tau_x, \theta, \tau) \iint s^*(u_x - \tfrac{1}{2}\tau_x, u - \tfrac{1}{2}\tau) e^{i\theta_x u_x + i\theta u}$$

$$\times\, s(u_x + \tfrac{1}{2}\tau_x, u + \tfrac{1}{2}\tau)\, du\, du_x. \tag{28}$$

The distribution is then

$$C(x, k, t, \omega) = \left(\frac{1}{2\pi}\right)^4 \iiiint M(\theta_x, \tau_x, \theta, \tau) e^{-i\theta_x x - i\tau_x k - i\theta t - i\tau\omega}$$

$$\times\, d\theta_x\, d\tau_x\, d\theta\, d\tau \tag{29}$$

$$= \left(\frac{1}{2\pi}\right)^4 \iiiiiint s^*(u_x - \tfrac{1}{2}\tau_x, u - \tfrac{1}{2}\tau) s(u_x + \tfrac{1}{2}\tau_x, u + \tfrac{1}{2}\tau)$$

$$\times\, \phi(\theta_x, \tau_x, \theta, \tau) \tag{30}$$

$$= e^{-i\theta_x x - i\tau_x k + i\theta_x u_x} e^{-i\theta t - i\tau\omega + i\theta u}\, du_x\, d\tau_x\, d\theta_x\, du\, d\tau\, d\theta. \tag{31}$$

For the case where $\phi = 1$, we have

$$W(x, k, t, \omega) = \left(\frac{1}{2\pi}\right)^2 \iint s^*(x - \tfrac{1}{2}\tau_x, t - \tfrac{1}{2}\tau)s(x + \tfrac{1}{2}\tau_x, t + \tfrac{1}{2}\tau) \times e^{-i\tau_x k - i\tau\omega}\, d\tau\, d\tau_x. \tag{32}$$

Two other useful forms are

$$W(x, k, t, \omega) = \left(\frac{1}{2\pi}\right)^2 \iint S^*(k + \tfrac{1}{2}\theta_x, \omega + \tfrac{1}{2}\theta)S(k - \tfrac{1}{2}\theta_x, \omega - \tfrac{1}{2}\theta) \times e^{-i\theta_x x - i\theta t}\, d\theta\, d\theta_x \tag{33}$$

and

$$W(x, k, t, \omega) = \left(\frac{1}{2\pi}\right)^2 \iint S^*(k + \tfrac{1}{2}\theta_x, t - \tfrac{1}{2}\tau)S(k - \tfrac{1}{2}\theta_x, t - \tfrac{1}{2}\tau) \times e^{-i\theta_x x - i\tau\omega}\, d\theta\, d\tau, \tag{34}$$

where the S's are given by (**16**) and (**17**).

For the sake of simplicity we will, in this chapter, restrict ourselves to (**32**), although most of the properties derived can be readily generalized to (**31**) by appropriate choice of kernel constraints. The advantages of using other distributions will be discussed elsewhere.

MARGINALS AND INSTANTANEOUS FREQUENCY

Marginals

It is readily verified that the two-dimensional marginals are satisfied,

$$\int C(x, k, t, \omega)\, d\omega\, dk = |s(x, t)|^2 \tag{35}$$

$$\int C(x, k, t, \omega)\, dt\, dx = |S(k, \omega)|^2 \tag{36}$$

$$\int C(x, k, t, \omega)\, d\omega\, dt = C(x, k; t) \tag{37}$$

$$\int C(x, k, t, \omega)\, dx\, dk = C(t, \omega; x). \tag{38}$$

Instantaneous Frequency and Instantaneous Spatial Frequency

Consider the first conditional moment of frequency,

$$\langle \omega \rangle_{x, k, t} = \int \omega C(x, k, t, \omega)\, d\omega \tag{39}$$

$$= \left(\frac{1}{2\pi}\right)^2 \int\int\int \omega s^*(x - \tfrac{1}{2}\tau_x, t - \tfrac{1}{2}\tau) s(x + \tfrac{1}{2}\tau_x, t + \tfrac{1}{2}\tau)$$

$$\times e^{-i\tau_x k - i\tau\omega} \, d\tau \, d\tau_x \, d\omega \tag{40}$$

$$= \frac{1}{2\pi i} \int \frac{\partial}{\partial \tau} s^*(x - \tfrac{1}{2}\tau_x, t - \tfrac{1}{2}\tau) s(x + \tfrac{1}{2}\tau_x, t + \tfrac{1}{2}\tau)$$

$$\times e^{-i\tau_x k}\Big|_{\tau=0} \, d\tau_x \tag{41}$$

$$= -\frac{1}{2\pi}\frac{1}{2i} \int \left[\frac{\partial s^*(x - \frac{1}{2}\tau_x, t)}{\partial t} s(x + \tfrac{1}{2}\tau_x, t) - s^*(x - \tfrac{1}{2}\tau_x, t) \right.$$

$$\left. \times \frac{\partial s(x + \frac{1}{2}\tau_x, t)}{\partial t} \right] e^{-i\tau_x k} \, d\tau_x \tag{42}$$

$$= -\frac{1}{2\pi}\frac{1}{2i} \int \left[A(x + \tfrac{1}{2}\tau_x, t) \frac{\partial A(x - \frac{1}{2}\tau_x, t)}{\partial t} - A(x - \tfrac{1}{2}\tau_x, t) \right.$$

$$\times \frac{\partial A(x + \frac{1}{2}\tau_x, t)}{\partial t} - iA(x + \tfrac{1}{2}\tau_x, t) A(x - \tfrac{1}{2}\tau_x, t) \frac{\partial}{\partial t}$$

$$\left. \times \{\varphi(x + \tfrac{1}{2}\tau_x, t) + \varphi(x - \tfrac{1}{2}\tau_x, t)\} \right]$$

$$\times e^{i\varphi[x + (1/2)\tau_x, t] - i\varphi[x - (1/2)\tau_x, t] - i\tau_x k} \, d\tau_x, \tag{43}$$

where in (**43**) we have expressed the signal in terms of its phase and amplitude

$$s(x, t) = A(x, t) e^{i\varphi(x, t)}. \tag{44}$$

The first conditional moment, $\langle \omega \rangle_{x, k, t}$, may be interpreted as the instantaneous frequency for a given spatial point and spatial frequency at a given time. If we further average over all spatial frequencies we obtain

$$\langle \omega \rangle_{x, t} = A^2(x, t) \frac{\partial}{\partial t} \varphi(x, t) \tag{45}$$

which is called the instantaneous frequency. The analogous quantity, $\langle k \rangle_{x, t, \omega}$, can be immediately written down by appropriate transcription.

SPECIAL CASES

Product Signals

Suppose the signal is of the form

$$s(x, t) = s_1(x) s_2(t); \tag{46}$$

then

$$C(x, k, t, \omega) = C_1(x, k)C_2(t, \omega), \tag{47}$$

where C_1 and C_2 are the two-dimensional densities given by (**18**) and (**19**).

Wave without Damping

For a wave of the form

$$s(x, t) = f(x - ct) \tag{48}$$

one has

$$C(x, k, t, \omega) = C_f(x - ct, k)\delta(\omega + ck), \tag{49}$$

where C_f is the two-dimensional distribution of $f(x)$ as given by (**19**). Therefore, as expected, for such waves, we have total correlation between frequency and spatial frequency.

Damping

Many damped waves have the form

$$s(x, t) = A(t)f(x - ct). \tag{50}$$

In this case we obtain

$$C(x, k, t, \omega) = C_f(x - ct, k)C_A(t, \omega + ck), \tag{51}$$

where C_f and C_A are respectively the two-dimensional densities of the $f(x)$ and $A(t)$.

WAVE PACKET MOTION

We now consider the first conditional moment of position. It is given by

$$\langle x \rangle_{k,t,\omega} = \int xC(x, k, t, \omega)\, dx. \tag{52}$$

Using (**34**) we have

$$\langle x \rangle_{k,t,\omega} = \left(\frac{1}{2\pi}\right)^2 \iiint xS^*(k + \tfrac{1}{2}\theta_x, t - \tfrac{1}{2}\tau)S(k - \tfrac{1}{2}\theta_x, t + \tfrac{1}{2}\tau) \times e^{-i\theta_x x - i\tau\omega}\, d\theta\, d\tau\, dx \tag{53}$$

$$= \frac{1}{2\pi i} \int \frac{\partial}{\partial \theta_x} S^*(k + \tfrac{1}{2}\theta_x, t - \tfrac{1}{2}\tau)S(k - \tfrac{1}{2}\theta_x, t + \tfrac{1}{2}\tau)e^{-i\tau\omega}|_{\theta_x=0}\, d\tau \tag{54}$$

$$= \frac{1}{2\pi}\frac{1}{2i} \int \left(\frac{\partial S^*(k, t - \frac{1}{2}\tau)}{\partial k} S(k, t + \tfrac{1}{2}\tau) - S^*(k, t - \tfrac{1}{2}\tau)\frac{\partial S(k, t + \frac{1}{2}\tau)}{\partial k}\right) \times e^{-i\tau\omega}\, d\tau. \tag{55}$$

We now take the special form

$$S(k, t) = F(k)e^{i\eta(k, t)} \tag{56}$$

with η real. This encompasses a large variety of propagating pulses. Substituting this form into (**55**) we obtain

$$\begin{aligned}\langle x \rangle_{k,t,\omega} &= \frac{1}{4\pi i}[FF^{*\prime} - F^*F'] \int e^{i\{\eta[k, t+(1/2)\tau] - \eta[k, t-(1/2)\tau]\} - i\tau\omega}\, d\tau \\ &\quad - \frac{1}{4\pi}|F|^2 \int \frac{\partial}{\partial k}\{\eta(k, t+\tfrac{1}{2}\tau) + \eta(k, t-\tfrac{1}{2}\tau)\} \\ &\quad \times e^{i\{\eta[k, t+(1/2)\tau] - \eta[k, t-(1/2)\tau]\} - i\tau\omega}\, d\tau \\ &= \frac{1}{2\pi}\eta_0'(k)|F(k)|^2 \int e^{i\{\eta[k, t+(1/2)\tau] - \eta[k, t-(1/2)\tau]\} - i\tau\omega}\, d\tau \\ &\quad - \frac{1}{4\pi}|F(k)|^2 \int \frac{\partial}{\partial k}\{\eta(k, t+\tfrac{1}{2}\tau) + \eta(k, t-\tfrac{1}{2}\tau)\} \\ &\quad \times e^{i\{\eta[k, t+(1/2)\tau] - \eta[k, t-(1/2)\tau]\} - i\tau\omega}\, d\tau,\end{aligned} \tag{57}$$

where in (**57**) we have written $F(k)$ in terms of its amplitude and phase

$$F(k) = |F(k)|e^{i\eta_0(k)}. \tag{58}$$

Specializing further, we consider

$$\eta(k, t) = \eta_1(k)t + \eta_2(k)t^2/2, \tag{59}$$

in which case we have

$$\begin{aligned}\langle x \rangle_{k,t,\omega} &= \frac{1}{2\pi}\eta_0'(k)|F(k)|^2 \int e^{i(\eta_1(k) + t\eta_2(k) - \omega)\tau}\, d\tau \\ &\quad - \frac{1}{4\pi}|F(k)|^2 \int [2t\eta_1'(k) + 2(t^2 + \tau^2/4)\eta_2'(k)]e^{i(\eta_1(k) + t\eta_2(k) - \omega)\tau}\, d\tau\end{aligned} \tag{60}$$

$$\begin{aligned}&= \eta_0'(k)|F(k)|^2\delta(\omega - \eta_1(k) - t\eta_2(k)) - |F(k)|^2[(t\eta_1'(k) + t^2\eta_2'(k)) \\ &\quad \times \delta(\omega - \eta_1(k) - t\eta_2(k)) + \tfrac{1}{4}\eta_2'\,\delta''(\omega - \eta_1(k) - t\eta_2(k))].\end{aligned} \tag{61}$$

Let us first discuss the linear term, that is when $\eta_2 = 0$. In that case

$$\langle x \rangle_{k,t,\omega} = [\eta_0'(k) - t\eta_1'(k)]\delta(\omega - \eta_1(k))|F(k)|^2. \tag{62}$$

This equation can be interpreted as follows. The average of the pulse at a certain frequency travels linearly with a velocity equal to $(-\eta_1'(k))$, but the only frequencies allowed are those given by the solutions of $\omega = \eta_1(k)$, that is the solutions of the dispersion relation. We can further integrate over all frequencies to obtain

$$\langle x \rangle_{k,t} = \int \langle x \rangle_{k,t,\omega}\, d\omega = (\eta_0'(k) - t\eta_1'(k))|F(k)|^2 \tag{63}$$

and if we further integrate over all spatial frequencies we obtain

$$\langle x\rangle_t = \int \langle x\rangle_{k,t}\, dk = \int \eta_0'(k)\,|F(k)|^2 - t\int \eta_1'(k)\,|F(k)|^2. \tag{64}$$

But

$$\int \eta_0'(k)\,|F(k)|^2 = \langle x\rangle_0 \tag{65}$$

and therefore we have that

$$\langle x\rangle_t = \langle x\rangle_0 - t\int \eta_1'(k)\,|F(k)|^2\, dk. \tag{66}$$

This result is very interesting and was first derived by D. Vakman (priv. comm.) in a different way. It shows that the average of a pulse travels with a constant velocity. The velocity is the spatial group delay averaged with the initial spatial spectrum.

Let us now address the question of how to obtain $\langle x\rangle_{t,\omega}$ which is given by

$$\langle x\rangle_{t,\omega} = \int \langle x\rangle_{k,t,\omega}\, dk = \int [\eta_0'(k) - t\eta_1'(k)]\,|F(k)|^2\,\delta(\omega - \eta_1(k))\, dk. \tag{67}$$

First consider the linear case and assume that the solution to the equation $\omega - \eta_1(k) = 0$ is $k = g(\omega)$. Then

$$\langle x\rangle_{t,\omega} = \left(\frac{\eta_0'(k)}{\eta_1'(k)} - t\right)|F(k)|^2\,|_{k=g(\omega)}. \tag{68}$$

If there are many solutions then we would have the sum of such terms, ranging over the particular solutions.

For the quadratic case we have

$$\langle x\rangle_{k,t} = (\eta_0'(k) - t\eta_1'(k) - t^2\eta_2'(k))\,|F(k)|^2 \tag{69}$$

and also

$$\langle x\rangle_t = \langle x\rangle_0 - \int (t\eta_1'(k) + t^2\eta_2'(k))\,|F(k)|^2\, dk. \tag{70}$$

WAVE EQUATIONS AND WAVE PACKET MOTION

We discuss wave packet motion independently of the issue of joint representations. For many wave equations, the solution is of the form

$$u(x,t) = \frac{1}{\sqrt{2\pi}}\int S(k,0)e^{ikx - i\omega(k)t}\, dk \tag{71}$$

$$= \frac{1}{\sqrt{2\pi}}\int S(k,t)e^{ikx}\, dk, \tag{72}$$

where $S(k, 0)$ is the initial spatial spectrum. The spectrum at a later time t is

$$S(k, t) = S(k, 0)e^{-i\omega(k)t}. \tag{73}$$

For these types of equations of evolution the magnitude of the spatial spectrum is time independent. Now consider the mean value of x and x^2 as a function of time. The position operator in the k representation is

$$\mathscr{X} = i\frac{\partial}{\partial k} \tag{74}$$

and hence

$$\langle x \rangle_t = \int S^*(k, t)\mathscr{X} S(k, t)\, dk \tag{75}$$

$$\langle x^2 \rangle_t = \int S^*(k, t)\mathscr{X}^2 S(k, t)\, dk \tag{76}$$

$$= \int |\mathscr{X} S(k, t)|^2\, dk. \tag{77}$$

If we define the spatial group velocity by

$$v_g(k) = \frac{d\omega(k)}{dk} \tag{78}$$

then

$$\mathscr{X} S(k, t) = i\frac{\partial}{\partial k} S(k, t) = \left(i\frac{\partial S(k, 0)}{\partial k} + tv_g S(k, 0)\right)e^{-i\omega(k)t} \tag{79}$$

and

$$\langle x \rangle_t = \int S^*(k, 0)\mathscr{X} S(k, 0)\, dk + t\int v_g |S(k, 0)|^2\, dk \tag{80}$$

$$= \langle x \rangle_0 + t\int v_g |S(k, 0)|^2\, dk \tag{81}$$

$$= \langle x \rangle_0 + t\langle v_g \rangle_0. \tag{82}$$

Similarly

$$\mathscr{X}^2 S(k, t) = i^2 \frac{\partial^2}{\partial k^2} S(k, t) \tag{83}$$

$$= \left[-\frac{\partial^2}{\partial k^2} S(k, 0) + it\left(\frac{dv_g}{dk} S(k, 0) + 2v_g \frac{\partial S}{\partial k}\right) + v_g^2 t^2 S(k, 0)\right] \times e^{-i\omega(k)t} \tag{84}$$

and after using the Hermitian properties of $\mathscr{X}$ we obtain

$$\langle x^2\rangle_t = \langle x^2\rangle_0 + it \int v_g\left(S^*(k,0)\,\frac{dS(k,0)}{dk} - S(k,0)\,\frac{dS^*(k,0)}{dk}\right) dk + t^2\langle v_g^2\rangle_0 \quad (85)$$

$$= \langle x^2\rangle_0 + it \int S^*(k,0)\left(v_g\,\frac{d}{dk} + \frac{d}{dk}\,v_g\right)S(k,0)\,dk + t^2\langle v_g^2\rangle_0 \quad (86)$$

$$= \langle x^2\rangle_0 + t\langle(v_g\mathscr{X} + \mathscr{X}v_g)\rangle_0 + t^2\langle v_g^2\rangle_0. \quad (87)$$

Now consider the standard deviation, that is the spread of the wave packet as a function of time,

$$\sigma_t^2(x) = \langle x^2\rangle_t - \langle x\rangle_t^2 \quad (88)$$

$$= \langle x^2\rangle_0 + t^2\langle v_g^2\rangle_0 + t\langle v_g\mathscr{X} + \mathscr{X}v_g\rangle_0 - [\langle x\rangle_0 + t\langle v_g\rangle_0]^2 \quad (89)$$

or

$$\sigma_t^2(x) = \sigma_0^2(x) + 2t\,\mathrm{Cov}\,(v_g x) + t^2\sigma_0^2(v_g) \quad (90)$$

where we have defined the covariance by

$$\mathrm{Cov}\,(v_g x) = \tfrac{1}{2}\langle v_g\mathscr{X} + \mathscr{X}v_g\rangle_0 - \langle v_g\rangle_0\langle x\rangle_0. \quad (91)$$

A Classical View

We now ask what type of ordinary density gives results which have the same functional form as derived. Suppose we have a density or probability distribution of x, $P(x)$ at $t = 0$ and suppose we make the transformation

$$y = x + v(x)t, \quad (92)$$

where $v(x)$ is an arbitrary function. The probability distribution for y is given by

$$P_y(y) = \int P(x)\delta(y - x - v(x)t)\,dx. \quad (93)$$

Now consider the first moment of y

$$\langle x\rangle_t = \langle y\rangle = \int yP_y(y)\,dx \quad (94)$$

$$= \int yP(x)\delta(y - x - v(x)t)\,dx\,dy \quad (95)$$

$$= \int (x + v(x)t)P(x) \quad (96)$$

$$= \langle x\rangle_0 + \langle v(x)\rangle_0\,t. \quad (97)$$

Similarly

$$\langle x^2 \rangle_t = \langle y^2 \rangle = \int y^2 P(x)\delta(y - x - v(x)t)\, dx\, dy \tag{98}$$

$$= \int (x + v(x)t)^2 P(x) = \langle x^2 \rangle_0 + \langle v^2(x) \rangle_0 t^2 + 2\langle xv(x) \rangle_0 t. \tag{99}$$

Also,

$$\sigma_t^2(x) = \langle x^2 \rangle_0 + 2\langle xv(x) \rangle_0 t + \langle v^2(x) \rangle_0 t^2 - (\langle x \rangle_0 + \langle v(x) \rangle_0 t)^2 \tag{100}$$

$$= \sigma_0^2(x) + 2t \text{ Cov } (v(x)x) + t^2 \sigma_0^2(v_g), \tag{101}$$

where now the covariance is defined in the standard classical manner

$$\text{Cov } (xv) = \langle xv(x) \rangle_0 - \langle v \rangle_0 \langle x \rangle_0 . \tag{102}$$

Thus we see the similarity of these equations with (**82**) and (**90**).

Generalization[d]

A more general form is to take

$$u(x, t) = \int L(k, 0) h(k, x) T(k, t)\, dk \tag{112}$$

[d]An intermediate generalization is to consider wave equations that yield

$$u(x, t) = \frac{1}{\sqrt{2\pi}} \int S(k, 0) e^{ikx - i\omega(k, t)}\, dk \tag{103}$$

$$= \frac{1}{\sqrt{2\pi}} \int S(k, t) e^{ikx}\, dk \tag{104}$$

with

$$S(k, t) = S(k, 0) e^{-i\omega(k, t)}. \tag{105}$$

$$= \int L(k, t)h(k, x)\,dk, \tag{113}$$

where

$$L(k, t) = L(k, 0)T(k, t), \tag{114}$$

where k now is not the wave number but the value of some physical parameter and where h and T are complete sets of eigenfunctions. This form comes about from various nonlinear equations and also linear equations with coefficients that depend on position and time. In particular, equations of the form where[e]

$$\sum_n a_n(x) \frac{\partial^n}{\partial x^n} u(x, t) = \sum_n b_n(t) \frac{\partial^n}{\partial t^n} u(x, t), \tag{115}$$

where

$$\sum_n a_n(x) \frac{\partial^n}{\partial x^n} h(k, x) = \varepsilon(k)h(k, x) \tag{116}$$

$$\sum_n b_n(t) \frac{\partial^n}{\partial t^n} T(l, t) = \gamma(l)T(l, t) \tag{117}$$

will yield solutions for (**115**) provided that

$$\varepsilon(k) = \gamma(l). \tag{118}$$

This is a solution to a number of nonlinear equations. We now define the time-dependent group delay by

$$v_g(k, t) = \frac{\partial \omega(k, t)}{\partial k} \tag{106}$$

and the same steps as before lead to

$$\langle x \rangle_t = \langle x \rangle_0 + \langle v_g(k, t) \rangle_0 \tag{107}$$

$$\langle x^2 \rangle_t = \langle x^2 \rangle_0 + t^2 \langle v_g^2(k, t) \rangle_0 + \langle (v_g(k, t)\mathscr{X} + \mathscr{X} v_g(k, t)) \rangle_0 \tag{108}$$

$$\sigma_t^2(x) = \sigma_0^2(x) + \sigma_0^2(v_g(k, t)) + 2\ \mathrm{Cov}\ (v_g(k, t)x), \tag{109}$$

where now the notation $\langle v_g \rangle_0$ means that the average is taken with the distribution at time zero

$$\langle v_g \rangle_0 = \int v_g(k, t)\,|S(k, 0)|^2\,dk \tag{110}$$

but nonetheless $\langle v_g \rangle_0$ is time dependent because of the explicit time dependence of v_g. Also,

$$\mathrm{Cov}\ (v_g x) = \tfrac{1}{2}\langle v_g(k, t)\mathscr{X} + \mathscr{X} v_g(k, t)\rangle_0 - \langle v_g(k, t)\rangle_0 \langle x \rangle_0. \tag{111}$$

[e]Basically we are extending the energy representation of quantum mechanics to more general equations than the Schrödinger equation.

This equation can be called a generalized dispersion relation. It may be solved for l as a function of k or for k as a function of l. The general solution is then

$$u(x, t) = \int L(k, 0)h(k, x)T(l(k), t)\, dk \tag{119}$$

with

$$L(k, 0) = \int u^*(x, 0)h(k, x)\, dx. \tag{120}$$

Henceforth, instead of $T(l(k), t)$ we simply write $T(k, t)$. Or, we could write

$$u(x, t) = \int L(k(l), 0)h(k(l), x)T(l, t)\, dl. \tag{121}$$

The mean position and position square are

$$\langle x \rangle_t = \int L^*(k, t)\mathscr{X} L(k, t)\, dk \tag{122}$$

$$= \int L(k, 0)T(k, t)\mathscr{X} L(k, 0)T(k, t)\, dk \tag{123}$$

$$\langle x^2 \rangle_t = \int L^*(k, t)\mathscr{X}^2 L(k, t)\, dk \tag{124}$$

$$= \int |\mathscr{X} L(k, t)|^2\, dk, \tag{125}$$

where the position operator $\mathscr{X}$ must be expressed in the k representation. The advantage of this form is that it is in the natural representation of the operators and that is why the initial density appears explicitly. The penalty to be paid is that we have to have the operator in that representation.

Example 1. We now do some examples whose solutions are well known and hence we may check the answers we obtain by the methods outlined above. Consider the Schrödinger equation

$$-\frac{\hbar^2}{2m}\frac{\partial^2\psi(x, t)}{\partial x^2} - Fx = i\hbar\frac{\partial\psi(x, t)}{\partial t}, \tag{126}$$

where F is a constant force. It is well known that this equation can be solved explicitly in the momentum representation and also in the position representation where Airy functions arise.[f] As is standard in quantum mechanics, we write

$$\left(-\frac{\hbar^2}{2m}\frac{\partial^2\psi(x,t)}{\partial x^2} - Fx\right) = h(E,x) \tag{132}$$

$$i\hbar\frac{\partial T(E,t)}{\partial t} = ET(E,t). \tag{133}$$

In this case we have taken $k = l = E$. The point we are trying to make is that the natural representation for this problem is the energy representation because in that representation it evolves simply in time. The position and momentum operators are[7,8]

$$\mathscr{X} = -\frac{1}{F}\left[E + \frac{F^2\hbar^2}{2m}\frac{\partial^2}{\partial E^2}\right]; \quad \mathscr{P} = -F\frac{\partial}{\partial E} \tag{134}$$

and also

$$T(E,t) = e^{-iEt/\hbar}. \tag{135}$$

Now suppose we have an initial wave function in the energy representation $\eta(E) = L(E, 0)$; then

$$\mathscr{X}\eta(E)T(E,t) = -\frac{1}{F}\left[E\eta(E) + \frac{d^2\eta(E)}{dE^2} - \frac{2it}{\hbar}\frac{d\eta(E)}{dE} - \frac{t^2}{\hbar^2}\eta(E)\right] \tag{136}$$

[f] In the momentum representation the time-independent Schrödinger equation is

$$\frac{p^2}{2m}u_E(p) - i\hbar F\frac{\partial u_E(p)}{\partial p} = Eu_E(p) \tag{127}$$

and the normalized solution is

$$u_E(p) = \frac{1}{\sqrt{2\pi\hbar F}}e^{(i/\hbar)(Ep - p^3/6m)}. \tag{128}$$

Hence, the momentum wave function evolves according to

$$\phi(p,t) = \int \eta(E)u_E(p)e^{-iEt/\hbar}\,dE \tag{129}$$

with

$$\eta(E) = \int \phi(p,0)u_E^*(p)\,dp. \tag{130}$$

This yields

$$\phi(p,t) = \phi(p - Ft, 0)\exp\left[-\frac{i}{6m\hbar F}(3F^2pt^2 - 3Fp^2t - F^3t^3)\right]. \tag{131}$$

Using this one can verify (**140**) and also derive the operators given by (**134**).

and hence

$$\langle x \rangle_t = \int \eta(E, t)\mathscr{X}\eta(E, t) \tag{137}$$

$$= -\frac{1}{F}\left[\int E\,|\eta(E)|^2 - \frac{F\hbar^2}{2m}\int \eta(E)\,\frac{d^2\eta(E)}{dE^2}\,dE + \frac{F\hbar^2}{2m}\right.$$

$$\left.\times \int\left(\frac{2it}{\hbar} + \frac{t^2}{\hbar^2}\,|\eta(E)|^2\right) dE\right] \tag{138}$$

$$= \langle x \rangle_0 + \frac{i\hbar t F}{m}\int \eta(E)\,\frac{d\eta(E)}{dE}\,dE + \frac{F}{2m}\,t^2 \tag{139}$$

$$= \langle x \rangle_0 + \frac{\langle p \rangle_0}{m}\,t + \frac{F}{2m}\,t^2 \tag{140}$$

which is a well-known result.

Example 2. Consider the Schrödinger equation for the harmonic oscillator

$$-\frac{\hbar^2}{2m}\,\frac{\partial^2\psi(x, t)}{\partial x^2} - kx^2 = i\hbar\,\frac{\partial\psi(x, t)}{\partial t}. \tag{141}$$

In this case the energy values are discrete and integrals become summations. An arbitrary wave function is given by

$$\psi(x, t) = \sum c_n(t)u_n(x) \tag{142}$$

with

$$c_n(t) = c_n(0)e^{-iE_nt/\hbar} = c_n(0)e^{-i(n+1/2)\omega t} \tag{143}$$

and where the u_n's are the harmonic oscillator eigenfunctions. The position operator, $\mathscr{X}$, in the harmonic energy representation is

$$\mathscr{X} = \frac{1}{\alpha}\sqrt{\frac{n+1}{2}}\,(1 + \Delta_n) + \frac{1}{\alpha}\sqrt{\frac{n}{2}}\,(1 + \Delta_n)^{-1}, \tag{144}$$

where $\alpha = m\omega/\hbar$, $\omega = \sqrt{k/m}$ and Δ is the differencing operator, $\Delta c_n = c_{n+1} - c_n$. The average value of position is now

$$\langle x \rangle_t = \sum_n c_n^*(t)\mathscr{X}c_n(t) \tag{145}$$

$$= \sum_n c_n^*(0)e^{i(n+1/2)\omega t}\mathscr{X}c_n(0)e^{-i(n+1/2)\omega t}. \tag{146}$$

Now

$$\mathscr{X}c_n(t) = \mathscr{X}c_n(0)e^{-i(n+1/2)t\omega} \tag{147}$$

$$= e^{-i(n+1/2)\omega t}[c_{n+1}(0)e^{-i\omega t} + c_{n-1}(0)e^{i\omega t}], \tag{148}$$

where we have used $(1 + \Delta_n)^m c_n = c_{n+m}$. Therefore

$$\langle x \rangle_t = \sum_n c_n^*(0) e^{i(n+1/2)\omega t} \mathscr{X} c_n(0) e^{-i(n+1/2)\omega t} \tag{149}$$

$$= \sum_n [c_n^*(0) c_{n+1}(0) e^{-i\omega t} + c_n^*(0) c_{n-1}(0) e^{i\omega t}] \tag{150}$$

which is also a well-known result.

SHORT SPACE–TIME FOURIER TRANSFORM

The spectrogram is a conceptually different approach to defining time-frequency distributions. However, it has been shown that it can be derived via (**11**) by choosing a particular kernel. It is interesting to consider what new features are presented in defining the spectrogram for the situation we have been considering.

To study the properties of the wave at time t and position x, we emphasize the signal at that space–time point and relatively suppress it at other points. To achieve this we multiply the signal by a window function, $h(x, t)$, centered at x, t, to produce a modified signal,

$$s_{x,t}(x', t') = s(x', t') h(x' - x, t' - t). \tag{151}$$

The modified signal is a function of four variables, the fixed time and space point we are interested in and the running space-time point x', t'. The window function is chosen to leave the signal more or less unaltered around the point x, t, but to suppress it at distant points.

The double Fourier transform will reflect the distribution of frequency and spatial frequency around the space–time point,

$$S_{x,t}(k, \omega) = \frac{1}{2\pi} \iint e^{-j\omega t' - jkx'} s_{x,t}(x', t')\, dx'\, dt' \tag{152}$$

$$= \frac{1}{2\pi} \iint e^{-j\omega t' - jkx'} s(x', t') h(x' - x, t' - t)\, dx'\, dt'. \tag{153}$$

The spectrogram is then

$$P_{SP}(x, t, k, \omega) = |S_{x,t}(k, \omega)|^2$$

$$= \left| \frac{1}{2\pi} \iint e^{-j\omega t' - jkx'} s(x', t') h(x' - x, t' - t) \right|^2 dx'\, dt'. \tag{154}$$

This definition of the two-dimensional spectrogram introduces new features over the one-dimensional case because we have two fundamental choices for the window function. First we can keep the window function as a two-dimensional function and second we can take the simpler product function form, $h(x, t) = h_1(x) h_2(t)$. The latter case produces relations very similar to the one-dimension case. But the more general case allows for a much greater variety of window functions. These types of spectrograms will be discussed elsewhere.

CONCLUSIONS

There is an interesting viewpoint that can be taken regarding these types of joint representations. To develop this idea we consider the calculation $\langle k\omega \rangle_{x,t}$:

$$\langle k\omega \rangle_{x,t} = \int k\omega C(x, k, t, \omega)\, dk\, d\omega \tag{155}$$

$$= \int k \langle \omega \rangle_{x,k,t}\, dk, \tag{156}$$

where $\langle \omega \rangle_{x,k,t}$ is given by (**43**). One obtains

$$\langle k\omega \rangle_{x,t} = A^2(x, t) \frac{\partial \phi(x, t)}{\partial t} \frac{\partial \phi(x, t)}{\partial x}. \tag{157}$$

Therefore we see that one can think of

$$\frac{\partial \phi(x, t)}{\partial t} \quad \text{and} \quad \frac{\partial \phi(x, t)}{\partial x}$$

as the frequency and spatial frequency in the x, t representation. Using this idea, answers can generally be written down immediately without recourse to calculations. For example, suppose we want $\langle k \rangle_{x,t}$; then we can immediately write

$$\langle k \rangle_{x,t} = A^2(x, t) \frac{\partial \phi(x, t)}{\partial x} \tag{158}$$

and so forth. Now the same viewpoint can be taken in the spectral domain. Suppose we want to calculate $\langle xt \rangle_{k,\omega}$. In the spectral domain x, t can be replaced by

$$\left(-\frac{\partial \psi(k, \omega)}{\partial \omega}\right) \quad \text{and} \quad \left(-\frac{\partial \psi(k, \omega)}{\partial k}\right),$$

where $\psi(k, \omega)$ is the spectral phase defined by

$$S(k, \omega) = B(k, \omega) e^{i\psi(k, \omega)}. \tag{159}$$

Therefore we can immediately write

$$\langle xt \rangle_{k,\omega} = B^2(k, \omega) \frac{\partial \psi(k, \omega)}{\partial \omega} \frac{\partial \psi(k, \omega)}{\partial k}. \tag{160}$$

Although our motive in discussing the above was to show a different viewpoint and to show that one can save considerable calculations, in some cases these relations also allow us to define the covariance between variables. The local covariance of frequency and spatial frequency is

$$\mathrm{Cov}_{x,t}(k\omega) = A^2(x, t)\left(\frac{\partial \phi(x, t)}{\partial t} \frac{\partial \phi(x, t)}{\partial x} - \frac{\partial \phi(x, t)}{\partial t} - \frac{\partial \phi(x, t)}{\partial x}\right) \tag{161}$$

and the global covariance is

$$\mathrm{Cov}\,(k\omega) = \int\!\!\int A^2(x, t)\left(\frac{\partial\phi(x, t)}{\partial t}\,\frac{\partial\phi(x, t)}{\partial x} - \frac{\partial\phi(x, t)}{\partial t} - \frac{\partial\phi(x, t)}{\partial x}\right) dx\, dt. \quad \mathbf{(162)}$$

Similar expressions can be written down for the covariance of position and time.

ACKNOWLEDGMENTS

I thank Gabriel Cristobal for discussions regarding spatial–spatial frequency representations, and David Vakman for discussions regarding propagating pulses.

REFERENCES

1. Cohen, L. 1989. Proc. IEEE **77:** 941 for a general review and list of references; Cohen, L. 1995. Time-Frequency Analysis. Prentice-Hall. Englewood Cliffs, N.J.
2. Cristobal, G., C. Gonzalo & J. Bescos. 1991. *In* Advances in Electronics and Electron Physics, Vol. 80: 309. Academic Press. New York.
3. Reed, T. & H. Wechsler. 1988. Signal Process. **14:** 95.
4. Jacobson, L. D. & H. Wechsler. 1988. Signal Process. **14:** 37.
5. Cohen, L. 1966. J. Math. Phys. **7:** 781.
6. Cohen, L. 1976. J. Math. Phys. **17:** 1863.
7. Cohen, L. 1966. Am. J. Phys. **34:** 684.
8. Cohen, L. 1996. Phys. Lett. **A212:** 315.

Time-Frequency Analysis of Variable Star Light Curves[a]

Z. KOLLÁTH[b] AND J. R. BUCHLER

Department of Physics
University of Florida
Gainesville, Florida 32611

INTRODUCTION

The determination of the period variations of stellar light curves is a classical problem in astronomy. The so-called O − C method has been widely used to search for any deviation from the strictly repetitive occurrence of the maxima or minima of the light curves. The plot of the observed (O) minus the calculated (C) time of the extrema—the O − C diagram—provides useful information about long-term period changes and even about the motion of the variable star in a binary system. This method, however, breaks down when the signal contains more than one frequency or it has strong modulations.

The different forms of frequency spectra (mainly the Fourier transform) became an important tool for the analysis of multiperiodic variations, and the Fourier decomposition of the periodic light curves of Cepheids and RR Lyrae stars turned into a powerful device for the comparison of observations and theoretical models. The astronomical data are usually gapped and unevenly sampled because of the rhythm of days and nights, the weather conditions and telescope time availability. These sampling properties introduce many complications for the period determination, like the aliasing due to the spectrum of the observing window. This is one of the reasons that the comparison of the Fourier spectra of different observational segments has been the principal method for investigating the change in the periods and amplitudes of oscillations for a long time.

The worldwide observational campaigns for short period stars (like white dwarf and δ Scuti stars) and the collection of mostly amateur observations of long period (RV Tauri, Mira, Semiregular) variables have made it possible to use more sophisticated tools of time-frequency analysis. In the last years the wavelet analysis with the Morlet kernel[1] has been applied to light variations. In the following we take into consideration the wavelet only as a tool for time-frequency description. For the application in scaling and finding self-similar behavior we refer to reference 2 and to Scargle in this volume.

The Morlet wavelet was first introduced in the astronomical literature by Goupil *et al.*,[3] where they applied it to the light curves of white dwarf stars. Later the method was tested for synthetic signals representing typical astronomical observations by Szatmáry *et al.*[4] Semiregular, Mira and W Virginis type variables were

[a]This work was supported by the NSF (Grant Nos. AST92-18068 and INT94-15868).
[b]On leave from Konkoly Observatory, Budapest, Hungary.

recently investigated with the method.[5–8] The wavelet analysis was also performed on the solar p-modes[9] and the variation of the solar cycle.[10] A similar method, the Gábor transform, was used to analyze frequency variation of an Ap star.[11]

While the Morlet wavelet has been the primary tool in variable star astronomy, other time-frequency descriptions, like the generalized Wigner distribution (see, e.g., references 12 and 13), have been successfully applied for engineering problems. The main purpose of this chapter is to compare the performance of these different methods on variable star data. First, we give an introduction to the Morlet wavelet and Gábor transforms. The time-frequency distributions are then briefly described and finally we present the analysis of observational data.

THE GÁBOR AND THE WAVELET TRANSFORMS

The classical tool of time-frequency analysis is the short-time Fourier transform (STFT), frequently referred to also as windowed Fourier transform. The signal $f(t)$ is weighted by a time-localized window function $h(t)$ and then Fourier transformed:

$$F(t, \nu) = \int_{-\infty}^{+\infty} f(\tau)h^*(\tau - t) \exp(-2i\pi\tau\nu)\, d\tau. \tag{1}$$

The STFT was introduced by Gábor[14] with the Gaussian analyzing window:

$$h(t) = \exp(-t^2/(2\sigma^2)). \tag{2}$$

We refer to this time-frequency representation as the *Gábor transform* (GT). (The general form of the STFT is also frequently called GT, independently of the choice of the window.) The spectrogram is the power spectrum version ($|F(t, \nu)|^2$) of the STFT.[12]

The *wavelet transform* (WT) of a time series is defined as:

$$S(t, a) = a^{-1/2} \int_{-\infty}^{+\infty} f(\tau)g^*\left(\frac{\tau - t}{a}\right) dt, \tag{3}$$

where $g(t)$ is the kernel of the wavelet transformation. The variable a corresponds to the scale (period) parameter. We use the following simplified Morlet wavelet kernel:

$$g(x) = e^{-x^2/2 + icx}, \tag{4}$$

where the parameter c controls the ratio of the time and frequency resolution. (Its usual value is 2π.) As in our previous works[8,15] we use a modified version of the wavelet transform, transforming it to a time-frequency representation instead of the time-scale version:

$$T(t, \nu) = a^{-1/2}S(t, a(\nu)), \tag{5}$$

where ν is the frequency. The $a(\nu) = c/(2\pi\nu)$ scaling gives the correct frequency of a periodic function at the maximum of the wavelet modulus, i.e., the absolute value of the $T(t, \nu)$ and this maximum value is proportional to the signal amplitude. Thus for $x(t) = \cos(\omega t + \phi)$ one can obtain $|T| = (\pi/2)^{1/2}e^{-1/2(a\omega - c)^2}$.

It is more practical to calculate the wavelet from the Fourier transform ($F(\theta)$) of the signal. With the Morlet kernel this is given by

$$T^{+}(t, \nu) = \frac{1}{\pi} \int_{0}^{+\infty} F(\theta) \exp\left(-\frac{1}{2}\frac{(\theta - \nu)^2}{\sigma^2(\nu)}\right) \exp(2i\pi\theta t)\, d\theta, \qquad (6)$$

where $\sigma(\nu) = \nu/c$. Here we have introduced T^{+} calculating the one-sided inverse Fourier transform. (We discuss later why it is advantageous.) From this equation it is clear that the Morlet wavelet is given by a bandpass filtering, with a Gaussian frequency transfer function centered at the analyzing frequency ν and with a ν dependent bandwidth. Since the value of the Gaussian at $\theta = 0$ is $\exp(-c^2/2)$, the overlap of the filter to the negative frequencies is negligible when c is greater than ≈ 3. In this case $T^{+}(t, \nu) \approx T(t, \nu)$. On the other hand, when c is very small, i.e., when the bandwidth of the filtering is larger than the bandwidth of the signal, T^{+} reduces to the *analytic signal* of the function $f(t)$. (The analytic signal is a complex function whose real part is the signal and its Fourier transform vanishes for the negative frequencies. For the properties of the analytic signal see, e.g., reference 12.) According to our definition, the negative frequency part of the Fourier transform of $T^{+}(t, \nu)$ vanishes for any fixed value of ν, i.e., T^{+} is the analytic signal of the bandpass filtered function.

The previously described properties of T^{+} remain the same, even when we replace σ by a constant in (**6**). Then the bandwidth of the filter is constant, i.e., we get a filtered Fourier transform, which is equivalent to the Gábor transform apart from a factor of $\exp(2i\pi\nu t)$ (which disappears when one plots the absolute value or the power). What is then the difference between the two forms? The wavelet locates the high-frequency components more precisely in time and is advantageous for example if one searches for sudden changes in the signal. On the other hand, the price we have to pay for the higher temporal resolution is a higher contamination in the higher frequency part of the spectrum (for higher frequencies the frequency-transfer function is wider). The WT is disadvantageous when the harmonics of the fundamental frequency (or the overtones of that mode) have much smaller amplitudes than the fundamental one. The situation is reversed for the GT.

For the numerical realization of both the GT and WT we used (**6**). A fast inverse Fourier transform is calculated for all values of ν. For the GT we replace $\sigma(\nu)$ by $\sigma(\nu_0)$, where ν_0 is a constant frequency. By this definition the Gábor transform is matched to the WT at the frequency ν_0, i.e., the resolution is the same in both time-frequency maps at ν_0.

In most applications that have been performed on variable star data, generally only the modulus of the wavelet or Gábor transform has been investigated. However the phase of the transform gives important information about the frequency evolution of the signal, and it can provide a better estimation of the local frequency. The instantaneous frequency is defined by the time derivative of the phase of the analytic signal. Similarly one can define the instantaneous frequency around an average frequency ν_0 by:

$$\nu_{\text{inst}} = \frac{\partial}{\partial b} \phi(\nu_0, b), \qquad (7)$$

where $\phi(v, b) = \arg(T^{+}(v, b))$. Here the frequency window is expected to be real. For a more rigorous derivation of the instantaneous frequency of STFT see reference 16.

TIME-FREQUENCY DISTRIBUTIONS

The generalized form of time-frequency distributions (GTFD) was introduced by Cohen:[13]

$$C(t, v) = \frac{1}{2\pi} \int\int\int \exp(-i\theta t - 2i\pi\tau v + i\theta u)\Phi(\theta, \tau) f^*\left(u - \frac{\tau}{2}\right) f\left(u + \frac{\tau}{2}\right) du\, d\tau\, d\theta, \tag{8}$$

where $f(t)$ is the (generally complex) signal and $\Phi(\theta, \tau)$ is the kernel of the distribution. With the simplest kernel ($\Phi(\theta, \tau) = 1$) the definition of the GTFD reduces to the Wigner distribution (WD):

$$W(t, v) = \int \exp(-2i\pi\tau v) f^*\left(t - \frac{\tau}{2}\right) f\left(t + \frac{\tau}{2}\right) d\tau. \tag{9}$$

The WD of the function $\exp(i\Omega t)$ is $W(t, v) = \delta(2\pi v - \Omega)$, giving a peak at the exact frequency only. However, since the transformation performs a nonlinear operation on the signal, multicomponent (or multiperiodic) signals have cross terms in the WD. Similarly while the WD of an impulse ($\delta(t - t_0)$) is well localized at $t = t_0$, the combination of time localized functions can give nonzero WD even for times when the signal vanishes. These properties of the WD can be very disturbing when one investigates strongly modulated multiperiodic time series. To avoid this problem one can insert a kernel with localizing properties into the generalized time-frequency distribution.

For our applications we use the kernel defined by Choi and Williams,[17] i.e., $\Phi(\theta, \tau) = \exp(-\theta^2\tau^2/\sigma)$, giving the following distribution:

$$C(t, v) = \frac{1}{2\pi^{1/2}} \int\int (\tau^2\sigma)^{-1/2} \exp(-\sigma(u - t)^2/\tau^2 - 2i\pi\tau v) f^*\left(u - \frac{\tau}{2}\right) f\left(u + \frac{\tau}{2}\right) du\, d\tau. \tag{10}$$

Our numerical implementation is based on the discretized and windowed version of the Choi–Williams distribution (see reference 17, equation 20). It introduces two more parameters, the lengths of the window in the time and frequency domain (M and N respectively). In our calculations N is given by the maximum frequency, i.e., we fix this parameter. The length of the temporal window is given by M times the sampling time.

There is a straightforward connection between the time-frequency distribution and the spectrogram (Gábor transform). If one replaces the kernel $\Phi(\theta, \tau)$ by the *ambiguity function* (the two-dimensional Fourier transform of the WD) of the *window* of the spectrogram, then the resulting time-frequency distribution is the spectrogram itself.[12] Since the ambiguity function of a Gaussian is a two-dimensional Gaussian, the Gábor transform is given by a time-frequency distribution with the kernel $\Phi(\theta, \tau) = c \exp(-\tau^2/(4\sigma^2) - \sigma^2\theta^2)$. From this it is clear that both the CWD and the spectrogram can be generated from the same time-frequency distribution by slightly different exponential kernels. Similarly, generalized time-scale energy distributions were introduced in reference 18, extending the wavelet transform.

APPLICATION TO VARIABLE STARS

In this section we present the time-frequency analyses of two variable star light curves. The raw data consist of the visual estimates of the brightness. To reduce the observational noise and produce an evenly sampled time sequence, we have first averaged the data in several-day-long bins, then interpolated by a smoothing spline (for this procedure see reference 15). The comparison of the Fourier transforms of the smoothed and unsmoothed data (together with the corresponding spectral windows) indicates that this preprocessing does not alter the signal at the frequencies of interest. The sampling time of the smoothed data was 2 days for Scuti and 5 days for T Umi.

For all data sets we plot the square root of the positive part of the Choi–Williams distribution. In this way we get the same amplitude scale as for the wavelet and Gábor transforms. (With a power scale the lower amplitude oscillations are hardly visible.)

T Ursae Minoris

T Ursae Minoris is a Mira star with a period of ≈ 300 days. The rapid decrease of the period of T Umi was reported by Gál and Szatmáry.[6] The period dropped from 314.5 days to 283.2 days during a 25-year period, i.e., ≈ 30 pulsational cycles. Their wavelet plot clearly shows the variation of the frequency, and the investigation of the O − C diagram confirms this finding. Here we give a comparison of the different methods on this data set.

The wavelet and the Gábor transforms of the light variation are presented on the top of FIGURE 1. While the two maps are the same around the average frequency of the variation (≈ 0.003 c/d)—guaranteed by the match of the window functions in that frequency—the Gábor transform resolves definitely better the harmonics of the pulsational frequency, i.e., the higher temporal resolution of the WT at higher frequencies is rather disadvantageous in this application. It is important, because the absolute change of the frequency is twice and three times larger at these harmonics. The Choi–Williams distribution (lower left box on FIG. 1) emphasizes even better the second and third harmonics. The parameters of the CWD are $\sigma = 10$ and $M = 128$. For comparison we present the instantaneous frequency (evaluated from the Gábor transform with (7)) in FIGURE 1 too. The solid line represents the frequency variation of the lowest frequency (f_0), while the dotted line marks one-half of the instantaneous frequency at $2f_0$. The second curve is noisier, due to the smaller amplitude, but the increasing trend in the frequency manifests itself in the same way. It has not been possible to calculate the same for the third frequency.

To be honest we have to note that the O − C diagram really provides the same information on the basic period variation. The additional feature of the time-

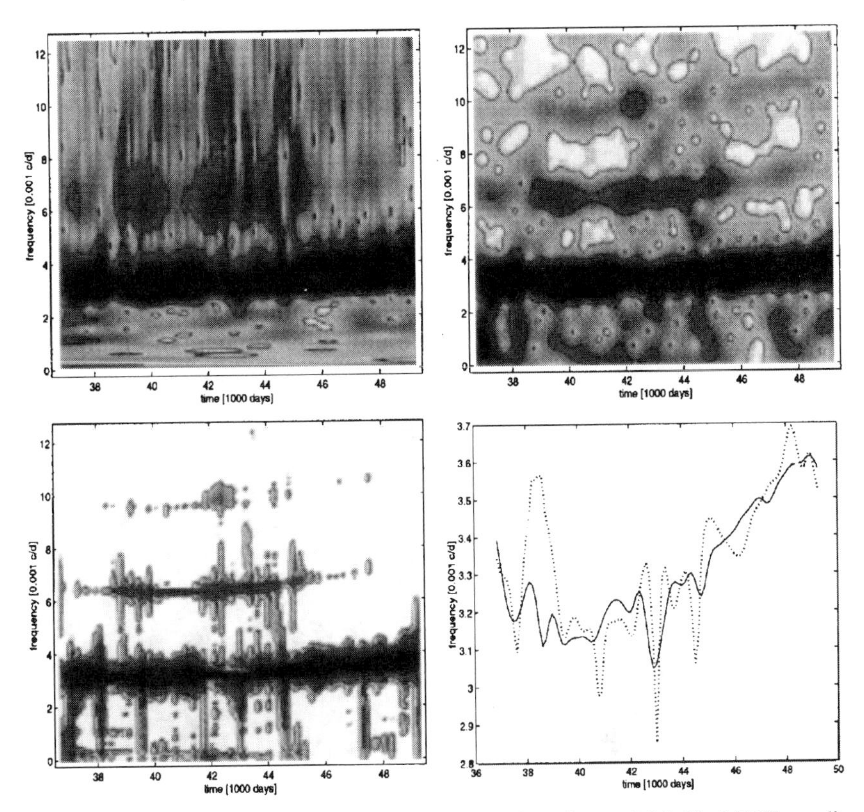

FIGURE 1. The wavelet map (**upper left**), Gábor transform (**upper right**), Choi–Williams distribution (**lower left**) and the instantaneous frequencies (**lower right**) of the light curve of T Umi.

frequency maps is to exhibit the temporal variations of the different amplitudes. For example after $t = 42{,}000$ the amplitude at $f \approx 0.003$ c/d drops while the amplitude of the harmonics increases for the same short period of time. This event correlates with the beginning of the period change. This can be a fingerprint of the physical process modifying the nature of the star (see reference 6 for the explanation as a possible helium shell flash).

R Scuti

R Scuti is an RV Tauri type variable star, with irregular light variations. An application of the global polynomial flow reconstruction method to the light curve has found that the brightness variations can be modelled by a four-dimensional flow or map[15] (see also Gouesbet *et al.* in this volume and reference 19).

The Fourier spectrum of the light curve displays two broad peaks near $\nu_1 \approx 0.069$ c/d and $\nu_2 \approx 0.0147$ c/d. These same frequencies show up in the linear stability

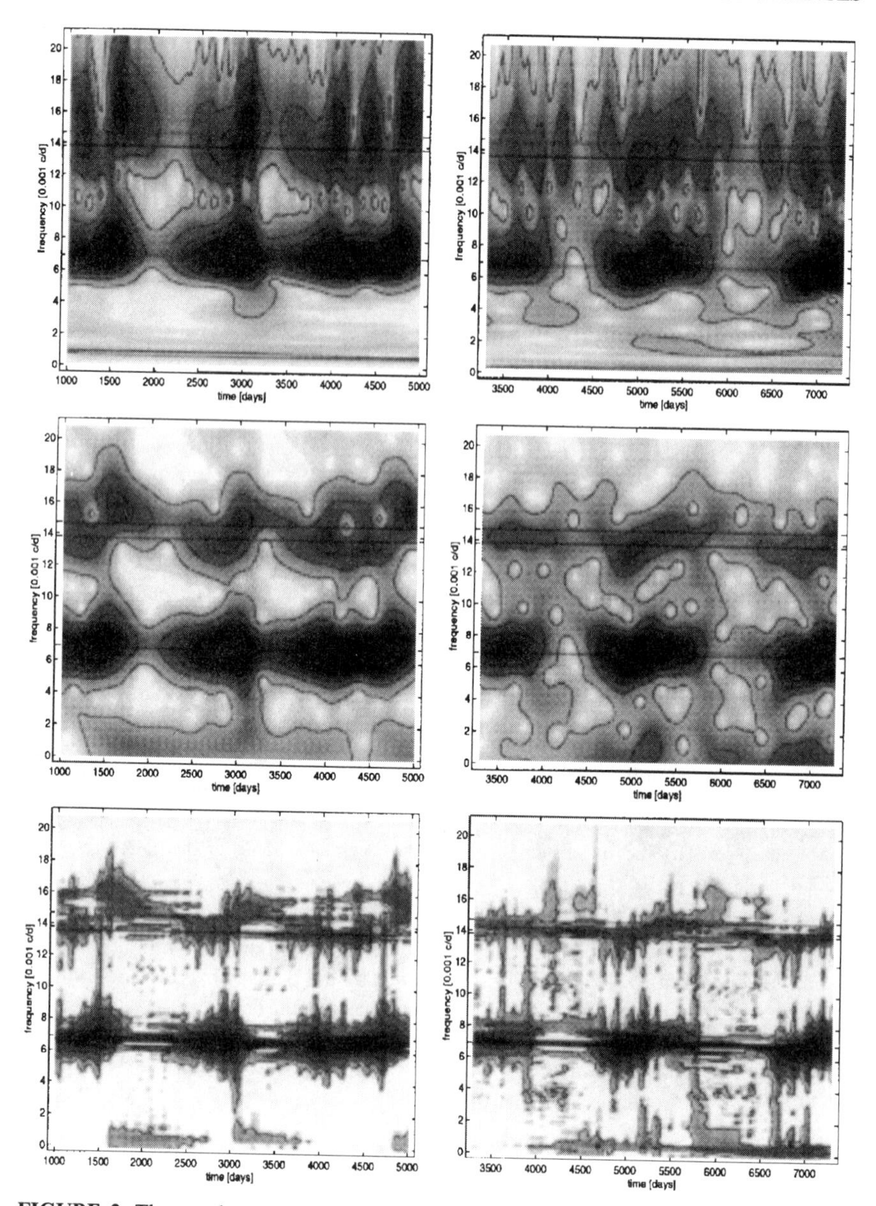

FIGURE 2. The wavelet map (**top**), Gábor transform (**center**) and Choi–Williams distribution (**bottom**) of the synthetic data (**left column**) and the light curve of R Sct (**right column**).

analysis of the fixed point of this map which has two spiral manifolds, one unstable and the other stable. In the 4D phase-space of the system one first sees a spiralling out of the trajectory along the ν_1 unstable directions. During this part of the pulsations one expects a strong harmonic content at $2\nu_1$ because nonlinear effects distort the pulsations away from sinusoidal. This phase is ended when nonlinear effects cause a reinjection into the ν_2 stable manifold, giving rise to oscillations with a decaying amplitude, and now with frequency ν_2. One can thus expect a switch back and forth between the neighboring frequencies $2\nu_1$ and ν_2, and this is clearly seen in the synthetic signals (i.e., in the signals produced by an iteration of the map). These features were clearly visible in the wavelet analysis of the *synthetic signal.*[5] However, the same analysis for the real observations of R Scuti by the WT was not informative at all. This is not astonishing because the light curve is contaminated by noise, and we have only a relatively short sample of the amplitude modulated signal, while from the clean synthetic signal one can choose an optimal segment. One of the motivations for this paper has precisely been a search for more sophisticated methods (or the combination of them) that enable one to display this behavior in the observational data.

Here we present a comparison of the analyses of the synthetic curve and of the R Scuti data by the three methods. We have calculated all the transforms for a grid of 128 frequency and 100 time values. The value of $c = 2\pi$ has been used in the wavelet analysis and we have set σ in the Gábor transform to match the resolution of the wavelet at $\nu = 0.007$ c/d. The parameters of the Choi–Williams distributions are: $\sigma = 2$ and $M = 128$. The gray scale plots of the distributions are presented in FIGURE 2. The values of ν_1, $2\nu_1$ and ν_2 are indicated by horizontal lines in all figures. We confirm that it is hard to collect any information other than the amplitude modulation of the signal from the plots of the wavelet modulus. Only the plot of the instantaneous frequency sheds some light on the frequency variation of the synthetic signal (see reference 15). On the other hand, the Gábor transform clearly shows the frequency shift for the synthetic signal, and there is some indication of it for the R Scuti data, as well. The Choi–Williams distribution in contrast seems to be the superior for both data sets. The frequency changes between $2\nu_1$ and ν_2 are clearly visualized in the plots.

It is very pleasing to us that the application of this type of time-frequency analysis (with the Choi–Williams distribution) provides a further test for a positive comparison of the synthetic data and the observed light curve of R Scuti.

CONCLUSIONS

Astronomers have essentially limited themselves to the use of the wavelet transform in their investigations of the time-frequency characteristics of variable light curves. Here we have compared the results that one obtains with the wavelet transform to those of other time-frequency methods, using both real light curves and a synthetic signal. From these tests we can conclude (1) that the Gábor transform provides much more informative results on the high frequency part of these data than the wavelet transform, but (2) that the time-frequency analysis with the Choi–Williams distribution is definitely superior to both methods, at least on these data.

We have also shown that the time-frequency analysis can provide us with an important tool for the comparison of chaotic data sets. Thus, in the case of the irregular star R Scuti our global flow reconstruction found a wavering between two neighboring frequencies. The time-frequency analysis was able to show that this subtle effect is actually also present in the noisy observational light curve data, thus further strengthening our conclusion that a low dimensional flow governs the dynamics of this star.

ACKNOWLEDGMENTS

We wish to thank Leon Cohen for helpful discussions, Károly Szatmáry for the light curve of T Umi and Janet Mattei for the AAVSO observational data.

REFERENCES

1. GROSSMANN, A., R. KRONLAD-MARTINET & J. MORLET. 1989. *In* Wavelets: Time-Frequency Methods and Phase Space. J. M. Combes, A. Grossmann & Ph. Tchamitchian, Eds.: 2. Springer-Verlag. New York/Berlin.
2. SCARGLE, J. D., T. STEIMAN-CAMERON, K. YOUNG, D. L. DONOHO, J. P. CRUTCHFIELD & J. IMURA. 1993. Astrophys. J. **411:** L91.
3. GOUPIL, M. J., M. AUVERGNE & A. BAGLIN. 1991. Astron. Astrophys. **250:** 89.
4. SZATMÁRY, K., J. VINKÔ & J. GÁL. 1994. Astron. Astrophys. Suppl. **108:** 377.
5. SZATMÁRY, K. & J. VINKÔ. 1992. Mon. Notices Roy. Ast. Soc. **256:** 321.
6. GÁL, J. & K. SZATMÁRY. 1995. Astron. Astrophys. **297:** 461.
7. SZATMÁRY, K., J. GÁL & L. L. KISS. 1996. Astron. Astrophys. **308:** 791.
8. KOLLÁTH, Z. & B. SZEIDL. 1993. Astron. Astrophys. **277:** 62.
9. BAUDIN, F., A. GABRIEL & D. GIEBERT. 1994. Astron. Astrophys. **285:** L29.
10. OCHADLICK, A. R., H. N. KRITIKOS & R. GIEGENGACK. 1993. Geophys. Res. Lett. **20:** 1471.
11. BOYD, P. T., P. H. CARTER, R. GILMORE & J. F. DOLAN. 1995. Astrophys. J. **445:** 861.
12. COHEN, L. 1994. Time-Frequency Analysis. Prentice-Hall. Englewood Cliffs, N.J.; also COHEN, L. 1989. Proc. IEEE **77:** 941.
13. COHEN, L. 1966. Math. Phys. **7:** 781.
14. GÁBOR, D. 1946. J. Inst. Elect. Eng. **93:** 429.
15. BUCHLER, J. R., Z. KOLLÁTH, T. SERRE & J. MATTEI. 1996. Astrophys. J. **462:** 489.
16. DELPRAT, N., B. ESCUDIÉ, P. GUILLEMAIN, R. KRONLAD-MARTINET, P. TCHAMITCHIAN & B. TORRÉSANI. 1992. IEEE Trans. Inf. Theory **38:** 644.
17. CHOI, H. I. & W. J. WILLIAMS. 1989. IEEE Trans. Acoustics Speech Signal Process. **37:** 862.
18. RIOUL, O. & P. FLANDRIN. 1992. IEEE Trans. Signal Process. **40:** 1746.
19. BUCHLER, J. R., Z. KOLLÁTH & T. SERRE. 1995. Ann. N.Y. Acad. Sci. **773:** 1.

Wavelets, Scaling, and Chaos[a]

JEFFREY D. SCARGLE

Space Science Division
NASA-Ames Research Center
Moffett Field, California 94035-1000

INTRODUCTION

Wavelets are the central idea of a broad framework for thinking about, displaying, and analyzing data—not just a new specialized technique. They are localized basis functions that provide an extremely flexible and efficient representation of time series. This overview concentrates on the use of wavelets to characterize features of time series connected with an underlying nonlinear dynamical process. Nonlinear dynamics often leads to scaling behavior. This connection is not yet well understood, and is the subject of much ongoing research. Wavelets are good tools for detecting, quantifying, and modeling scaling behavior.

WAVELET ANALYSIS OF DIGITAL SIGNALS

Wavelets comprise a complete set of basis functions. Their use in time series analysis, as with Fourier analysis, stems from the fact that linear combinations of them can represent suitably behaved functions in general. The special properties of wavelets summarized in this section make them very useful for certain applications.

Overview

Wavelets are *localized*, i.e., the wavelet functions are nonzero in subintervals of the total sampled time interval. This feature makes them especially useful for signals that contain sharply defined, localized features like jumps, jerks, and bumps, but their frequency (or, equivalently, inverse time scale) properties make them useful as well for representing $1/f$ noise, chirps, fractals, and other multi-scale or self-similar structures.

Another important feature of wavelets is that there is a close relationship between the magnitudes of the wavelet coefficients and the smoothness of the corresponding function—a relationship not shared by Fourier analysis. A somewhat formal theorem, discussed by Meyer,[1] basically states that it is possible to generate an arbitrarily messy function (full of discontinuities and other nasty things) by diminishing the amplitudes and adjusting the phases of the Fourier coefficients of a smooth, continuous function. Hence one is not guaranteed of success when trying to smooth a function by diminishing some of its Fourier components. In contrast, it

[a]This paper is based in part on work supported by grants from NASA's Astrophysics Data Program.

can be shown that with wavelet coefficients there is such a guarantee. This feature is responsible for the existence of wavelet-based smoothing algorithms with very nice properties. Briefly, decreasing the size of some of the wavelet coefficients in a systematic way results in a nicely noise-free version of the sampled data (see below). Further, many or even most of the wavelet coefficients can be thrown away completely without materially affecting the representation of the function (or image). Such data compression is proving quite valuable commercially and has been used by the FBI to store digital fingerprint images in one-twentieth the storage capability.

Wavelets allow analysis of not just fluctuations on a specific scale, but also the relationships between those that occur on different scales. Hence they are great tools for exploring scaling behavior in physical processes—such as $1/f$ noise.

Lastly, wavelet analysis, and its generalization *multi-scale analysis*, should not be regarded as just specific data analysis tools. They are part of an overall change in the way one looks at data in the first place. One invents display techniques for highlighting behaviors on different scales and visualizing their relationships. One analyzes images into the sum of a smoothed and a detailed component. The following theme pervades much of wavelet analysis: one decomposes data into components on a hierarchy of scales, operates on the wavelet coefficients at each scale individually, and then re-synthesizes the data. Further, one thinks in terms of not just the harmonic content of a signal, but more of the scale of its structures in the time domain; and one studies non-stationarities such as sharp ("localized") features and time-evolution of structural characteristics. One seeks (and finds) methods to remove random observational noise from data, without rounding off the sharp edges present in the signal—"denoising without smoothing." And one finds data analysis and visualization techniques that are suited to the multi-scale behaviors characteristic of $1/f$ noise and chaotic dynamical systems.

This chapter is not meant to introduce the reader to technical aspects of wavelets, or to their actual use in data analysis. There is now a large literature on the theory and applications of wavelets; good sources include books by Daubechies,[2] Meyer[1,3,4] and Chui;[5,6] see also Graps[7] and the World Wide Web sites given later. Much of the published research is oriented toward abstract functional analysis, but the monograph by Meyer[4] discusses the goals of wavelets in data analysis and provides an historical overview, a treatment of time-frequency methods, plus four chapters on applications (computer vision, fractals, turbulence, and the study of distant galaxies).

Instead, this review will concentrate on top-level descriptions of some of the multi-scale data analysis techniques that have been developed in the last few years.

Visualization and Analysis of Temporal Structure: The Scalogram

Many applications of wavelets involve displaying a signal's wavelet coefficients as a function of the independent variables representing temporal (or sometimes spatial) location and scale. This simple technique can reveal many of the systematics and specifics of non-stationarities. It can also be thought of as a simple time-frequency diagram.

The papers by Goupil *et al.*[8,9] contain pictures of such three-dimensional displays. The term *scalogram* (e.g., Rioul and Flandrin[10]) refers to the absolute value of the wavelet coefficients, although this quantity is sometimes called the *wavelet modulus*. One often sees variants of this basic display, most often using the logarithm of the coefficients, because of the large dynamic range of the components of most signals, and because scaling relations are most often in the form of power laws. Inspection of the scalogram (or of the wavelet coefficients themselves) is generally useful when one needs to view frequency/scale and location information at the same time.

Detection and Characterization of 1/f Noise: The Scalegram

The *scalegram* is the wavelet analog of the power spectrum. Both are of use when one is interested in the components of a signal as a function of frequency (or its inverse, scale) but does not care about location. The scalegram can be thought of as the scalogram projected (integrated) onto the scale axis. To avoid possible confusion of terminology, the scalegram is sometimes called the *wavelet spectrum.*

The scalegram is defined as the absolute square of the wavelet coefficients, averaged over all values of the location parameter allowed at a given scale. The resulting function of the scale parameter contains much the same information as the power spectrum; indeed, it is very much like the power spectrum plotted as a function of $\log(1/\omega)$. For data with non-normal errors, it may be preferable to define the scalegram as the mean absolute value of the wavelet transform, as discussed by Norris *et al.*[11] This makes for a more robust estimator.[12]

In almost all practical situations, one has observations of a signal corrupted by observational noise. There is a very simple way to make an estimate of the scalegram that corrects for the presence of noise—as long as one can make simplifying assumptions about the noise, namely that it is independently distributed. Two noise models are frequently used in astronomy: a signal plus normally distributed observational errors, and a signal determined by counting photons. In the second case, not only is the noise not normally distributed, but it is not even additive. The model is that the observed samples are random variables with mean value equal to the true signal, and with a Poisson distribution. A straightforward computation of the scalegram of both models yields the result that the noise simply adds a constant (independent of scale) to the true scalegram. The constant is just the variance of the observational noise in the first case, and the mean counting rate in the second. This relation means that in practice it is quite easy to estimate the true scalegram, although of course this estimate is uncertain if the signal-to-noise ratio is low. Scargle, Steiman-Cameron, Young, Donoho, Crutchfield, and Imamura used this method to obtain scalegrams of the variability of the x-ray source Scorpius X-1; see reference 13 for definitions and computational details on the scalegram. They used the scalegram to diagnose scaling behavior in the time series, and this in turn led to a model of the nonlinear dynamics of the accretion process that predicts that this and other low-mass x-ray binary star systems display a special kind of nonlinear deterministic behavior called *transient chaos.*

Wavelet Denoising Methods

Donoho and coworkers[14–26] have proposed specific algorithms, based on wavelets, for a number of estimation and time series analysis problems. An underlying theme of this work is a technique called *thresholding or wavelet shrinkage*, i.e., computing the wavelet coefficients of a noisy signal; establishing a threshold; setting to zero those wavelet coefficients which, in absolute value, are below the threshold; and then inverse wavelet transforming. This reconstruction of the original signal typically has the background noise removed in a way that can be optimal, as long as the statistics of the noise are well behaved.

A good introduction to the methodology is reference 19 in the context of estimating unknown functions embedded in noise (see references 21, 22 and 24); more technical matters are discussed in reference 23. The use of wavelets to estimate probability distributions from data is discussed in reference 25. The papers[20,26] contain nice overviews of denoising, inverse problems, wavelet packet denoising, segmented multi-resolutions, and nonlinear multi-resolutions.

Among the mathematical results in these papers are proofs that the basic thresholding technique is amazingly good both in a practical sense (as depicted with synthetic examples) and in a theoretical sense (in that the rate at which the variance of the estimator improves with sample size is greater than for conventional methods—nearly as great as for an ideal estimator). In particular, the soft-threshold reconstruction method yields an estimate that is at least as smooth as the unknown function, and the corresponding estimator comes nearly as close in mean square to the unknown function as any measurable estimator can.[14] These procedures generally provide estimates of noise-corrupted signals that are accurate, without the smearing of sharp features (steps, edges, spikes, etc.) that accompanies most conventional techniques.

Bendjoya *et al.*[27] have developed a variant of thresholding based on recognizing significant patterns in the wavelet coefficients. Their approach starts with an ordinary wavelet thresholding. In a second pass, previously rejected coefficients are restored if they relate in specified ways to those retained in the first pass. For example, coefficients are restored if they exceed a second, lower, threshold and are contiguous to the first-pass coefficients in scale-location space. The net result is the ability to more effectively denoise signals with structures that correspond to the criteria used in this pattern recognition process. I think we are just beginning to see a flood of innovative variants of the basic concept of wavelet shrinkage denoising.

Pursuit Methods

Wavelets have directly or indirectly led to methods in which overcomplete sets of functions (some of which are wavelets or wavelet-like) are used in place of complete, orthogonal bases.

The general framework in which these methods operate is as follows. Given time series data

$$X_n = X(t_n),\ n = 1, 2, \ldots, N, \tag{1}$$

and a set, D, of functions,

$$D = \{\psi_k(t)\},\ k = 1, 2, 3, \ldots, \tag{2}$$

the idea is to represent the data as a linear combination of functions from D, i.e., of the form

$$X_n = \sum_k c_k \psi_k(t_n), \quad n = 1, 2, \ldots, N; \quad \psi_k \in D, \tag{3}$$

where the c_k are constant coefficients, the $\psi_k(t)$ are *atoms*, the set D is a *dictionary* of atoms, and the relation (3) is called an *atomic representation*.

It is usually demanded that D be such that any function can be accurately represented by a linear combination of its atoms; but the dictionary may be *overcomplete*, so that this representation is not unique. Among the perhaps many possible linear combinations that represent the data equally well, one is typically selected by some auxiliary condition—that the number of terms in the expansion be as small as possible. Examples of atomic dictionaries include wavelets themselves, wavelet packets[28,29] and cosine packets, and Gábor atoms[30]—sine waves modulated by a Gaussian envelope.

Atoms are intended to be elementary building blocks, well suited for the construction of relevant signal structures. They are often simple and such that they are localized in time and their Fourier transform is localized in frequency—in which case they are called *time-frequency atoms.* One way to make a useful dictionary of such atoms is to pick a function $g(t)$, and then construct copies of it shifted, $g(t - l)$, and scaled, $g(t/s)$, in time. This is exactly how wavelets are constructed from the *mother wavelet,*[2] since the functions can be grouped into hierarchical subsets of different "scales of variation." This arrangement, called *multi-resolution analysis* (MRA), was invented in 1986 by Mallat and Meyer (see references 2 and 3). MRA systematizes the connections between one scale and the next, and leads to recursive schemes for computing the N-point wavelet transform in $N \log N$ time instead of N^2.

How does one pick the best representation in these cases in which there is not a unique one? If the dictionary were complete and orthonormal, the procedure is of course straightforward and unique. Coifman and Wickerhauser[29] developed a method called *wavelet packet pursuit,* which allows one to find the basis which generates the representation of a given data stream that has the largest entropy. In practice, this maximum entropy condition provides a basis that allows a clean separation between signal and noise, and has other useful properties. Mallat and Zhang[30] introduced a method, called *matching pursuit,* that basically seeks the atoms which have the largest inner product with the data, using a recursive technique that has some similarities to the *CLEAN* technique well-known in radio astronomy.[31]

Donoho and his student Shaobing Chen have developed a method that is perhaps the most general of all—*Basis Pursuit.*[32] The idea here is to seek not the best atom at a given step, by optimizing a match between atoms and the data, but to seek the best basis by optimizing a match between the dictionary as a whole and the data. An example of such a global property is the question "How good a representation can be obtained with a small number of terms?," one that is ignored in matching pursuit. All of these procedures are implemented in the publically available **Wavelab** software package—see the final section.

Time-Frequency Methods

Donoho and coworkers have developed a number of algorithms for carrying out a generalized multi-scale analysis of time series, with the goal of producing time-frequency distributions. The basic idea involves the construction of a library of multi-scale representations of the data; one member in the library is ordinary wavelet analysis, one is close to ordinary Fourier analysis, and the others are mixtures of the two. The scheme involves selecting the member of the library which best represents the data at hand—by minimizing a measure of the entropy of the representation. The (generally) mixed representation found in this way is then converted into a time-frequency distribution.

This technique appears to produce time-frequency, or "phase space plots" that are very good at detecting variable periodic or quasi-periodic components in the time series. In addition, the method completely avoids some unpleasant side-effects of the classical approach to time-frequency distributions, such as the well-known interference terms between multiple periodic components.[33]

This section ends by briefly describing some techniques that, while not strictly using wavelets, are close to their spirit. For example, as mentioned above, *multi-resolution analysis* refers to methods that separate and deal with time series fluctuations on different time-scales and time-locations. Sliding window or *short-time Fourier transform* (STF) analysis is a way to do this (see reference 2 for a discussion of windowed, or short-time, Fourier analysis and its relation to wavelets). A variant of the STF transform, called the Gábor transform, is being used by a group based at NASA-Goddard[34,35] to investigate astronomical systems, both observational and theoretical, containing harmonic signals with time-varying frequencies. The goal of their techniques is to provide insight into non-stationary evolution of chaotic and other nonlinear physical systems. They studied numerical data from computations of a single star scattering gravitationally off a binary system. The Gábor transform was used to generate time-frequency plots exhibiting the dual-frequency behavior of this three-body system, and to show the evolution of the frequencies over time. In addition, they analyzed Hubble telescope photometric data of the star HD 605435 and found evidence that changes in the frequencies are too abrupt to be due to a beat phenomenon, as proposed by other workers.

Detection and Characterization of Discontinuities: The Continuous Wavelet Transform

The term *continuous wavelet transform* in the literature refers to the transform evaluated at a continuum of scales (not just discrete scales related to each other by factors of 2^s, where s is an integer)—and does not refer to the discrete values of the independent variable. It is very useful in locating sudden jumps in time series, and then measuring the order of the discontinuity. In a very interesting paper,

Alexandrescu *et al.*[36] showed how both the location and order of singularities can be measured by locating ridge lines of a map of the wavelet coefficients.

The idea is to inspect the scalogram based on the continuous wavelet transform. Of course, one computes the transform at a discrete set of values of the scale and location parameters—the basic point is that the transform is evaluated at a number of scales *between* the powers of two used in the discrete transform. This is much like oversampling the Fourier transform, although in practice it seems to yield much more information.

If a singularity is present in the data, a three-dimensional plot of the scalogram has a "ridge" of power located at the singularity and growing in the direction of decreasing scale. The order of the singularity is determined by the logarithmic slope of the ridge line. This method works surprisingly well even in the presence of substantial noise, as Alexandrescu and coworkers demonstrate by analyzing synthetic data with various "jerks" present.

Wavelet Transform of Unevenly Spaced Data

For well-known reasons, astronomical time series are often unevenly spaced. Techniques have been developed to compute Fourier transforms, correlation functions,[37] and power spectra,[38] for data with arbitrary sampling in time. The goal is not only to find computational techniques for estimating the corresponding quantities from unevenly spaced time series, but to correct to the extent possible for the unevenness of the sampling.

Because of the simplicity of the Haar wavelet transform, it is relatively straightforward to compute it from data that have arbitrary sampling. Since the Haar wavelet is constant over dyadic intervals, the wavelet coefficient can be computed by counting the number of points in each such interval.

As with Fourier analysis of unevenly spaced data, it is not obvious how many levels in the scale hierarchy to include. An upper limit is set by the condition that the finest scale be as small as the smallest sample interval in the data. This is overkill if there are one or only a few very small intervals. The condition that the smallest interval be on the order of the mean sampling interval is probably reasonable in practice, although one may wish to add a few additional levels in order to capture information on these fine scales, if the signal-to-noise ratio warrants. The only other real problem is the treatment of dyadic intervals that contain no samples (e.g., due to pure chance if the sampling is random). In the computation of the scalegram, one should average the squared wavelet coefficients corresponding only to intervals which contain samples.

Translation-Invariant Wavelet Transforms

There is a practical problem with wavelets, directly resulting from their localized nature. Suppose a signal has one or more localized features. The coefficients of the wavelets whose scale is on the order of the scale of the feature, and whose location overlaps the feature, can clearly be quite sensitive to the precise position of the wavelet relative to the feature. This means that results of various data analytic procedures, such as denoising, will depend on the location of the wavelets—which is usually arbitrarily related to the structures in the signal.

Coifman and Donoho[39] have developed a special translation invariant wavelet transform that, in essence but not directly, considers all possible rotations of the data by one sample (with wraparound), thus in a sense averaging over the effects connected with special phase relationships. The **WaveLab** software contains a directory of tools for implementing this extension of the ordinary wavelet transform in several ways.

SCALING BEHAVIOR AND NONLINEAR DYNAMICS

What is the relevance of scaling behavior, and the use of wavelets to diagnose it, to nonlinear dynamics? In my view, this is an important outstanding question. Some connections have been discussed by various workers, but the essence of the connection between dynamics and multi-scale behavior has been elucidated only in a few special cases.

One thread that many have followed is that dimensionality (fractal dimension, correlation dimension, etc.) seems to relate to self-similarity and scaling behavior on the one hand, and to multi-scale spatial and temporal dynamical behavior on the other. Internal connections between different mathematical aspects of scaling—such as the power spectrum, the autocorrelation function, and dimensions—have been made (e.g., references 40 and 41). However, none of the concepts related to dimensionality are unique to, or diagnostic of, chaotic behavior; indeed, purely random processes can have any dimension between 1 and infinity.

The things that are specifically diagnostic of the dynamics, such as Lyapunov exponents, have not to my knowledge been related directly to scaling behavior. Almost certainly the exponential growth connected with positive Lyapunov exponents and chaotic evolution in time have something to do with the appearance of structures on a wide range of scales. Furthermore, complex systems—such as those represented as many simple components whose mutual interactions lead to high complexity—have the degrees of freedom that are needed for scaling behavior over a wide scale-range.[42] Other workers have pursued the connection between *self-organized critical states* and scaling behavior (e.g., references 43–45). Still others have focused on fractal Brownian motion and its connection with self-similarity.[46–48] Ramanathan and Zeitouni[49] derived a theorem characterizing fractional Brownian motion by the covariance structure of its wavelet transform, and Arneodo *et al.*[50] clarified the nature of the wavelet transform of multifractals.

Wavelet methods are more and more being used for the solution of differential equations describing physical processes that occur over a wide range of scales—both temporal and spatial scales. The techniques include expansion of the dependent variables in wavelets as functions of the independent variables, as well as determination of grid sizes and positions using multi-resolution analysis. Bendjoya and Slezak[51] discuss the use of wavelets in the solution of partial differential equations in the characterization of turbulence and fractals. Studies of turbulence are increasingly making use of wavelet methods (see Chapter 10 of reference 4), for obvious reasons. It is interesting that wavelets play an important dual role in science, because of their importance in both data analysis and in theoretical computations. The key connection, of course, is the scaling behavior of the physical variables.

The previously mentioned work on Scorpius X-1[13] was an example of this, in the sense that wavelet analysis of the time series data ultimately led to a nonlinear model for the accretion process that involves diffusion (as well as accretion and local instabilities). While this work did not invoke wavelet methods for solution of differential equations *per se*, it is nevertheless true that diffusion equations are among those that are being solved using wavelet techniques. The resulting model, the *Dripping Handrail*,[52] was later applied by other authors to other accretion systems.[53,54]

A discussion of some related concepts is to be found in the book by Schroeder.[55]

SUMMARY OF APPLICATIONS OF WAVELETS IN ASTRONOMY

We are beginning to see many applications of wavelets to astronomical data analysis. I have no doubt that as astronomers come to appreciate the great utility of wavelet techniques in smoothing, noise reduction, characterization, modeling, and data compression problems, the applications will grow, and within a few years multi-resolution techniques will be quite familiar, and wavelets will actually be more used than Fourier techniques. This section contains brief descriptions of applications of wavelets to the analysis of *dynamical processes* in astronomy—where this term is used in a rather liberal way. A more general review of wavelets in astronomy will appear elsewhere.[56]

Some of these applications make use of the multiscale nature of wavelets to study various scaling issues in spatial distributions. For example, Gill and Henriksen,[57] using a two-dimensional analog of the scalegram, detected and characterized some scaling relations in the spatial distribution of molecular emission in the star-forming region L1551. Escalera and Mazure[58] used two- and three-dimensional wavelet transforms to study and compare subclustering in toy models of, and actual data on, the distribution of galaxies in space. Meyer (see Chapter 11 of reference 4) describes the use of wavelets in the study of the distribution of galaxies.[59,60]

Other studies have concentrated on time-scaling relations, using time series data from various astronomical objects. Goupil *et al.*[8,9] have used the scalogram to study the variability in white dwarfs, and in the ZZ Ceti star G191 16. They find evidence for nonlinear behavior in various stars.[61] At the same time they were able to diagnose the photometric quality of a given set of observations, by using the same wavelet methods on the observations of comparison stars. Another example of application to time series is the modeling of the flickering of Scorpius X-1, by the author and coworkers, described earlier (see also reference 13).

With a very different goal, Norris and coworkers[11] studied the light curves at a large number of Gamma-ray bursts, in an effort to detect the cosmic time-dilation that would be present if these objects are at cosmological distances. In any model of the Universe, other than those which ascribe the observed redshifts to an unknown mechanism that causes photons to lose energy as they propagate over large distances, redshifted objects should appear to be varying more slowly than nearby objects. Since the redshifts of the Gamma-ray burst sources are unknown, these authors looked for a correlation between some measure of time scale and the apparent brightness—arguing that on average the apparent brightness will be inversely correlated with distance. The scalegram (referred to there as the *wavelet spectrum*) was used to define one such measure of time scale. The sought-after correlation was found using several techniques, suggesting that these objects lie at large distances (redshifts on the order of unity).

In this work, wavelets were also used to denoise the raw data, using slight modifications of the Donoho wavelet shrinkage. Another signal estimation problem in which the basic goal is improved estimation of a signal with certain assumed properties, embedded in noise, is the analysis of Voyager image data. One-dimensional scans across the Encke gap ringlet in Saturn's ring system were analyzed using a novel technique (mentioned above), by Bendjoya *et al.*[27] This work yielded some evidence for small substructures in the ringlet that otherwise are lost in the noise.

An interesting problem, in connection with quasar gravitational lenses and other astronomical phenomena, is the estimation of the time delay between two noisy time series. Hjorth *et al.*[62] have studied the gravitationally lensed quasar 0957 + 561A,B. They used a multi-resolution analysis corresponding to Daubechies wavelets to analyze several optical time series for the two components. After linear interpolation, they used various aspects of the wavelet decomposition of the resulting evenly spaced data to assess the effects of patterns in the sampling—combined with the interpolation itself. They examined a kind of cross-correlation of the wavelet amplitudes; the resulting indications of a detectable time delay were not definitive, but were not inconsistent with other results.

Vigouroux and Delache[63] have studied the relative efficiency with which time series data on the apparent diameter of the sun are representable with Fourier and wavelet methods. They conclude that real solar data, as well as data synthesized to have similar properties, are more compactly and accurately represented with wavelet than Fourier expansions.

This review deals mainly with one-dimensional time series, and ignores the extremely active research on wavelets for image processing. This work is scientifically, technically and commercially very important, but for the most part not oriented toward understanding of dynamical systems. Exceptions to this include the study of the distributions of: galaxies,[53] x-rays within galaxy clusters,[64] and asteroid orbital parameters.[65]

WAVELAB—SOFTWARE TOOLS FOR WAVELETS AND OTHER MULTI-RESOLUTION ANALYSIS

There is a tremendous amount of information about wavelets available on the Internet and the World Wide Web (WWW).

Tex/LaTex and other versions of most of the papers by Donoho and coworkers are readily available on the Internet, via an anonymous ftp. Do an ftp to the node "playfair.stanford.edu" as "anonymous," enter your Internet address in lieu of a password, change to the directory ./pub/donoho, (type "cd pub" and "cd donoho"), and then "get" followed by the name of the file you want. The WWW address for this material is:

http://playfair.stanford.edu/~donoho/.

In addition, a complete wavelet analysis toolkit called **WaveLab**, implementing all the wavelet operations discussed here and many more, can be obtained at no cost from this WWW site:

http://playfair.stanford.EDU:80/~wavelab/.

This is a very complete system of **MatLab** (copyright, *The MathWorks, Inc.*) scripts that implement both the basic wavelet and related transforms, and the more advanced techniques described in the first section of this chapter. There is full documentation, a set of tutorials, and a section of "Toons," short for *cartoons.* These scripts reproduce from scratch the figures in many of the Stanford group's papers describing the theoretical research underlying the algorithms in **WaveLab**. These scripts recreate the figures in the publication in complete detail. By studying these scripts and experimenting with the data, the reader can learn all details of the process that led to the figure. With conventional paper publications, at best, the reader can understand the basic idea of a figure or, at worst, be left with unanswered questions and unresolved ambiguities.

This approach is part of a discipline called *Reproducible Research*, the idea of which is to provide the reader full access to all details (data, equations, code, etc.) needed to completely reproduce all the results normally presented only in summary form in scientific publications. Two demonstration projects have established the basic philosophy and feasibility of Reproducible Research. The pioneer is Jon Claerbout, of the Stanford University Geophysics Department. His ideas and accomplishments can be inspected on a document titled "Reproducible electronic documents".[66] This project began by achieving a large degree of reproducibility without using the Internet. Their first foray into reproducible research was a textbook in which every figure caption contained the name of a computer program. The programs themselves, organized by chapter and section of the book, were provided on an accompanying magnetic medium—which in addition contained all necessary numerical and geophysical data. Their project has since evolved through phases when publication was through the media of CD-ROMs and the Internet.

The *Wavelet Digest* is a very good general source of information on wavelets, with pointers to papers and meetings, plus a question and answer forum: http://www.wavelet.org/wavelet/index.html/. Other general WWW sites are

http://www.amara.com/current/wavelet.html

and

http://www.mat.sbg.ac.at/~uhl/wav.html/.

Fionn Murtagh has an extensive collection of papers on astronomical applications of wavelets, multi-resolution, noise suppression, filtering, image restoration and compression at

http://http.hq.eso.org/~fmurtagh/wavelets.html/.

ACKNOWLEDGMENTS

I wish to thank Dave Donoho and Ian Johnstone, of Stanford University, who have made many valuable suggestions, as have their students Shaobing Chen, Jon Buckheit, Xiaoming Huo, and Thomas Pok-Yin Yu. Zhang Zhifeng provided help with the matching pursuit software, and Minh Duong-van made valuable comments about chaotic dynamics.

REFERENCES

1. MEYER, Y. 1992. Wavelets and Operators. Cambridge University Press. Cambridge, U.K. English translation of 1990 book.
2. DAUBECHIES, I. 1992. Ten Lectures on Wavelets. Society for Industrial and Applied Mathematics. Philadelphia.
3. MEYER, Y. 1992. Wavelets and Applications, Proc. Int. Conf. Marseille, France, May 1989. Springer-Verlag. New York.
4. MEYER, Y. 1993. Wavelets: Algorithms and Applications. SIAM: Philadelphia. Translation from French by R. Ryan.
5. CHUI, C. K. 1992. An Introduction to Wavelets. Academic Press. Boston.
6. CHUI, C. K. 1992. Wavelets: A Tutorial in Theory and Applications. Academic Press. Boston.
7. GRAPS, A. 1995. IEEE Signal Process. Mag.; also:

   http://www.best.com/agraps/current/wavelet.html/.

8. GOUPIL, M. J., M. AUVERGNE & A. BAGLIN. 1990. A Wavelet Analysis of the ZZ Ceti Star G191 16. Proc. 7th Euro. Workshop on White Dwarfs. G. Vauclair, Ed. NATA. Toulouse, France.
9. GOUPIL, M. J., M. AUVERGNE & A. BAGLIN. 1991. Astron. Astrophys. **250:** 89.
10. RIOUL, O. & P. FLANDRIN. 1992. IEEE Trans. Signal Process. **40:** 1746.
11. NORRIS, J. P., R. J. NEMIROFF, J. D. SCARGLE, C. KOUVELIOTOU, G. J. FISHMAN, C. A. MEEGAN, W. S. PACIESAS & J. T. BONNELL. 1994. Ap. J. **424:** 540.
12. CLAERBOUT, J. F. & F. MUIR. 1973. Geophysics **38:** 826.
13. SCARGLE, J., T. STEIMAN-CAMERON, K. YOUNG, D. DONOHO, J. CRUTCHFIELD & I. IMAMURA. 1993. Ap. J. Lett. **411:** L91.
14. DONOHO, D. L. 1992. Stanford Statistics Department Tech. Rep. No. 409. This and the other papers by Donoho, Johnstone, and others at Stanford, Palo Alto, CA, are available as compressed postscript documents on the WWW at

    http://playfair.stanford.edu/subjects/wavelets.html/.

15. DONOHO, D. L. 1992. Stanford Statistics Department Tech. Rep. No. 408. Palo Alto, CA.
16. DONOHO, D. L. 1992. Stanford Statistics Department Tech. Rep. No. 410. Palo Alto, CA.
17. DONOHO, D. L. 1994. *In* Recent Advances in Wavelet Analysis. L. Schumaker & G. Webb, Eds. Academic Press. New York.
18. DONOHO, D. L. 1992. Stanford Statistics Department Tech. Rep. No. 403. Palo Alto, CA.
19. DONOHO, D. L. 1993. Stanford Statistics Department Tech. Rep. No. 425. Palo Alto, CA.

20. DONOHO, D. L. 1993. Stanford Statistics Department Tech. Rep. No. 437. Palo Alto, CA.
21. DONOHO, D. L. & I. M. JOHNSTONE. 1992. Stanford Statistics Department Tech. Rep. No. 400. Palo Alto, CA.
22. DONOHO, D. L. & I. M. JOHNSTONE. 1993. Stanford Statistics Department Tech. Rep. No. 425. Palo Alto, CA.
23. DONOHO, D. L. & I. M. JOHNSTONE. 1992. Stanford Statistics Department Tech. Rep. No. 401. Palo Alto, CA.
24. DONOHO, D. L. & I. M. JOHNSTONE. 1992. Stanford Statistics Department Tech. Rep. No. 402. Palo Alto, CA.
25. DONOHO, D. L., I. M. JOHNSTONE, G. KERKYACHARIAN & D. PICARD. 1993. Stanford Statistics Department Tech. Rep. No. 426. Palo Alto, CA.
26. DONOHO, D. L., I. M. JOHNSTONE, G. KERKYACHARIAN & D. PICARD. 1993. Stanford Statistics Department Tech. Rep. No. 419. Palo Alto, CA.
27. BENDJOYA, PH., J.-M. E. PETIT & F. SPAHN. 1993. Icarus **105:** 385.
28. WICKERHAUSER, M. V. 1994. Adapted Wavelet Analysis, from Theory to Software. A. K. Peters. Wellesley, Mass.
29. COIFMAN, R. & M. V. WICKERHAUSER. 1992. IEEE Trans Inform. Theory **38:** 713.
30. MALLAT, S. & Z. ZHANG. 1993. IEEE Trans. Signal Process. **41:** 3397.
31. HOGBOM, J. A. 1974. Astron. Astrophys. Suppl. **15:** 417.
32. CHEN, STANFORD & D. L. DONOHO. 1995. Tech. Rep.
33. COHEN, L. 1989. Proc. IEEE **77:** 941.
34. BOYD, P. T., P. H. CARTER, R. GILMORE & J. F. DOLAN. 1994. Ap. J. **445:** 861.
35. BOYD, P. T. & S. L. W. MCMILLAN. 1994. Chaos **3:** 507.
36. ALEXANDRESCU, M., D. GILBERT, G. HULOT, J.-L. LE MOUËL & G. SARACCO. 1995. J. Geophys. Res. **100:** 12557.
37. SCARGLE, J. 1989. Astrophys. J. **343:** 874.
38. SCARGLE, J. 1982. Astrophys. J. **263:** 835.
39. COIFMAN, R. R. & D. L. DONOHO. 1995. *In* Wavelets and Statistics. A. Antoniadia, Ed. Springer-Verlag Lecture Notes. Springer-Verlag. New York.
40. OSBORNE, A. & A. PROVENZALE. 1989. Physica **D35:** 357.
41. DUONG-VAN, M. & M. D. FEIT. 1996. 1/*f*-Power Spectrum as a Consequence of Self-Similar (Fractal) System of High Complexity (Large Fractal Dimension). Preprint.
42. PERSKY, N. & S. SOLOMON. 1996. Phys. Rev. E **54:** 4399.
43. LU, E. T. & R. J. HAMILTON. 1991. Ap. J. Lett. **380:** L89.
44. LU, E. T. 1995. Phys. Rev. Lett. **74:** 2511.
45. MINESHIGE, S., N. B. OUCHI & H. NISHIMORI. 1994. Publ. Astron. Soc. Japan **46:** 97.
46. MANDELBROT, B. 1983. The Fractal Geometry of Nature. W. H. Freeman. New York.
47. MANDELBROT, B. 1989. Pure Appl. Geophys. **131:** 5.
48. MANDELBROT, B. 1990. Physica **A163:** 306.
49. RAMANATHAN, J. & O. ZEITOUNI. 1991. IEEE Trans. Inform. Theory **37:** 1156.
50. ARNEODO, A., G. GRASSEAU & M. HOLSCHNEIDER. 1988. Phys. Rev. Lett. **61:** 2281.
51. BENDJOYA, P. & E. SLEZAK. 1993. Celest. Mech. Dynam. Astron. **56:** 231.
52. YOUNG, K. & J. SCARGLE. 1996. Ap. J. **468:** 617.
53. IMAMURA, J., J. MIDDLEDITCH, J. SCARGLE, T. STEIMAN-CAMERON, L. A. WHITLOCK, M. WOLFF & K. WOOD. 1993. Ap. J. **419:** 793.
54. STEIMAN-CAMERON, T., K. YOUNG, J. SCARGLE, J. CRUTCHFIELD, J. IMAMURA, M. WOLFF & K. WOOD. 1993. Ap. J. **435:** 775.
55. SCHROEDER, M. 1991. Fractals, Chaos, Power Laws; Minutes from an Infinite Paradise. W. H. Freeman. New York.
56. SCARGLE, J. 1996. Wavelet Methods in Astronomical Time Series Analysis. Proc. Con. Applications of Time Series Analysis in Astronomy and Meteorology, Padua, Italy, 3–10 September, 1993. Chapman and Hall. To be published.
57. GILL, A. & R. HENRIKSEN. 1990. Ap. J. **365:** L27.
58. ESCALERA, E. & A. MAZURE. 1992. Ap. J. **388:** 23.
59. SLEZAK, E., A. BIJAOUI & G. MARS. 1990. Astron. Astrophys. **227:** 301.
60. ESCALERA, E., E. SLEZAK & A. MAZURE. 1992. Astron. Astrophys. **269:** 379.
61. GOUPIL, M. J., M. AUVERGNE & A. BAGLIN. 1988. Astron. Astrophys. **196:** L13.

62. HJORTH, P., L. VILLEMOES, J. TEUBER & R. FLORENTIN-NIELSEN. 1992. Astron. Astrophys. **255:** L20.
63. VIGOUROUX, A. & PH. DELACHE. 1993. Astron. Astrophys. **278:** 607.
64. SLEZAK, E., F. DURRET & D. GERBAL. 1994. Astron. J. **108:** 1996.
65. BENDJOYA, P., E. SLEZAK & C. FROESCHLE. 1991. Astron. Astrophys. **251:** 312.
66. SCHWAB, M. & J. CLAERBOUT. 1995. Reproducible electronic documents, located on the WWW at

http://sepwww.stanford.edu/redoc/.

Transitional Dynamics of Chaotic Orbits[a]

HENRY E. KANDRUP[b] AND BARBARA L. ECKSTEIN

Department of Astronomy
University of Florida
Gainesville, Florida 32611

INTRODUCTION AND MOTIVATION

One major enterprise in galactic dynamics is the construction of model galaxies, idealized as time-independent solutions to the collisionless Boltzmann equation, i.e., what mathematicians term the gravitational Vlasov–Poisson system. In one way or another, this entails specifying a mass density $\rho(\mathbf{r})$, and an associated potential $\Phi(\mathbf{r})$, the forms of which are determined by a one-particle distribution function $f(\mathbf{r}, \mathbf{v})$. Much work in this vein has assumed, at least tacitly, that the potential Φ should be integrable or near-integrable, so that all the orbits in Φ are regular and there is no global stochasticity. However, despite earlier admonitions by a number of different authors, notably Schwarzschild,[1] there is no compelling reason, either theoretical or observational, to restrict attention to potentials that admit only regular orbits. Indeed, a synthesis of observational data and numerical investigations of orbits in fixed potentials and the results of N-body simulations suggest strongly that, in certain settings, chaos may be unavoidable.

The past decade or so has seen growing evidence that even such seemingly "simple" objects as elliptical galaxies are irregularly shaped, more complicated than the axisymmetric forms that were assumed in the 1970s. The existence of twisted isophotes (cf. reference 2) is usually interpreted as evidence that one is not seeing a spheroidal luminosity distribution in projection. Moreover, residuals in fits of the projected brightness distribution to a cos 2θ law,[3] especially pronounced in high density environments,[4] often imply a three-dimensional configuration that is more complicated in shape than a triaxial ellipsoid. High resolution photometry of nearby galaxies also suggests that many galaxies may have cuspy centers, whereby, even on scales as small as a few parsecs, the density diverges, rather than approaching a constant value (cf. references 5 and 6 and numerous citations therein).

These observations are significant in view of numerical experiments which suggest that, generically, breaking symmetries tends to increase the overall abundance of chaos. If, e.g., one takes an ellipsoidal potential admitting relatively little chaos and introduces small perturbations, idealized as $P_l(\cos\theta)$ corrections consistent in form and amplitude with the observed residuals, one can observe a huge increase in the total amount of chaos, as measured by the Kolmogorov entropy.[7] A

[a]HEK was supported in part by National Science Foundation Grant No. PHY92-03333. BLE was supported by NASA through the Florida Space Grant Consortium. Some of the numerical calculations reported herein were facilitated by computer time provided by *IBM* through the Northeast Regional Data Center (Florida).

[b]Also: Institute for Fundamental Theory, University of Florida and Department of Physics, University of Florida.

similar increase can also arise if one introduces a central perturbation to mimic a supermassive black hole and/or cusp.[7,8] Indeed, recent attempts by Merritt and Fridman[9] to construct cuspy triaxial models using a variant of Schwarzschild's[1] method indicate that, if the cusp is too steep, it may be impossible to model a cuspy triaxial system without incorporating a significant number of chaotic orbits. Despite recent work indicating that, when viewed in the full many-particle phase space, the gravitational N-body problem is exponentially unstable towards small changes in initial conditions (cf. reference 10 and citations therein), the evidence for large-scale chaos in N-body simulations is somewhat less compelling. However, there *is* at least limited evidence for chaotic orbits in simulations of barred spiral galaxies, presumably resulting from resonance overlap.[11–13]

Assuming that galactic models can admit a significant amount of chaos, there is a clear need for useful short time characterizations of chaotic orbit segments. If one waits long enough, it may be true that, in terms of their average properties, all orbits in a connected chaotic phase space region are identical. However, when considering galaxies one is confronted with systems that are relatively young in natural units, at most ~ 100 crossing times t_{cr} in age. The important point then is that, when examining the forms of different chaotic orbits even from the same phase space region over astrophysically relevant timescales $< 100t_{cr}$, there is an enormous degree of diversity, far too much to be explained, or even characterized, by quantities like Lyapunov exponents, which are only defined in an asymptotic $t \to \infty$ limit. Sometimes a chaotic orbit segment will look wildly irregular; other times it will be essentially indistinguishable from a regular orbit. Moreover, when tracking the evolution of a single chaotic orbit, one can observe apparent transitional behavior, whereby the orbit seemingly evolves from one "class" of chaotic behavior to another.[14–17]

But why are these sorts of phenomena important to galactic dynamicists interested in constructing models of galaxies? To construct galactic models using some variant of Schwarzschild's method, selecting orbit ensembles which reproduce Φ self-consistently, one needs to identify the building blocks from which the model is to be made. For regular orbits this is easy since the orbits are all multiply periodic. The situation is different for chaotic orbits which are aperiodic and, consequently, space filling. However, one can still identify a natural building block for chaotic orbits, namely the *invariant distribution* or *invariant measure.* Specifically, for any phase space region Γ which is connected in the sense that a single chaotic orbit will eventually probe the entire region, one can identify a unique invariant measure $N(\Gamma)$, which corresponds to a probability density that, when evolved into the future using the equations of motion appropriate for Φ, remains invariant.[18] Because of its invariance under time translation, $N(\Gamma)$ can be identified as the natural building block for the construction of a time-independent equilibrium.

Given the identification of $N(\Gamma)$ as the natural building block for a self-consistent model, one would like to understand as completely as possible its fundamental properties. However, earlier investigations of such invariant distributions have led to the recognition that, oftentimes, $N(\Gamma)$ seems to comprise different "classes" of orbit segments,[16,19] reflecting, e.g., topological semi-obstructions associated with cantori[20] or an Arnold web.[21] If, e.g., one considers chaotic orbits in a time-independent, two-dimensional potential, the observed transitions in behavior and the apparent division of orbits into two or more seemingly distinct classes are both associated with cantori, which correspond physically to a lower dimensional surface in the phase space that contains a cantor set of holes.

In this connection, it is also important to identify the characteristic timescale associated with transitions between one class of chaos and another. This will, e.g., provide potentially useful information about the efficiency with which different parts of phase space can "communicate" as a result of intrinsic diffusion (cf. references 22 and 23). Moreover, it is important to understand how low amplitude periodic disturbances and low amplitude friction and noise can accelerate this phase space transport by serving as a source of extrinsic and/or modulational diffusion.[18] Idealizing a galaxy as a time-independent solution to the collisionless Boltzmann equation involves a neglect of various physical effects which one knows must be present, e.g., irregularities associated with discrete substructures or internal pulsations and the effects of an external environment. However, numerical investigations, both for maps (cf. reference 24) and for differential equations,[19,25,26] have shown that even very weak irregularities, modeled as friction and noise or as time-periodic disturbances, can dramatically accelerate diffusion through various semi-obstructions. Indeed, in some cases they can essentially obliterate the distinction between different chaotic orbit classes.

The objective of the research reported here was the development and implementation of various tools of analysis that can be exploited fruitfully to probe complex chaotic phase spaces and, especially, to identify transitions between different chaotic orbit classes.

The first of these is short time, or local, Lyapunov exponents, $\chi(\Delta t)$, constructed as finite time analogues of the usual Lyapunov exponents. Short time Lyapunov exponents were first introduced in a mathematically useful way by Grassberger *et al.*[27] and Sepúlveda *et al.*,[28] and have been applied subsequently to a number of problems of astronomical interest, including kinematic dynamos (cf. references 29 and 30). The idea of computing a distribution of short time Lyapunov exponents, $N[\chi(\Delta t)]$, for an ensemble of orbits, e.g., a sampling of an invariant distribution, was apparently first considered[31] in the context of galactic dynamics.

The work on short time Lyapunov exponents has led to three significant conclusions:

1. Chaotic orbit segments that are more regular in visual appearance tend systematically to be less unstable and, consequently, to have smaller short time Lyapunov exponents.
2. Distributions of short time exponents, $N[\chi(\Delta t)]$, generated from a sampling of an invariant distribution can exhibit a complex structure with multiple peaks, reflecting the fact that the orbit segments that were sampled divide into several different classes. However, whether $N[\chi(\Delta t)]$ is singly or multiply peaked is a strong function of the sampling time Δt. In particular, an analysis of the form of $N[\chi]$ as a function of Δt provides nontrivial information about the characteristic timescale associated with transitions between different classes of chaotic orbits.
3. Even for a given class of chaotic orbit, χ exhibits significant variability on short timescales. Indeed, if a single orbit is divided into segments of length Δt and $\chi(\Delta t)$ computed for each segment, the resulting set of segments, viewed as

time series, typically exhibits roughly self-similar behavior when the sampling interval Δt is changed.

The second tool of analysis involves computing Fourier transforms of chaotic orbit segments. The point here is that qualitative changes in the behavior of quantities like $\mathbf{r}(t)$ are also manifested in the transformed $\mathbf{r}(\omega)$. If a single chaotic orbit is integrated for a long time, it will eventually probe an invariant distribution and, as such, be characterized by an invariant Fourier spectrum, the form of which is independent of the detailed choice of initial conditions.[17] However, different segments of that single orbit will have spectra that look very different. A chaotic segment that appears visually to be nearly regular will have a Fourier spectrum that closely resembles that of a regular orbit, characterized by a few sharp peaks, whereas a segment that appears wildly irregular will exhibit broader band power. The objective here is to implement the idea[19] that, in a precise mathematical sense, chaotic orbits can be viewed as complex superpositions of regular orbits, a notion well-motivated physically by the interpretation of chaos as arising from resonance overlap.

The following section describes the use of distributions of short time Lyapunov exponents as tools for analyzing complex chaotic phase spaces, emphasizing the shape of the distribution as a function of the sampling interval. The next section discusses the results of a multi-resolution analysis of sequences of short time Lyapunov exponents, summarizing, in particular, the evidence for self-similar structure. A further section uses Fourier transforms to identify a precise sense in which some chaotic orbit segments are more irregular, and hence complex, than others. The work described there demonstrates in particular that there is a strong direct correlation between the "complexity" of an orbit segment, as probed by its Fourier spectrum, and the value of its largest short time Lyapunov exponent. The final section extends that analysis to quantify the sense in which a given chaotic orbit, nearly regular or wildly chaotic, tends to comprise "pieces" of different regular orbits with different "topological" properties.

SHORT TIME LYAPUNOV EXPONENTS AS DIAGNOSTICS FOR PHASE SPACE TRANSPORT

One distinguishing characteristic of chaos, which can be taken as a definition, is the fact that chaotic orbits exhibit an exponentially sensitive dependence on initial conditions. This observation leads naturally to the notion of Lyapunov exponents, which probe the average rate of exponential instability in a $t \rightarrow \infty$ limit. Specifically, given an initial phase space perturbation $\delta Z(0)$, the Lyapunov exponent is defined as (cf. reference 32)

$$\chi \equiv \lim_{t \rightarrow \infty} \lim_{\delta Z(0) \rightarrow 0} \frac{1}{t} \ln \left[\frac{\|\delta Z(t)\|}{\|\delta Z(0)\|} \right], \tag{1}$$

where $\|.\|$ represents a suitable norm. An orbit is said to be chaotic if at least one

Lyapunov exponent is positive. This corresponds to the statement that there is at least one phase space direction in which the orbit is exponentially unstable.

Such Lyapunov exponents provide important characterizations of the average properties of chaotic orbits. However, as noted already, different segments of a single chaotic orbit can exhibit very different qualitative behavior. It seems natural, therefore, to ask whether these changes correlate with the overall degree of exponential instability over shorter time intervals. This question can be addressed by computing short time Lyapunov exponents $\chi(\Delta t)$,[27,28] which are defined in the obvious way as finite time analogues of the asymptotic χ, i.e.,

$$\chi(\Delta t) \equiv \lim_{\delta Z(0)\to 0} \frac{1}{\Delta t} \ln \left[\frac{\|\delta Z(\Delta t)\|}{\|\delta Z(0)\|} \right]. \tag{2}$$

Associated with any given initial condition evolving in an N-dimensional potential, there will be $2N$ independent exponents, one for each of the phase space dimensions, although, for the special case of a time-independent Hamiltonian, two of the $t \to \infty$ exponents must vanish and the others must come in pairs, $\pm\chi$ (cf. reference 18). It is possible to compute explicitly the values of all the exponents by considering an orthogonal set of initial perturbations and then implementing a Gramm–Schmidt orthogonalization procedure (cf. reference 33). However, in many cases the most important question involves the magnitude of the most positive exponent, which will dominate the exponential instability. At least for late times, this largest exponent can be obtained relatively easily by introducing a random $\delta Z(0)$ and evolving that initial perturbation into the future, the point being that, eventually, any exponential growth will be dominated by the largest exponent, independent of either $\delta Z(0)$ or the choice of norm.

By contrast, when considering a shorter time exponent, the $\chi(\Delta t)$ associated with some initial condition *can* depend significantly on both the initial $\delta Z(0)$ and the norm. Suppose, however, that one chooses the obvious Euclidean norm, so that, e.g., for a two-dimensional system, $\|\Delta Z\|^2 \equiv |\delta x|^2 + |\delta y|^2 + |\delta p_x|^2 + |\delta p_y|^2$. It is then clear that, on time scales $\gg T_\chi \sim \chi_{max}^{-1}$, with χ_{max} the largest Lyapunov exponent, the $\chi(\Delta t)$ computed from a random $\delta Z(0)$ should provide a good approximation to the largest short time exponent for the interval $0 < t < \Delta t$. In what follows, attention will focus on the largest short time exponent, generated in precisely this way.

The remainder of this section summarizes a series of numerical experiments which involved (1) sampling an invariant distribution to generate an ensemble of chaotic initial conditions, (2) computing short time Lyapunov exponents $\chi(\Delta t)$ for each orbit in the ensemble, (3) combining these to generate distributions $N[\chi(\Delta t)]$, and then (4) examining and interpreting the shape of $N[\chi(\Delta t)]$ as a function of Δt. The conclusions derive from a detailed examination of three different potentials which, despite having very different qualitative properties, yield similar results, both qualitatively and semi-quantitatively. Two of these are well known to nonlinear dynamicists, namely the sixth-order truncation of the three-particle Toda lattice potential[34] and the so-called dihedral, or $D-4$, potential.[35] The third, more realistic from the standpoint of galactic dynamics, is the so-called *KAMB* potential,[19] which is constructed as the sum of an axisymmetric Plummer potential and a nonaxisymmetric Plummer potential. Specifically,

$$V(x, y) = -\frac{1}{(c^2 + x^2 + y^2)^{1/2}} - \frac{m}{(c^2 + x^2 + a^2y^2)^{1/2}}, \tag{3}$$

with $m = 0.3$, $a = \sqrt{0.1} \approx 0.316$, and $c = 1.0$. For typical energies in this potential, a typical crossing time corresponds to $t \sim 10$, so that the Hubble time can be identified conveniently as $t_H = 1024$. Because all three potentials yield similar results, attention here will focus on results derived for the *KAMB* potential.

FIGURES 1, 2, and 3 exhibit distributions of short time Lyapunov exponents, $N[\chi(\Delta t)]$, generated from samplings of invariant distributions in the *KAMB* potential at three different energies, $E = -0.3$, $E = -0.55$, and $E = -0.6$. In each case eight different sampling intervals were used, ranging from $t = 256$ to $t = 32{,}768$, i.e., from roughly $\frac{1}{4}t_H$ to $32t_H$. For each energy, there is only one large connected phase space region for chaotic orbits, and hence a unique invariant distribution. The $E = -0.3$ hypersurface is typical of a phase space region comprising primarily chaotic orbits. Regular islands do exist in the stochastic sea, but they are relatively small. Cantori are seemingly unimportant and there is relatively little trapping of chaotic orbits near regular regions. By contrast, the $E = -0.55$ hypersurface contains large regular regions, so that cantori are very important. In this case, trapping is relatively common and can persist for times $t > t_H$. The $E = -0.6$ hypersurface represents an extreme case where, presumably because the holes in the confining cantori are very small, trapping can persist for periods as long as $100t_H$ or more.

For the energies exhibited in these illustrations, as well as for other energies and other potentials, on sufficiently short timescales, $<10-20t_{cr}$ or so, the distribution $N[\chi(\Delta t)]$ is relatively broad and essentially featureless, being characterized by a single peak. However, as the sampling time Δt increases, the distribution becomes

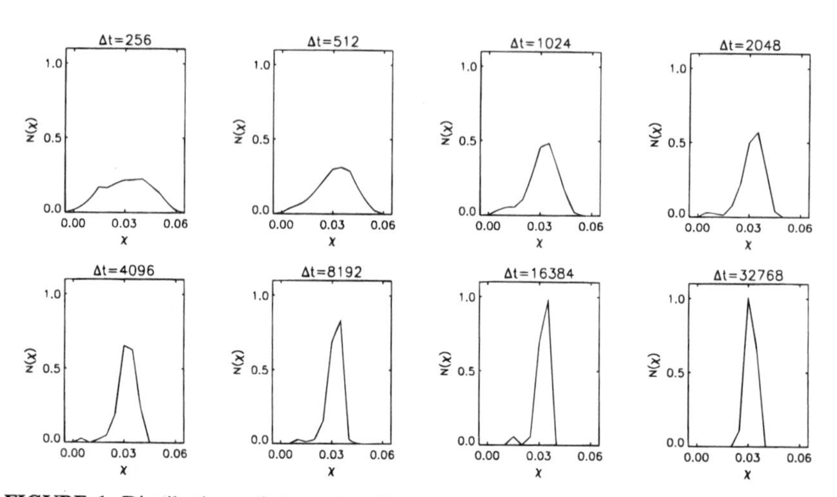

FIGURE 1. Distributions of short time Lyapunov exponents, $N[\chi(\Delta t)]$, generated from a sampling of the invariant measure of chaotic orbits in the *KAMB* potential with energy $E = -0.3$, allowing for eight different sampling intervals Δt. The curves are normalized to have equal integrated area.

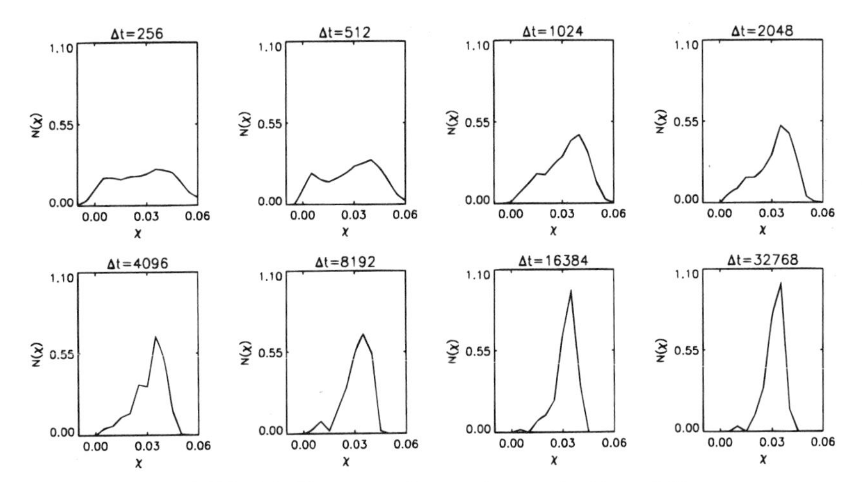

FIGURE 2. The same as FIGURE 1, except for $E = -0.55$.

narrower and, more significantly, nontrivial structures can arise. Three general possibilities exist:

1. If the phase space is almost completely chaotic, one observes typically a single peaked distribution, seemingly well approximated as a Gaussian, which becomes progressively narrower as t increases.
2. If the phase space admits a large measure of regular orbits, one can observe instead a multi-modal distribution with more than one peak, seemingly well approximated as a sum of two or more Gaussians.
3. Often one observes an intermediate behavior, where $N[\chi]$ is singly peaked but is noticeably different from a Gaussian because of the presence of a low-χ tail.

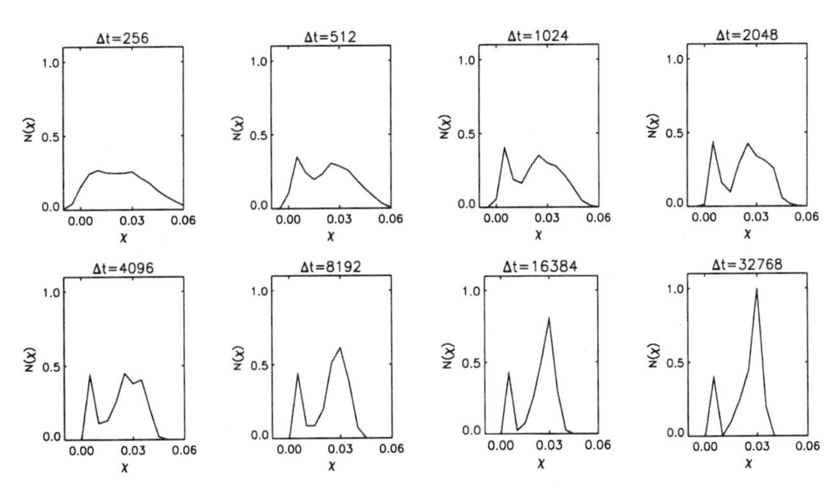

FIGURE 3. The same as FIGURE 1, except for $E = -0.6$.

For sufficiently long sampling times Δt any nontrivial structures must disappear since, in the $t \to \infty$ limit, the distribution must converge towards an infinitely sharp distribution concentrated at the value of the true Lyapunov exponent. Note also that, because the initial conditions were generated as a sampling of the invariant distribution, the mean χ associated with $N[\chi]$ should, modulo finite number statistics, be independent of Δt. Because of the Ergodic Theorem, the statistical properties of a collection of N orbit segments each integrated for a total time Δt should be equivalent to the properties of a single segment integrated for a total time $N \times \Delta t$. As discussed elsewhere,[31] the late time χ has physical meaning even on short times, representing the average degree of exponential instability for a collection of orbits that samples the invariant distribution.

The observed changes in the form of $N[\chi]$ with increasing sampling time Δt are easily understood in terms of different classes of chaos and transitions between these classes. The basic points in this are (1) that a multi-peaked distribution for time Δt reflects the *de facto* existence, on a time scale Δt, of several distinct chaotic classes, and (2) that chaotic orbits which are more regular in visual appearance tend systematically to have smaller short time Lyapunov exponents.

When examining chaotic orbit segments over very short time scales, it is difficult to distinguish confined chaotic orbits, which are trapped near regular islands, from unconfined chaotic orbits, which travel essentially unimpeded through the deeps of the stochastic sea. This is reflected by the fact that, for sufficiently short times Δt, the same value of $\chi(\Delta t)$ can be found for both confined and unconfined chaotic orbits. Alternatively, on sufficiently short time scales two orbits that eventually appear relatively similar in appearance, e.g., both confined near the same regular island, can have very different values for $\chi(\Delta t)$.

Over longer time scales, however, it becomes increasingly apparent whether the orbit segment in question is unconfined or whether instead it is trapped near some regular island. This is manifested by the fact that distributions of short time Lyapunov exponents computed for ensembles containing both confined and unconfined segments evidence several different populations which, when combined together, yield a composite multi-peaked distribution. For two-dimensional systems, such multi-peaked distributions are most common for energies where regular islands have a significant measure. This reflects the fact that the distinctions between different populations are enforced by cantori, fractured *KAM* tori associated with the breakdown of integrability. At energies where the regular regions are smaller, cantori become less important and trapping is less common.

Despite the existence of cantori and/or other semi-obstructions, over sufficiently long time scales individual orbit segments will eventually switch from one orbit class to another so that the distinctions between different orbit classes are erased. This is reflected in the fact that, for sufficiently large Δt, any multi-peaked structure eventually disappears, so that $N[\chi(\Delta t)]$ assumes a form appropriate for a single population. The obvious point, then, is that the time scale Δt on which these structures disappear provides a reasonable measure of the characteristic time scale on which chaotic orbits breach cantori or other obstructions to probe different phase space regions.

This characteristic time Δt can vary enormously for different energies. At energies where regular orbits are relatively uncommon, the holes in cantori tend to be

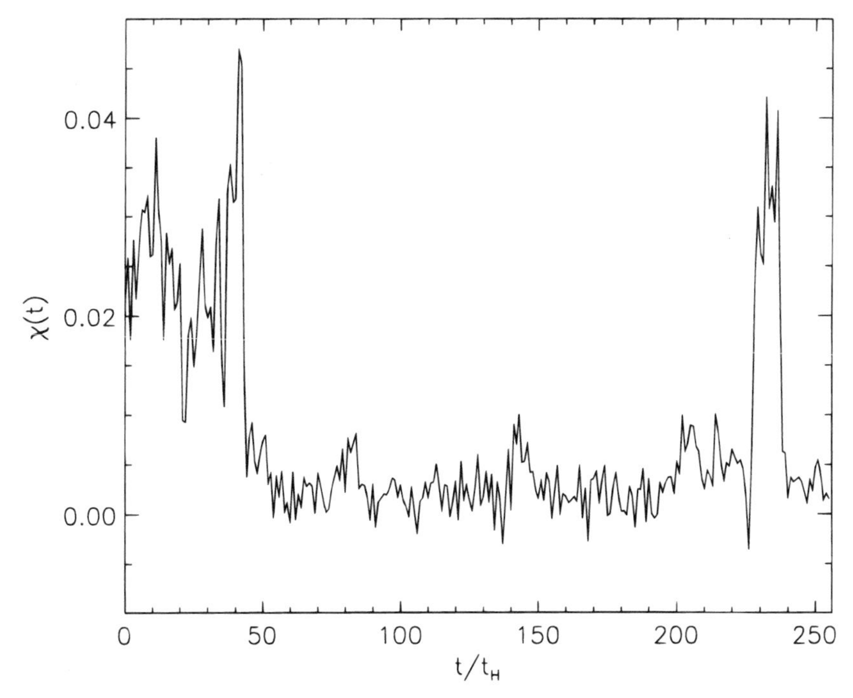

FIGURE 4. A time series of short time Lyapunov exponents $\chi(t)$, generated by partitioning a single long time integration with $E = -0.6$ into segments of length $t = 1024 \equiv t_H$.

large so that transitions occur relatively rapidly. However, at energies where there is little global stochasticity the holes become much smaller, so that the time scale associated with transitions becomes significantly larger. Thus, e.g., for energies $E \sim -0.3$ in the *KAMB* potential, little if any structure is observed for sampling times > 2–$4t_H$, whereas, for energies as low as $E \sim -0.6$, one observes structures on scales of 100–$200t_H$ or more.

To at least a limited degree, short time Lyapunov exponents can also serve as diagnostics as to whether and when an individual orbit segment has made a transition from one class of chaotic behavior to another. Thus, e.g., a knowledge of $\chi(\Delta t)$ for some interval $\Delta t \sim t_H$ can provide information about whether, overall, the segment was characterized by wildly chaotic or more nearly regular behavior. However, on much shorter time scales the range of possible values of χ for a given class is sufficiently large that short time exponents are not useful in distinguishing between confined and unconfined behavior. Only after a time as long as ~ 10–$20t_{cr}$ can one make clear distinctions between confined and unconfined behavior and, consequently, one cannot use short time exponents to identify the time when such a transition has occurred more accurately than to within 10–$20t_{cr}$.

One final question remains. When an orbit is trapped near a regular region, is the value of χ simply very small, or could it in fact vanish identically? Algorithms

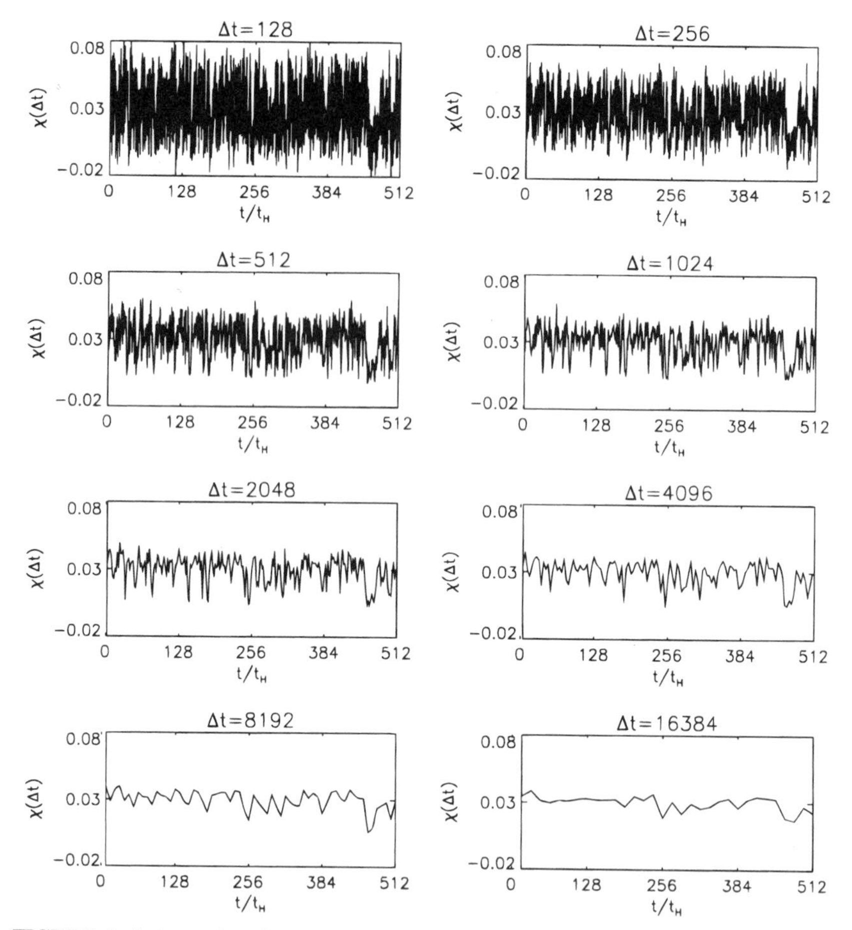

FIGURE 5. A time series of short time Lyapunov exponents $\chi(\Delta)$, generated by partitioning a single long time integration with $E = -0.55$ into segments of variable length Δt.

like those used in the work described here make this question difficult to answer definitively. However, there is strong evidence that χ is not vanishingly small. If the true value of χ is in fact zero, one would anticipate that a computation of $\chi(t)$ for progressively longer times will yield a χ that eventually decays to zero as a power law in t.[18] However, this means that if $\chi(t)$, evaluated for a total time $t = N\Delta t$, is partitioned to extract a collection of N different χ's for successive intervals Δt, there should be no obvious sense in which, systematically, the $\chi(\Delta t)$'s become progressively smaller in value. FIGURE 4 exhibits data for a single chaotic orbit with $E = -0.6$ which was integrated for a total time $256t_H$ and then analyzed to extract a collection of $\chi(\Delta t)$'s with $\Delta t = 1024 = 1t_H$. It is evident that, despite being trapped near a regular island for nearly $200t_H$, from about $50t_H$ to $225t_H$, there is absolutely no sense in which the exponent decays systematically towards zero.

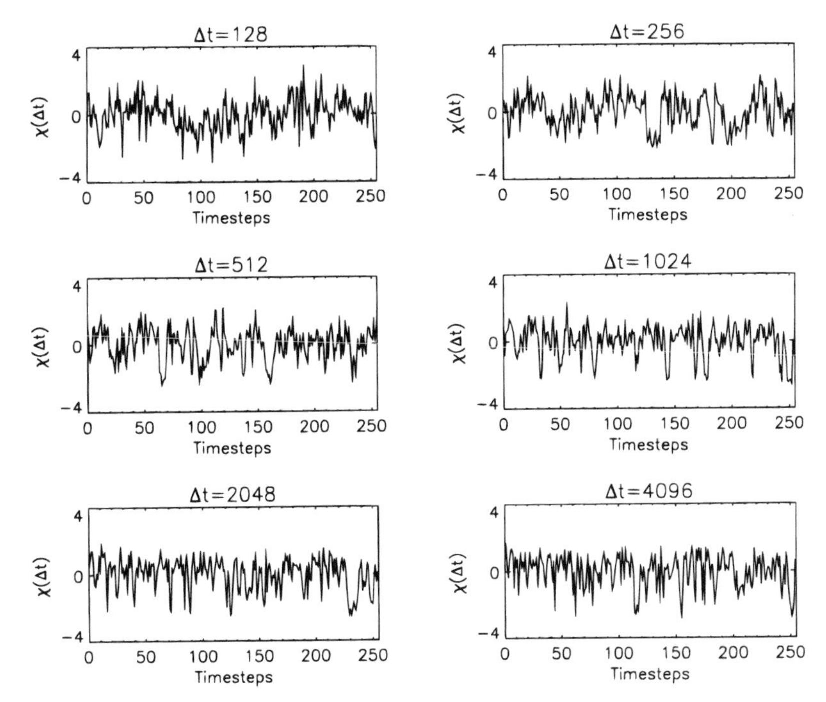

FIGURE 6. Data from FIGURE 5, now reanalyzed to exhibit 256 timesteps of variable length Δt.

MULTI-RESOLUTION ANALYSIS OF SHORT TIME LYAPUNOV EXPONENTS

As noted already, statistical properties extracted from a long time integration of a single chaotic initial condition should be identical to the statistical properties of an ensemble of chaotic segments generated from a sampling of the invariant measure Γ. This implies, however, that to analyze the structure of some connected phase space region as a function of sampling time Δt, it suffices computationally to study the properties of a single orbit that was integrated for a very long time and then partitioned into a number of smaller segments. Thus, in particular, to extract statistical properties of short time Lyapunov exponents, one can (1) compute $\chi(T)$ for a single long time interval $T = 2^n \delta t$, with δt some fixed interval at which data are recorded, (2) partition the integration into a collection of segments of length $\Delta t = 2^k \delta t$, with $k < n$, and then (3) extract for each segment i an exponent $\chi(\Delta t_i)$ via the obvious prescription

$$\chi(\Delta t_i) \equiv \frac{(t_i + \Delta t)\chi(t_i + \Delta t) - t_i\,\chi(t_i)}{\Delta t}. \tag{4}$$

The resulting string of 2^{n-k} short time Lyapunov exponents can be viewed as a time series, and it is natural to probe the amount of structure in this time series as a function of k or Δt. This is closely related to the subject matter of the previous section since the distribution of short time exponents $\chi(\Delta t)$ associated with this time series should, modulo finite number statistics, agree completely with the distribution $N[\chi(\Delta t)]$ generated from a sampling of the invariant measure.

The obvious question then is: How does the amount of structure in such a time series change as a function of Δt? Qualitatively, it is evident that, as Δt becomes larger, the time series generated for a fixed interval T becomes smoother and deviations from the mean become smaller in magnitude. This is, for example, illustrated in FIGURE 5, which exhibits data for a chaotic initial condition in the $KAMB$ potential with $E = -0.55$ integrated for a time $T = 512t_H$ and then partitioned into segments varying in lengths from $\frac{1}{8}t_H$ to $16t_H$.

However, probing a fixed number of sampling intervals, rather than a fixed integration time T, typically leads instead to roughly self-similar structure. This is easily seen by computing, for a fixed number of time steps, a renormalized time series

$$\chi_\sigma(\Delta t_i) = \sigma(\Delta t) \times (\chi(\Delta t_i) - \bar{\chi}), \tag{5}$$

with $\bar{\chi}$ the mean value of χ and $\sigma(\Delta t)$ the corresponding dispersion, and then studying the form of this time series as a function of the sampling time Δt. FIGURE 6 exhibits the results of such an analysis for the same orbit that was used to generate FIGURE 5. Each panel in this illustration exhibits a total of 256 χ's, the sampling time Δt varying from $\frac{1}{8}t_H$ to $4t_H$, for a total time T between $32t_H$ and $1024t_H$. It is apparent that the different panels exhibit a comparable amount of structure, consistent with an assumption of approximate self-similarity.

Self-similarity is hardly surprising for a time series that corresponds to a distribution of short time Lyapunov exponents, $N[\chi(\Delta t)]$, which remains approximately Gaussian for a range of different sampling intervals. In this case, the distribution N itself is self-similar, remaining invariant if one implements a scaling of the form (**5**) in terms of the time-dependent dispersion $\sigma(\Delta t)$.

One useful diagnostic in probing the degree of structure in the time series is how the dispersion σ scales with the sampling time Δt. In earlier work,[16,31] it was observed that, for a relatively small range of sampling intervals, i.e., $10t_H < \Delta t < 100t_H$ or so, the dispersion σ can often be approximated by a power law $\sigma \propto (\Delta t)^{-p}$, with p a positive constant $\leq \frac{1}{2}$. The experiments reported here tested the accuracy of this scaling for a much larger range of sampling times.

The basic conclusion derived from these experiments is that for potentials, energies, and sampling times corresponding to a single peaked, near-Gaussian distribution of short time exponents, such a power law fit is in fact quite good. Thus, a log-log plot of σ versus Δt can be approximately linear for two or more decades of sampling times, with a best fit exponent $p \sim 0.3$–0.4. However, when the distribution $N[\chi(\Delta t)]$ is more complicated, e.g., characterized by two or more local maxima, a log-log plot can instead show statistically significant curvature, with σ decreasing only much more slowly with increasing Δt. Physically, this reflects the fact that even if the dispersions associated with individual components of the total distribution are decreasing relatively rapidly with Δt, the dispersion for the composite distribution remains large because of the finite gap between the different peaks.

This general behavior is illustrated in FIGURE 7, which exhibits $\sigma(\Delta t)$ for chaotic orbits in the *KAMB* potential with three different energies, namely $E = -0.3$ (solid curve), $E = -0.55$ (dashed curve), and $E = -0.6$ (dot-dashed curve). The solid curve for $E = -0.3$, an energy characterized (cf. FIG. 1) by singly peaked distributions $N[\chi]$, is approximately linear, with a best fit slope $p = 0.31 \pm 0.03$. For short sampling times Δt, the other two curves are also approximately linear, although they acquire significant curvature as Δt increases. In particular, for $E = -0.6$ the dispersion $\sigma(\Delta t)$ seems to be approaching a constant value, independent of Δt, reflecting the large gap between the two local maxima in the distribution at $\chi \sim 0.005$ and 0.03. The curve for $E = -0.55$ exhibits an intermediate behavior, with the slope decreasing slightly for $2.5 < \log \Delta t < 3.5$ but then increasing again.

It is tempting to analyze these time series to extract quantities like the information and/or correlation dimensions,[18] which serve as natural probes of self-similar structures. However, this is not well motivated physically since, at least for the algorithm used in this paper, a consideration of the $\Delta t \rightarrow 0$ limit does not seem well motivated. However, it *is* worth recognizing that to the extent that the distribution of exponents is approximately Gaussian, and that the dispersion σ is well fit by a power law $\sigma \propto (\Delta t)^{-p}$, all the generalized dimensions assume that the values $D_q = p$.

In this context, it should be recognized that, as discussed elsewhere,[31] the value $p = \frac{1}{2}$ has a special significance: for large Δt, $N[\chi(\Delta t)]$ can be viewed as a convolution of a large number of distributions $N[\chi(\delta t)]$ for shorter time intervals δt. If, however, the values of $\chi(\delta t)$ for different δt intervals are completely uncorrelated, the Central Limits Theorem guarantees that the long time distribution will converge towards a Gaussian with a dispersion scaling as $k^{-1/2}$, with k the number of shorter

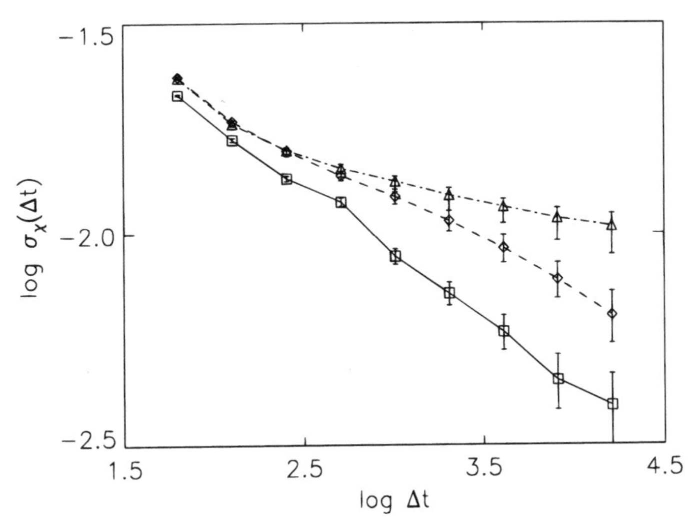

FIGURE 7. The dispersion $\sigma(\Delta t)$ associated with samplings of the invariant distribution for energies $E = -0.3$ (*solid curve*), $E = -0.55$ (*dashed curve*), and $E = -0.6$ (*dot-dashed curve*). The data are presented in a log-log plot.

time intervals. The fact that this number grows linearly with Δt thus implies a scaling $\sigma \propto (\Delta t)^{-1/2}$. A smaller value of p reflects the fact that successive intervals are *not* completely uncorrelated and, as such, provides another probe of the extent to which the degree of exponential instability at two times t and $t + \Delta t$ are, or are not, completely decoupled.

THE COMPLEXITY OF CHAOTIC ORBIT SEGMENTS

The objective of this and the following section is twofold, namely (1) to exhibit an algorithm which can be used to quantify the extent to which a given chaotic orbit segment is, or is not, "nearly regular," and (2) to suggest a concrete diagnostic which can be used to quantify the sense in which, at least visually, chaotic orbit segments seem often to comprise "pieces" of different regular orbits.

Based on a visual inspection of numerous orbits, most galactic dynamicists would probably agree that chaotic orbits are typically more complex overall than regular orbits, and that wildly chaotic orbit segments are more complex than chaotic orbit segments that are nearly regular. One way to quantify this intuition is by exploiting notions from the theory of *algorithmic complexity* (cf. reference 36), so as to implement the idea that more "bits" of information are required to characterize a wildly chaotic orbit segment of given length than a more nearly regular segment of equal length. The key realization is that the reason that a regular orbit looks simple, rather than complex, is that it is periodic. In particular, the Fourier spectrum associated with a regular segment will have virtually all the power concentrated at or near a few special frequencies. However, this would suggest that one natural way to quantify the extent to which a chaotic segment is, or is not, nearly regular is to determine how much of the total power is concentrated at a few special frequencies. A wildly chaotic orbit would be expected to exhibit a much broader band spectrum than a nearly regular chaotic orbit.

Another setting in which to formulate essentially the same idea is in the context of *compression*, which arises naturally in a number of different settings in astronomy and space science (cf. reference 37). The idea here is to determine the extent to which a chaotic orbit segment, generated as a time series of N equally spaced points, can be reconstructed if one (1) expands the time series in terms of a set of N basis functions, (2) sets some large number of the expansion coefficients equal to zero, and then (3) inverts the truncated basis function expansion. Specifically, one can ask: What is the minimum number j of basis coefficients required so that an approximate orbital time series reconstructed using only j basis coefficients will be "close to" the original time series with respect to some natural norm? Thus, e.g., one could demand that the L^2 distance between the true and reconstructed time series be small compared with the L^2 size of the original time series, i.e.,

$$\|x_T - x_R\| \equiv \left(\sum_{i=1}^{N} [x_T(t_i) - x_R(t_i)]^2 \right)^{1/2} < \varepsilon \left(\sum_{i=1}^{N} [x_T(t_i)]^2 \right)^{1/2} \equiv \varepsilon \|x_T\|, \qquad (6)$$

for some small ε.

If one decomposes the original time series in a Fourier basis, there is a direct connection between the former tact, which views the orbit in k-space, and the latter, which considers instead a suitably reconstructed time series. Assuming that the discrete Fourier expansion involves a complete orthonormal basis, Parseval's theorem (cf. reference 38) implies that, for either the true or the reconstructed orbit, $\sum_i |x(\omega_i)|^2 = \sum_i |x(t_i)|^2$. However, it is easy to see that if one selects a set of j frequencies $\{\omega_j\}$ such that

$$\sum_{i=j+1}^{N} |x(\omega_i)|^2 < \varepsilon^2 \sum_{i=1}^{N} |x(\omega_i)|^2, \tag{7}$$

the reconstructed orbit will satisfy (**6**).

The possibility of classifying orbits in terms of complexity, as probed by the degree to which power is concentrated at a relatively few frequencies, was tested through an examination of a number of chaotic orbits in the *KAMB* potential. This involved computing trajectories for a variety of initial conditions for a total time $T = 128 \times 1024$, sampling at fixed intervals $\delta t = 1/2$, partitioning the total orbit into 128 segments of equal length, and then computing the Fourier spectrum for the resulting segments using a standard *FFT* routine.[39] Given a knowledge of $|x(\omega_i)|^2$ and $|y(\omega_i)|^2$, one can then determine the minimum numbers of frequencies, $N_x(f)$ and $N_y(f)$, that must be retained to capture a fixed percentage f of the total x- and y-powers. The overall complexity of the chaotic orbit segment at a given level f was defined as the sum $N(f) \equiv N_x(f) + N_y(f)$. The threshold percentage was varied considerably, from 50 to 99.75%. A variety of different energies were considered, ranging from low energies $E \sim -0.6$, where phase space transport is relatively slow and a chaotic orbit can look nearly regular for a very long time, to higher energies $E \sim -0.3$, where changes in visual appearance occur on much shorter time scales.

Characterizations of the overall degree of complexity based on this prescription are in accord with physical expectations in the sense that those orbits which appear visually to be the most regular are identified as the least complex and those orbits which appear visually to be the most irregular are identified as the most complex. Moreover, this diagnostic is robust in the sense that, for a relatively broad range of values, different choices of f yield relatively similar results. Thus, e.g., if one orders the 128 segments of a given long time integration in terms of $N(60)$ and $N(95)$ and computes a rank correlation for the resulting lists, one typically obtains a number $\mathcal{R} \sim 0.7$–0.8. If, however, f is too large, say 98–99% or more, this measure of complexity yields results that are less compelling: When one raises f to numbers very close to 100%, corresponding to a nearly perfect reconstruction, one is confronted with the problem that, even for very nearly regular orbits, the peaks have a finite width. Viewed in terms of a diagnostic like $N(99.9)$, nearly regular and wildly chaotic orbits do not differ greatly.

It should be stressed that the success of this classification scheme relies on the fact that one is using a Fourier basis, which is the optimal basis for extracting intrinsic periodicities. A classification based on a significantly different set of basis functions would yield results that are very different. Thus, e.g., when expanding in a Daubechies-20 wavelet basis (cf. reference 40), one finds instead that, independent of

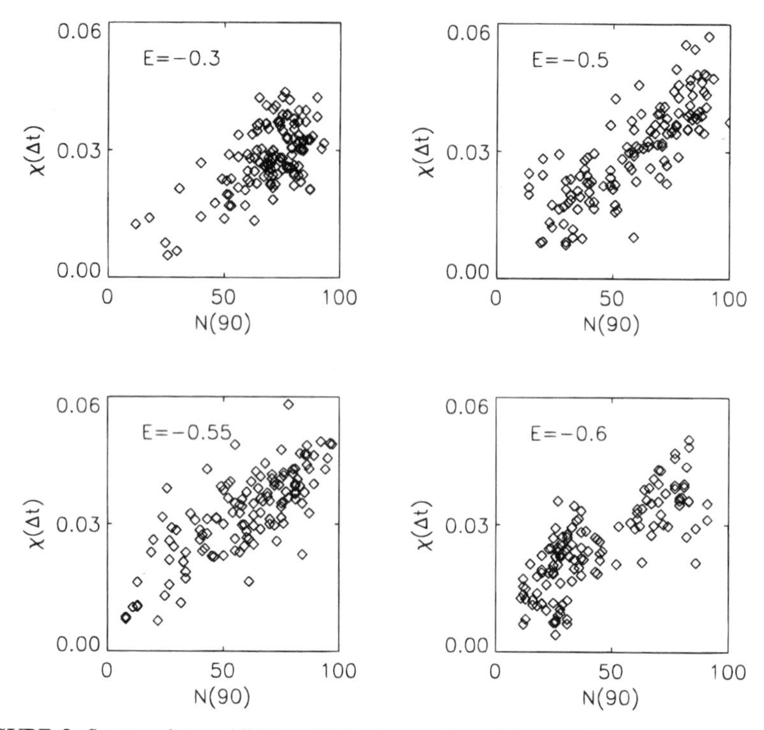

FIGURE 8. Scatter plots exhibiting $N(90)$, the number of frequencies required to capture 90% of the power in an orbit segment, and the short time Lyapunov exponent $\chi(\Delta t)$. Each panel exhibits data for 128 segments of length t_H generated from a single long time integration of time $T = 128t_H$.

the threshold f, comparable numbers of coefficients are required to yield approximate reconstructions of both regular and chaotic orbits.

One final question remains: To what extent does a characterization of orbit segments in terms of complexity agree with a characterization based on short time Lyapunov exponents? The answer is that there is a strong direct correlation between these two characterizations and that this correlation is especially strong for energies where orbits exhibit significant qualitative changes in behavior. This is illustrated in FIGURE 8, which exhibits scatter plots of $\chi(\Delta t)$ versus $N(90)$ for four different long time integrations in the *KAMB* potential with $E = -0.6$, -0.55, -0.5, and -0.3.

To quantify this trend, one can compute a rank correlation relating $\chi(\Delta t)$ and $N(f)$ for $f \sim 90$–95. The resulting $\mathcal{R}$ typically ranges from a value ~ 0.8 at lower energies to ~ 0.5 at much higher energies. As illustrated in FIGURE 9, rank correlations can also be used to exhibit visually the sense in which orderings in terms of values of N and χ agree at least approximately. The illustrations in FIGURE 9 were generated from a single long time integration with $E = -0.5$ by ranking each of 128

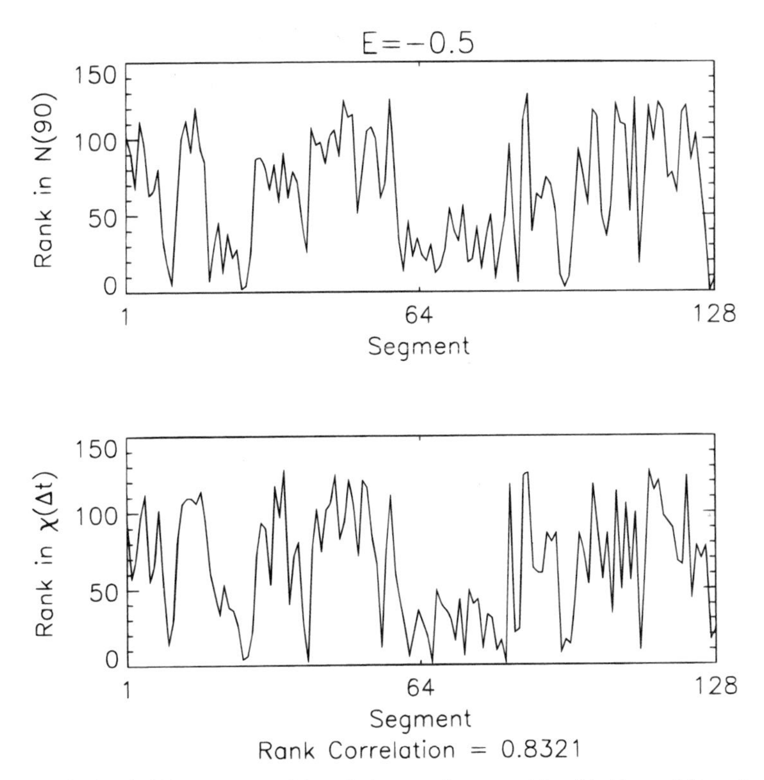

FIGURE 9. **(Top)** 128 segments of length $\Delta t = t_H$ for an orbit with $E = -0.5$, ranked in terms of $N(90)$, the number of frequencies required to capture 90% of the total power. Large $N(90)$ corresponds to a high rank. **(Bottom)** The same segments ranked in terms of the short time Lyapunov exponent $\chi(\Delta t)$.

segments in terms of $N(90)$ and $\chi(\Delta t)$, thereby generating numbers between 1 and 128, and then plotting the rankings as a function of segment number. The obvious point is that the curve in the top panel, which exhibits a ranking based on $N(90)$, is very similar in appearance to the curve in the bottom panel, which is based on $\chi(\Delta t)$.

CHAOTIC ORBITS AS SUPERPOSITIONS OF REGULAR ORBITS

Power spectra associated with different types of regular orbits are distinctly different. Consider, e.g., a time-independent two-dimensional potential where, generically, there are three basic types of regular orbits, namely loops, boxes, and bananas. Simple loop orbits have x- and y-powers, $|x(\omega_i)|^2$ and $|y(\omega_i)|^2$ which are both concentrated at the same two frequencies, ω_1 and ω_2. By contrast simple box orbits elongated in the y-direction have x-power concentrated at two frequencies ω_1 and ω_3 and y-power concentrated at an intermediate frequency ω_2 (cf. reference 41). Spectra for banana orbits are similar to those for box orbits except that the lower frequency $\omega_1 \to 0$.

Chaotic orbit segments that look "nearly regular" have power spectra which, not surprisingly, are very similar to the spectra of the regular orbits which they resemble. By contrast, complex "wildly chaotic" segments have more complicated spectra exhibiting a larger number of peaks. However, even for very complex chaotic segments one can often identify visual features strikingly reminiscent of two or more different regular orbits, features which correlate directly with various peaks of the power spectrum. For this reason, as motivated more carefully in reference 17, it seems natural to try to view any chaotic orbit segment, nearly regular or otherwise, as comprising "pieces" of one or more different regular orbits plus some low amplitude leftover junk. In particular, one might like to identify an algorithm which, given the power spectrum associated with some given orbit segment, allows one to quantify the degree to which the segment is "nearly loopy" or "nearly a box."

One natural starting point for such an algorithm is the recognition that, unlike boxes and bananas, loop orbits have power spectra $|x(\omega_i)|^2$ and $|y(\omega_i)|^2$ which both peak at the same ω_1 and ω_2. Specifically, one might proceed by (1) identifying peaks in the Fourier spectra, (2) calculating how much of the total power is concentrated at or near these peaks, and then (3) determining the extent to which the peak frequencies for the x- and y-powers are in fact the same. In the first instance, this can be done working directly with the raw power spectrum. However, one might choose instead to "smooth" the jagged peaks with a filter and analyze the resulting smoothed spectrum.

The most difficult task in implementing any such algorithm appears to be defining what constitutes a peak and, related to this, identifying which features in the spectrum should be interpreted as low amplitude "junk". Ideally one should probably implement a prescription which tailors the definition of peaks and valleys to reflect various properties of the specific orbit segment. However, one can achieve useful results without any such refinements by exploiting a hard threshold filter (cf. reference 40), identifying the "non-junky" frequencies as being those for which the x- and/or y-power exceeds some limiting value. Thus, in particular, working with x and y individually, one can specify a threshold amplitude corresponding to some fixed fraction z of the maximum value assumed by the power spectrum or, perhaps, a fraction of the average of several of the highest amplitudes.

The following prescription was used as a concrete test of these general ideas:

1. Define separate thresholds for the x- and y-powers, corresponding to a specified fraction z of the peak amplitudes A_x and A_y achieved by $|x(\omega_i)|^2$ and $|y(\omega_i)|^2$.
2. Use hard thresholding to select all frequencies ω_i where $|x(\omega_i)|^2$ and/or $|y(\omega_i)|^2$ exceeds, respectively, $z \times A_x$ or $z \times A_y$.
3. Determine the fraction f of the frequencies selected in this fashion which have both $|x(\omega_i)|^2$ and $|y(\omega_i)|^2$ larger than the threshold value.

For an unfiltered signal, reasonable results were obtained for z of order 5%. For filtering algorithms which smooth the highest peaks, a somewhat larger threshold z typically works better.

The implementation of this algorithm for regular loop orbits, where the x- and y-powers are concentrated at the same frequencies, yields relatively large values of f, approaching unity. Alternatively, when applied to a regular box orbit, much smaller values are obtained, typically approaching $f \approx 0$. For chaotic orbit segments that closely resemble regular orbits, the same qualitative conclusions prevail. However, "wildly chaotic" segments can exhibit a more complicated behavior.

If, as is sometimes the case, the chaotic segment appears visually to comprise "pieces" of several different box orbits, one again observes a relatively small value of f; and, similarly, if the orbit looks as if it is constructed from several different loop orbits, f will again be relatively large. In general, however, wildly chaotic segments tend to appear visually to be part box and part loop, and for such segments, one typically finds a value of f that is significantly displaced from both 0 and 1. Choosing a relatively large value for z tends to make the f associated with a true loop approach the ideal value $f = 1$ more closely, whereas choosing a smaller value of z tends to yield a smaller value for f. As such, if one were interested simply in determining whether some given regular orbit is a box or a tube, the selection of a relatively large value of x would seem appropriate. However, a small value of z yields a more sensitive diagnostic for distinguishing between different degrees of "near boxy" and "near loopy" behavior.

Representative results obtained using this algorithm may be derived from an inspection of FIGURE 10, which illustrates four different chaotic segments of length $t = 1024$, each extracted from the same chaotic orbit with $E = -0.5$ evolved in the *KAMB* potential. The first two segments, illustrated in (a) and (b), closely resemble

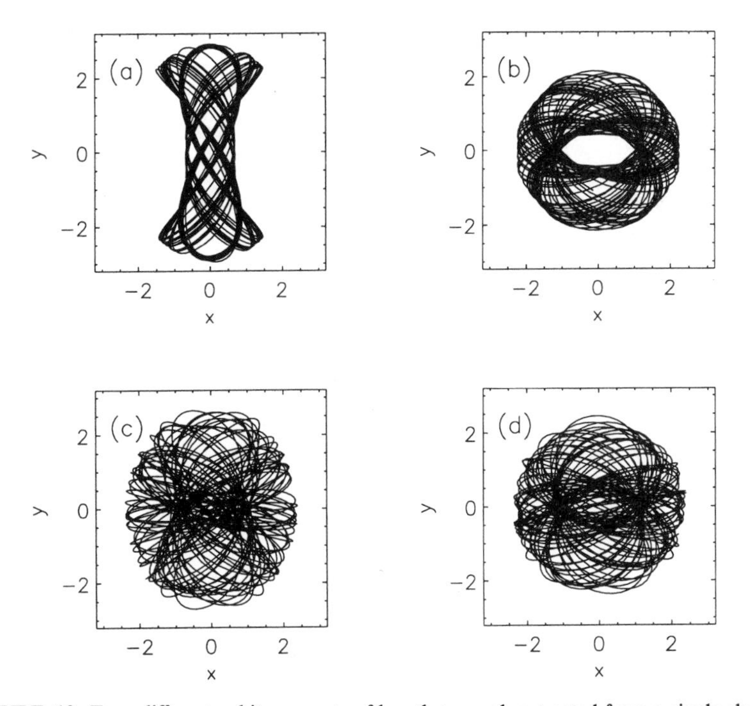

FIGURE 10. Four different orbit segments of length t_H, each extracted from a single chaotic orbit with $E = -0.5$.

regular orbits, one a box and the other a loop. The remaining two exhibit a more complex, intermediate behavior, with the segment in (c) seemingly more boxy than that in (d). Analysis with a hard threshold $z = 0.01$ assigns respectively to the four different segments the values $f = 1.0$, $f = 0.0$, $f = 0.541$ and $f = 0.266$, all numbers in seeming agreement with physical expectations.

This simplistic algorithm is far from optimal. However, its relative success is significant in at least two respects: (1) The fact that any such algorithm can be implemented at all corroborates the physical intuition that chaotic segments in a real sense comprise "pieces" of different regular orbits. (2) There would seem every reason to believe that some other more sophisticated algorithm could make this intuition even more precise.

In conclusion, it is important to stress exactly what information a Fourier spectrum can—and cannot—provide about a chaotic orbit segment. Determining the number of frequencies that contain some specified fraction $\mathcal{P}$ of the total power provides a quantitative characterization of the extent to which any given orbit segment can be viewed as "nearly regular." Similarly, determining the extent to which the power for different components of the orbit, e.g., $x(t)$ and $y(t)$, is or is not concentrated at the same frequencies can be interpreted as providing a quantitative characterization of the extent to which that segment is predominantly "boxy" or "loopy." However, the Fourier spectrum associated with some chaotic orbit segment *cannot* be used to predict the form of the chaotic orbit at much later times.

The reason for this is clear. The given segment is a piece of a chaotic orbit which, when integrated for very long times, will eventually exhibit power over a broad range of frequencies. Relatively short segments of a single chaotic orbit often look strikingly regular, with spectra closely resembling spectra appropriate for true regular orbits. However, different segments of a single chaotic orbit which all look regular will not in general all look like the same regular orbit: one segment may look like a regular box orbit, but another like a regular loop. Two different chaotic segments could resemble two significantly different loops. Different nearly regular segments sample different narrow portions of a single broad band spectrum, the invariant spectrum of chaos associated with the long time integration of a fully chaotic orbit.

ACKNOWLEDGMENTS

The authors acknowledge useful discussions with Robert Abernathy, Brendan Bradley, Ilya Pogorelov, and Christos Siopis.

REFERENCES

1. SCHWARZSCHILD, M. 1979. Astrophys. J. **232:** 236.
2. PELETIER, R. F., R. L. DAVIES, G. D. ILLINGWORTH, L. E. DAVIS & M. C. CAWSON. 1990. Astron. J. **100:** 1091.
3. BENDER, R., P. SURMA, S. DOBEREINER, C. MOLLENHOF & R. MADEJSKY. 1989. Astron. Astrophys. **217:** 35.
4. ZEPF, S. E. & B. C. WHITMORE. 1993. Astrophys. J. **418:** 72.

5. Lauer, T. R., E. A. Ajhar, Y.-I. Byun, A. Dressler, S. M. Faber, C. Grillmair, J. Kormendy, D. Richstone & S. Tremaine. 1995. Astron. J. **110:** 2622.
6. Kormendy, J. & D. Richstone. 1995. Ann. Rev. Astron. Astrophys. **33:** 581.
7. Udry, S. & D. Pfenniger. 1988. Astron. Astrophys. **198:** 35.
8. Hasan, H. & C. Norman. 1990. Astrophys. J. **361:** 69.
9. Merritt, D. & T. Fridman. 1996. Astrophys. J. **460:** 136.
10. Kandrup, H. E., M. E. Mahon & H. Smith. 1994. Astrophys. J. **428:** 458.
11. Athanassoula, E., O. Bienayme, L. Martinet & D. Pfenniger. 1983. Astron. Astrophys. **127:** 349.
12. Sparke, L. S. & J. A. Sellwood. 1987. Mon. Not. R. Astr. Soc. **225:** 653.
13. Pfenniger, D. & D. Friedli. 1991. Astron. Astrophys. **252:** 75.
14. Contopoulos, G. 1971. Astron. J. **76:** 147.
15. Aubry, S. & G. Andre. 1978. *In* Solitons and Condensed Matter Physics. A. R. Bishop and T. Schneider, Eds.: 264. Springer-Verlag. Berlin.
16. Mahon, M. E., R. A. Abernathy, B. O. Bradley & H. E. Kandrup. 1995. Mon. Not. R. Astr. Soc. **275:** 443.
17. Kandrup, H. E. & B. O. Bradley. 1996. Preprint. University of Florida.
18. Lichtenberg, A. J. & M. A. Lieberman. 1992. Regular and Chaotic Dynamics. Springer-Verlag. Berlin.
19. Kandrup, H. E., R. A. Abernathy & B. O. Bradley. 1995. Phys. Rev. E **51:** 5287.
20. Mather, J. N. 1982. Topology **21:** 45.
21. Arnold, V. I. 1964. Russian Math. Surv. **18:** 85.
22. MacKay, R. S., J. D. Meiss & I. C. Percival. 1984. Phys. Rev. Lett. **52:** 697.
23. MacKay, R. S., J. D. Meiss & I. C. Percival. 1984. Physica D **13:** 55.
24. Lieberman, M. A. & A. J. Lichtenberg. 1972. Phys. Rev. A **5**: 1852.
25. Habib, S., H. E. Kandrup & M. E. Mahon. 1996. Phys. Rev. E **53:** 5473.
26. Habib, S., H. E. Kandrup & M. E. Mahon. 1997. Astrophys. J. **480:** in press.
27. Grassberger, P., R. Badii & A. Poloti. 1988. J. Stat. Phys. **51:** 135.
28. Sepúlveda, M. A., R. Badii & E. Pollak. 1989. Phys. Rev. Lett. **63:** 1226.
29. Varosi, F., T. M. Antonsen & E. Ott. 1991. Phys. Fluids A **3**: 1017.
30. Finn, J. M., J. D. Hanson, I. Kan & E. Ott. 1991. Phys. Fluids B **3:** 1250.
31. Kandrup, H. E. & M. E. Mahon. 1994. Astron. Astrophys. **290:** 762.
32. Bennetin, G., L. Galgani & J.-M. Strelcyn. 1976. Phys. Rev. A **14:** 2338.
33. Wolf, A., J. Swift, H. Swinney & J. Vastano. 1985. Physica D **16:** 285.
34. Toda, M. 1967. J. Phys. Soc. Japan **22:** 431.
35. Armbruster, D., J. Guckenheimer & S. Kim. 1989. Phys. Lett. A **140:** 416.
36. Bennett, C. 1994. *In* Physical Origins of Time Asymmetry. J. J. Halliwell, J. Pérez-Mercader and W. H. Zurek, Eds.: 33. Cambridge University Press. Cambridge, U.K.
37. Deprit, A. & W. Poplarchek. 1975. Celestial Mech. **11:** 53.
38. Reed, M. & B. Simon. 1972. Functional Analysis. Academic Press. New York.
39. Press, W. H., B. P. Flannery, S. A. Teukolsky & W. T. Vetterling. 1992. Numerical Recipes, 2nd edn. Cambridge University Press. Cambridge, U.K.
40. Wickerhauser, M. V. 1994. Adapted Wavelet Analysis from Theory to Software: 454. Peters. Wellesley, MA.
41. Binney, J. J. & D. Spergel. 1982. Astrophys. J. **252:** 308.

Information, Language, and Pixon-Based Image Reconstruction[a]

R. C. PUETTER

Center for Astrophysics and Space Sciences
University of California, San Diego
La Jolla, California 92093-0111

IMAGE RECONSTRUCTION AND THE INVERSE PROBLEM

This chapter reviews pixon-based image reconstruction. Pixons (or *informatons*—their extension to generalized data sets) represent an optimized, information theory-based coordinate system for parameter estimation. Pixon-based methods are now becoming recognized as one of the most successful methods for practical image reconstruction, with performance exceeding those of other commonly used techniques such as Maximum Likelihood and Maximum Entropy methods. Like many of these techniques, pixon-based reconstruction can be viewed within the Bayesian estimation framework. However, an important departure of pixon-based methods from other Bayesian schemes is the use of information theory concepts, specifically Algorithmic Information Content, to optimize the Bayesian prior and constrain the reconstruction. Much of this material has been presented before, e.g., references 1–11. The chapter summarizes these discussions and presents a short, practical tutorial on implementation of these methods and presents a number of applications of the method to practical problems.

BAYESIAN METHODS

For the general inverse problem, the method of Bayesian estimation seeks to infer the most probable set of "parameters-of-interest" from a set of measured or otherwise given parameters. In the case of image restoration, this normally means estimating a higher resolution (or undistorted) image from a blurry (or distorted) image. In order to use probabilistic methods, one typically examines the joint probability distribution of the data, D, the underlying, unblurred image which is to be estimated, I, and the model, M, which describes the detailed relationship of the data to the image. This joint probability distribution, $p(D, I, M)$, can be factored

[a]This research was supported by the NSF, NASA, the California Association for Research in Astronomy, and Cal Space.

using conditional probabilities in the following manner:

$$p(D, I, M) = p(D \mid I, M)p(I, M), \tag{1}$$

$$= p(D \mid I, M)p(I \mid M)p(M), \tag{2}$$

$$= p(I \mid D, M)p(D, M), \tag{3}$$

$$= p(I \mid D, M)p(D \mid M)p(M), \tag{4}$$

$$= p(M \mid D, I)p(D, I), \tag{5}$$

$$= p(M \mid D, I)p(D \mid I)p(I), \tag{6}$$

where $p(X \mid Y)$ is the probability of X given that the value of Y is known. By equating various terms, these equations give rise to the formulae

$$p(I \mid D, M) = \frac{p(D \mid I, M)p(I, M)}{p(D \mid M)} \propto p(D \mid I, M)p(I \mid M), \tag{7}$$

$$p(I, M \mid D) = \frac{p(D \mid I, M)p(I \mid M)p(M)}{p(D)} \propto p(D \mid I, M)p(I \mid M)p(M). \tag{8}$$

The formula of (7) is the typical starting place for Bayesian methods. This equation essentially assumes that the appropriate model, M, is known and will not be varied during the reconstruction. This assumption is made explicit in the proportionality of equation in (7) where the term $p(D \mid M)$ is dropped (it is assumed to be constant—D and M are not varied during the reconstruction). Our preferred formulation, however, is given by the formula of (8). Here we do not assume that all of the parameters associated with the model belong to the so-called "nuisance parameters", i.e., parameters which might require estimation, but which are of no interest to the scientist. In pixon-based reconstruction, the model parameters associated with the local smoothness scale of the image are extremely important.

The significance of the terms on the far right-hand side of (7) and (8) is readily understood. The term $p(D \mid I, M)$ is a goodness-of-fit (GOF) quantity, measuring the likelihood of the data given a particular image and model. The terms, $p(I \mid M)$ and $p(M)$, are "priors", and incorporate our prior knowledge (or expectations) about the measurement situation or the nature of suitable selections for I and M. The term "prior" is used since each of these terms make no reference to the data, D, and hence can be decided on *a priori*, i.e., before the act of making the measurement.

In GOF (or maximum likelihood) image reconstruction, $p(I \mid M)$ and $p(M)$ are assumed to be constant, i.e., there is no prior bias concerning the image or parts of the model that might be varied. The standard choice for $p(D \mid I, M)$ is to use $p(D \mid I, M) = (2\pi\sigma^2)^{n_{pixels}/2} \exp(-\chi^2/2)$, i.e., the joint probability distribution for n_{pixels} independent pixels with normally distributed (Gaussian) noise. This assumes that (1) within the chosen space of image/model pairs, a statistically acceptable fit to the data is obtainable, and (2) that near this solution, image/model space uniformly fills data space so that the resulting probability distribution of data realizations is dominated by the probability density of the Gaussian distributed measurement noise, i.e.,

$$p(\text{Blurred Image} + \text{Noise} \mid I, M) = p(\text{Blurred Image} \mid \text{Noise}, I, M)p(\text{Noise} \mid I, M) \tag{9}$$

$$\cong const \times p(\text{Noise} \mid I, M). \tag{10}$$

The "constant" $p(\text{Blurred Image} \mid \text{Noise}, I, M)$ is then generally dropped from the expression. This is generally not too bad an assumption, but it is by no means guaranteed, especially if the chosen space of image/models is selected to be restrictive in some way.

In Maximum Entropy (ME) image reconstruction, the image prior is based upon "phase space volume" or counting arguments and the prior is expressed as $p(I \mid M) = \exp(\alpha S)$, where S is the entropy of the image and α is an adjustable constant that is used to weight the relative importance of the GOF and image prior. Again, usually the model is considered to be known and held fixed, i.e., $p(M) = constant$, and the usual assumption is to choose $p(D \mid I, M) = (2\pi\sigma^2)^{n_{\text{pixels}}/2} \cdot \exp(-\chi^2/2)$. While many specific formulations for α and S appear in the literature,[12–15] all ME methods capitalize on the virtues of incorporating prior knowledge of the likelihood of the image into the image restoration algorithm.

BAYESIAN PRIORS, EFFICIENT MODELS, AND OCCAM'S RAZOR

While most scientists have a well-developed intuition regarding the merits of achieving a good fit to the data, i.e., optimizing the GOF criteria, a well-developed intuition concerning the Bayesian priors and the effects of model selection on the likelihood of a particular image/model pair is often lacking. Investigation into the general effects of model selection, however, turns up the interesting fact that simple, highly predictive models are highly favored. In fact, such considerations provide a very practical and real basis for the principle of *Occam's Razor*, i.e., the predilection of scientists to choose the simpler of two models that provide equally good fits to the data. One of the best discussions of the Bayesian embodiment of Occam's Razor is that presented by MacKay.[16] Our discussion shall parallel MacKay's discussion.

In order to judge the relative merits of two hypotheses, H_1 and H_2, for explaining the data set D, Bayesians would form the relative probability ratio:

$$\frac{p(H_1 \mid D)}{p(H_2 \mid D)} = \frac{p(H_1)}{p(H_2)} \frac{p(D \mid H_1)}{p(D \mid H_2)}. \tag{11}$$

The first ratio involves hypothesis priors, i.e., hypothesis probability densities with no functional dependence on the data. These can be used to express our prior conceptions of the relative likelihood of hypothesis H_1 relative to H_2. As pointed out by MacKay, this corresponds to the usual motivations behind Occam's Razor. However, it is unnecessary to make any prior assumptions about the relative probabilities $p(H_1)$ and $p(H_2)$. Equation (**11**) will still give rise to an Occam's Razor-like effect. This is because simpler hypotheses, say H_1, make more precise predictions. Complex hypotheses embrace a much larger data space and hence spread their predictive probability $p(D \mid H_2)$ more thinly over this larger space. Thus, in the case in which both hypotheses H_1 and H_2 are equally compatible with the data, the simpler hypothesis H_1 will be favored by (**11**), without having to express any subjective dislike for one hypothesis relative to the other. Of course, even if one assigns an equal prior probability to certain aspects of the hypothesis, e.g., the model in image reconstruction, one can usually make sensible arguments for appropriate priors for the other aspects of the hypothesis, e.g., the image prior. For example, in the image reconstruction case, the hypothesis for a given data set is the pair (I, M). The prior is thus $p(I, M) = p(I \mid M)p(M)$. So even if one expresses no prior bias for a given model, e.g., $p(M) = constant$, one might easily assign a sensible prior probability distribution for $p(I \mid M)$, e.g., a ME prior. Such an assignment also acts in the cause of simple models, since any sample state drawn from a simple model typically has a higher probability simply because it represents a larger fraction of all possible states. Thus at every turn probability theory favors Occam's Razor, i.e., the selection of the simplest possible image/model (hypothesis) to explain the data. Thus in order to perform the very best image reconstruction, we need to simplify the image/model as much as possible. This will both optimize the Bayesian prior and provide the "highest possible peak" in the GOF term.

SUCCESSES AND FAILURES OF CLASSICAL METHODS

The very fact that image restoration/reconstruction has been performed for many years is testimony to its basic success as a method. For PSF blurring problems, its primary advantage is its ability to provide enhanced resolution. The first attempts at PSF deconvolution employed Fourier methods. However such methods only provide suitable performance if noise associated with the measurement is absent or inconsequential. Non-linear methods provide a significant improvement over Fourier deconvolution methods in moderate to low SNR situations by providing greater control over noise propagation. This is done by insisting (through the GOF criterion) that the modeled data (e.g., the data predicted by the inferred image and model) match the measured data within the uncertainty specified by the noise. This is quite different from Fourier methods, where no such demand is made. Of course, insisting that the predicted data "fit" the measured data within the noise limits does not guarantee that the inferred image is without significantly larger noise. Indeed, each image reconstruction problem can have a significant "null space" in which noise can be hidden—i.e., maps to zero in data space.

Common problems with GOF (or Maximum Likelihood) and ME methods are overfitting of the data, the production of signal correlated residuals, and spurious sources. These problems are related to using an inappropriate number of degrees-of-freedom (DOFs) to fit the data and consequently are largely overcome by pixon-based methods. Briefly, however, signal correlated residuals are produced by a "diluted" GOF criterion. Suppose we have imaged a galaxy in a large CCD frame. Assume that the CCD has dimensions 1024×1024 pixels. Further suppose that the galaxy only fills the central 100×100 pixels. If the astronomer were to use the entire data frame in his reconstruction and chose to represent the image as a rectangular grid of numbers with the same spatial frequency as the data (as is commonly done), then the reconstruction would use roughly 10^6 DOFs. If a standard GOF criterion were used, then one would stop the iterative procedure when there were less than 10^3 residuals larger than 3σ (0.1% of the residuals are 3σ or larger in a population of Gaussian distributed variables). Since iterative procedures that minimize GOF functions normally spend most of their time fitting the bright sources, they will be working on adjusting the bright source levels when the stopping criterion is met, resulting in the vast majority of the large residuals lying under the bright sources.

The origin of spurious sources in standard image reconstruction techniques is even easier to understand. Using the example above, it should be clear to most readers that the 10^6 DOFs used in the reconstruction are probably far more than are needed to describe the information present in the data. In fact, the data probably are inadequate to constrain the values of this many DOFs. Hence the vast majority of these DOFs are free to produce whatever bumps and wiggles they like, with the only requirement being that after they are smoothed by the PSF they average to zero within the noise. Hence their amplitude can be very large as long as they have a spatial scale small relative to the PSF and are arranged in such a manner that hills and valleys tend to cancel each other out. Since this large number of unconstrained DOFs represents a huge phase space and the typical member has numerous bumps and wiggles, spurious sources are guaranteed.

One of the main motivations for developing Maximum Entropy methods was to help overcome the spurious source problem. After all, the ME prior acts to make unsmooth images undesirable. However, in their zeal to flatten spurious sources, ME methods can over-flatten the true sources. This affects the ability of ME methods to perform statistically unbiased photometry. In fact, while ME methods are significantly more immune to spurious sources (they more effectively reject situations with large adjacent hills and valleys than GOF methods), they still tend to systematically underestimate source brightnesses.[17,18]

Most of the problems facing standard GOF and ME methods are overcome by pixon-based methods. Pixon-based methods are a generalization of ME methods and perform the Bayesian estimation problem in a more natural language. The pixon language is information-based. Consequently it concisely and accurately describes the key variables in the problem and avoids the use of too many DOFs in the reconstruction. Because the number of DOFs is tuned to the number required to accurately describe the image, problems with stopping the reconstruction because a "diluted" GOF criterion is used are avoided. In addition, spurious source production is avoided since the pixon-basis automatically rejects spurious sources as being unneeded in the reconstruction. The generalization of the pixon to arbitrary data sets is the *informaton*, i.e., a quanta of information, and has a wide range of application to a variety of problems. To see how concepts from information theory can be used as a powerful constraint for the image reconstruction problem, we turn next to a discussion of information itself.

TYPES OF INFORMATION

The concepts of information and entropy have been discussed in numerous works. While the thermodynamic concepts of entropy have a very specific meaning and apply to the state of an ensemble of particles, the concepts of information have a much broader scope. Nonetheless, the laws of statistical physics can be derived directly from information theory concepts,[19–21] and information and entropy (or negentropy) are often talked about as the same quantities. The most common definition of information, now known as Shannon information, was introduced in 1948[22,23] and specifies the average information contained in a string of symbols transmitted across a communication line. In this definition, the average information per symbol is given by

$$H = -\sum_{i=1}^{n} p_i \log_2 p_i, \tag{12}$$

where n is the total number of possible symbols, and p_i is the probability of occurrence of the ith symbol. There are many reasons why the Shannon information is a suitable definition for information. One of the most convincing arguments is given by the *Noiseless Source Coding Theorem* (see, e.g., reference 24), which states that for an ergodic source with an alphabet of n characters, that as the size of the transmitted string approaches infinite length, it is possible to construct a unique random length binary code for the string that has an average length in bits per encoded symbol equal to, but no less than, the Shannon entropy.

While the *Noiseless Source Coding Theorem* in and of itself speaks strongly for Shannon entropy as a sensible information measure, there are still other motivations. For example, it can be proven that the Shannon information is the unique and self-consistent measure which satisfies conditions which would reasonably be expected of a measure of information.[21,25,26] Using Khinchin's formulation of the properties uniquely defining an information measure, it is found that only the function H satisfies the conditions that (1) H takes its largest value when all the p_i are equal, (2) H does not change its value when additional impossible events are added—i.e., events for which $p_i = 0$, and (3) the information in two not necessarily independent events is given by the information of the first event plus the expectation value of the additional information provided by the second type of event after the first event has occurred.

Readers who have had classical training in physics will be interested that the formula of (**12**) is essentially equivalent to the physical definition of entropy, $\sigma = k \cdot \ln W$ (modulo a constant to convert logarithm types), where k is Boltzmann's constant and W is the total number of available states. In fact, (**12**) is the Boltzmann (or Gibbs) definition of entropy. Kittel[27] (see his page 118) gives a particularly fluid derivation of the Boltzmann definition of entropy starting from the Boltzmann factor.

While we have now provided motivation for the merits of Shannon information, of what use is this concept for solving the inverse problem in general or image reconstruction in particular. To be sure, ME methods make use of the entropy or Shannon information definition. However, Shannon information does not get to the heart of the problem. This definition of information does not allow us to describe the *information content* of an individual image or data set. As has been pointed out by several authors,[28,29] Shannon information is an ensemble notion. It is a measure of ignorance concerning possible realizations of events which have a given *a priori* probability distribution, and measures the statistical rarity or "surprise value" of a particular realization.[30] It does not deal with the information content expressed by a given realization. For this we need a different concept, i.e., that of algorithmic information content (AIC), algorithmic randomness, or algorithmic complexity.[31–33] As commonly used in information or computer science, the AIC of a string of characters is defined to be the size of the minimum computer program required for a universal computer to produce the specified string as its output. (A computer U is universal if for any other computer M there is a prefix program μ such that the program μp makes U perform the same tasks as performed by computer M running the program p, i.e., the program μ makes computer U simulate computer M.) As suggested by the name, AIC describes the information content of the specific item under scrutiny. In fact it does this in a very practical way, i.e., algorithmically. It provides a prescription for how the data can be reproduced. It should also be clear that the AIC represents the optimal compression of the data. There is no shorter description that will allow a complete reconstruction of the original information. In fact, it is this optimal compression of the information that will allow us to optimally extract an image in the image reconstruction problem in a practical manner.

THE ROLE OF LANGUAGE IN AIC AND BAYESIAN ESTIMATION

The term "randomness" comes into the definition of AIC since random strings of characters have maximum complexity or information content. This is because if the pattern of the string is non-random, there is some way of describing the string which is typically shorter than listing the string itself. For example, in strings of 1s and 0s, a string with 10^9 copies of the number 1 would take a lot of paper to write down, but can be described with a single sentence. Similarly this same string with a 0 placed in the millionth place can also easily be described without wasting paper. However, a totally random string of 10^9 digits cannot be so easily described, and if no rule is known (and this is what we mean by random), only the actual string will impart all of the information. It is well known that quantification of the AIC is a function of the "richness" of the language used to describe the character string or data set. In the example used above, we demonstrated how the English language could be used to briefly describe large strings of digits. Other languages, e.g., the binary digits 0 and 1, decimal digits, or even hexidecimal digits, do not provide such a terse description.

The above discussion makes it quite obvious that rich, i.e., complex, languages allow a terse description of a data set—the AIC is low for rich languages. This seems like an obvious advantage. It says that the data can be compressed to a higher degree and that the information can be more precisely located and identified. Similar concepts occur in Bayesian estimation, e.g., the concept of Occam's Razor. Clearly, in the Bayesian estimation problem, it is highly beneficial to develop concise, simple hypotheses (images or image/model pairs) to explain the data. Such simple hypotheses give rise to optimized priors [$p(I\mid M)$ or $p(I\mid M)p(M)$]. Consequently, they produce optimal reconstructions and a more likely M.A.P. image or image/model. Thus it is seen that minimizing the AIC of the image/model, optimally compressing the information in the data, and performing highly optimized Bayesian image reconstruction are directly related. Hence the goal for improved image reconstruction is clear. One must improve the language in which the reconstruction is performed, or equivalently the language in which the image is described. The language must be rich so that the description can be concise, the AIC low, and the Bayesian prior optimized. The fact that the description will be concise suggests that the language for describing the image must be "natural" in some sense, and we turn now to a discussion of possible languages for image description.

LANGUAGES FOR IMAGES

There are many "languages" for images in current use. The most common of these are usually not even thought of as languages since they seem so natural. For example, in electronic or digital imaging we essentially expect to see every image in the form of a rectangular pixel array. This is so natural that we do not even consider the consequences of this language for the image. However there are consequences. The pixels are almost always rectangular and so is the picture format. However this may not be optimal for image reconstruction. Certainly in the case of astronomical imaging, many objects are not square, e.g., stars. Also images of star clusters or galaxies, etc., seldom uniformly fill a rectangular picture to its corners. Hence these clearly are not an optimal format or pixel shape for astronomical situations. There are usually too many pixels in the image and many pixels are used to describe single objects such as stars.

To motivate a discussion of what constitutes an optimal language for image description, let us first ask the question: "What is the nature of picture information?" This, of course, can have many answers, and the answer changes, depending on the object at which the photographer points his camera. We are interested, here, in "generic" images. So let us first ask the general question: "What do we really expect when we take a picture?" Thinking about this in the broadest sense, all we can really say is that we expect the picture to contain some limited amount of information. We expect, in particular, that the manufacturer of the film or the designer of the camera will have made the grain or pixel size fine enough to do service to the picture, i.e., not to leave out any important detail. We expect interesting features, perhaps people, with open spaces between them. In other words, we expect that at each point in the image there is a finest spatial scale of interest and that there is no information content below this scale. Indeed, this is how photographic grain sizes are chosen and why data are not sampled finer than the Nyquist frequency when pixel elements are at a premium.

How does one capture this prior expectation in mathematic form and incorporate it into a set of image basis functions? Again, the key comes from thinking about photographic grain sizes or Nyquist sampling. We would do just as well at recording the picture information with large photographic grains in portions of the image with coarse structure. We need only have fine grains when we need to record fine spatial structure. This means that the picture information can be dealt with by using variable sized cells, with the cell sizes set so as to capture the spatial information present. In other words, we are proposing that a multi-resolution language is an appropriate choice, i.e., will be concise and hence fulfill the goals of Occam's Razor.

A PRACTICAL MULTI-RESOLUTION PIXON LANGUAGE FOR IMAGE DESCRIPTION

Above, we argued that a multi-resolution image description language would be suitable (i.e., concise) for describing generic images. The success of multi-resolution languages is not surprising. In fact, these ideas are familiar to most scientists and are fundamental to simple, well-known concepts, e.g., the use of fine grain photographic film to capture fine detail. So our multi-resolution pixon language will use the idea that generic images can be concisely described by using fewer degrees-of-freedom (DOFs) per unit area in portions of the image which are smooth and a greater density of degrees-of-freedom where there is greater detail. Each of these degrees-of-freedom might then be likened to a single photographic grain or a generalized pixel. The value of this pixel represents the average brightness in a given region. This, in fact, is the origin of the name "pixon". Each pixon is a single DOF used to describe the image in a particular region. The "pix" part of the name recognizes its pixel heritage, while the "on" suffix recognizes its more fundamental nature (the pixon is fundamental to the image, not to the instrument that took the picture). However, since any scheme which controls the local density of DOFs used to describe the local image information is suitable for constraining the image reconstruction and optimizing the Bayesian prior, rather than using image signal contained in cells with hard boundaries, we have chosen to use a local correlation scale formulation to control the DOF density. We call these "fuzzy pixons". To formalize the definition, at each point in the image, $\vec{x}$, we write the image, I, as

$$I(\vec{x}) = (K \otimes I_{pseudo})(\vec{x}) \tag{13}$$

$$(K \otimes I_{pseudo})(\vec{x}) = \int dV_{\vec{y}} K(\vec{x}, \vec{y}, \delta(\vec{x})) I_{pseudo}(\vec{y}) \tag{14}$$

$$K(\vec{x}, \vec{y}, \delta(\vec{x})) = K\left(\frac{\|\vec{x} - \vec{y}\|}{\delta(\vec{x})}\right); \text{ for radially symmetric pixons,} \tag{15}$$

where (13) through (15) show that the image is a local convolution of a "pseudo-image" with a blurring function with a given local scale, $\delta(\vec{x})$. Note that the local scale varies with position $\vec{x}$ in the image and the integration in (14) is carried out over volume in pseudo-image space. We have also indicated in (15) that one suitable functional form for the pixon kernel function, K, might be a radially symmetric function that depends only on the distance between the kernel center and the image position relative to the local scale. We have found this functional form to work quite well for centrally peaked kernel functions with a finite footprint. We normally use truncated paraboloids, i.e.,

$$K(\vec{x}, \vec{y}, \delta(\vec{x})) = \begin{cases} \left(1 - \dfrac{\|\vec{x} - \vec{y}\|^2}{\delta(\vec{x})^2}\right) \Big/ \displaystyle\int dV_{\vec{y}} \left(1 - \dfrac{\|\vec{x} - \vec{y}\|^2}{\delta(\vec{x})^2}\right); & \|\vec{x} - \vec{y}\| \le \delta(\vec{x}) \\ 0 & ; \|\vec{x} - \vec{y}\| > \delta(\vec{x}) \end{cases} \tag{16}$$

Of course, AIC concepts suggest that a richer language, e.g., elliptical pixons, would yield a more concise image description, which Occam's Razor then says should have a more optimized prior. In fact, we have used elliptical pixons recently to perform some image restorations. Nonetheless, for generic images it seems clear that we have reached a point of diminishing returns, and it is unlikely that pixon kernels more complicated than ellipses are warranted for generic images.

SOLVING FOR THE M.A.P. IMAGE/MODEL PAIR

Now that we have described a suitable language in which to solve the problem, we shall move on to the details of finding the M.A.P. (Maximum *A Posteriori*) image/model pair, i.e., the image/model pair that maximizes $p(I, M|D)$. To maximize $p(I, M|D)$, we need to maximize the product of $p(D|I, M)$ and $p(I, M)$. As mentioned above, the common choice for $p(D|I, M)$ in the case of pixelized data with independent Gaussian noise is $p(D|I, M) = (2\pi\sigma^2)^{n/2} \exp(-\chi^2/2)$, where χ^2 is the standard chi-squared value [i.e., $\chi^2 = \sum_1^n (x_i - \langle x_i \rangle)^2/\sigma^2$], σ is the standard deviation of the noise, and n is the number of independent measure parameters (here the number of pixels). While this functional form for the GOF criterion is almost always adopted, the various choices for $p(I, M)$ are far from standard. The choice of ME enthusiasts is something like

$$p(I|M) = \frac{N!}{n^N \prod_{i=1}^{n} N_i!} = e^S \tag{17}$$

$$p(M) = constant, \tag{18}$$

where n is the number of pixels, N is the total number of counts in the image, N_i is the number of counts in pixel i, and S is the entropy. This choice is made on the basis of counting arguments. The arguments are basically sound, but operate within a fixed language, e.g., the pixel basis for the image. As discussed previously, a more appropriate choice for the image basis would be multi-resolution pixons. This more critically models the data and provides a superior prior.

For our pixon prior we could use the same prior as used by ME workers with the pixons substituted for the pixels. The *a priori* probability arguments for the prior of (**17**)–(**18**) remain valid. It is just that we now recognize that the pixons are a more appropriate coordinate system and that we will use vastly fewer pixons (i.e., DOFs) to describe the image than there are pixels in the data. To make the formulae explicit, we define the pseudo-image on a pseudo-grid (we loosely refer to this as the "pixel grid" when talking about the pseudo-image) which is as least as fine as the data pixel grid (we use this finer "pixel" grid for the image too) and then use the following substitutions:

$$p(I|M) = \frac{N!}{n^N \prod_{\substack{i=1,\\ i \in Image}}^{n_{pixons}} N_i!} \tag{19}$$

$$N = \sum_{\substack{i=1,\\ i \in Image}}^{n_{pixons}} N_i = \sum_{\substack{j=1,\\ j \in Image}}^{n_{pixels}} N_j \tag{20}$$

$$N_i = \int dV_{\vec{y}} k(\vec{x}, \vec{y}, \delta(\vec{x})) I_{pseudo}(\vec{y}) \quad (21)$$

$$\sum_{\substack{i=1 \\ i \in Image}}^{n_{pixons}} N_i ! = \prod_{\substack{j=1 \\ j \in Image}}^{n_{pixels}} (N_j !)^{p_j} \quad (22)$$

$$p(\vec{x}) = 1 \Big/ \int dV_{\vec{y}} k(\vec{x}, \vec{y}, \delta(\vec{x})) \quad (23)$$

$$p(M) = constant, \quad (24)$$

where N_i is the number of counts in pixon i, N_j is the number of counts in pixel j, p is the pixon density, i.e., the number of pixons per pseudo-pixel (in image space), and $k(\vec{x}, \vec{y}, \delta(\vec{x}))$ is the pixon shape kernel normalized to unity at $\vec{y} = \vec{x}$. [Note that while the formula of (**19**) can be easily justified, it is not the only justifiable expression. Other expressions might be more appropriate in certain situations.]

To obtain the M.A.P. image/model pair, one can now proceed directly by minimizing the product of $(2\pi\sigma^2)^{n_{pixels}/2} \exp(-\chi^2/2)$ and (**19**) with respect to the local scales, $\{\delta(\vec{x}_j)\}$—the pixon map, and the pseudo-image values, $\{I_{pseudo,\, j}\}$. However, this is not what we do in practice. Instead, we divide the problem into a sequential chain of two repeated steps: (1) optimization of the pseudo-image with the local scales held fixed; (2) optimization of the local scales with the pseudo-image held fixed. The sequence is then repeated until convergence—see FIGURE 1. To formally carry out this procedure, in step (1) we should find the M.A.P. pseudo-image, i.e., the pseudo-image that maximizes $p(I \mid D, M)$—note that we are using the notation here that the local scales belong to M, while the pseudo-image values are associated with I. In step (2) we would then find the M.A.P. model, i.e., the scales that maximize $p(M \mid D, I)$.

While the above procedure is quite simple, we have made still further simplifications. In neither step do we evaluate the prior. We simply evaluate the GOF term $p(D \mid I, M)$. So in step (1) we essentially find the Maximum Likelihood pseudo-image with a given pixon map. In step (2) we must take into account some knowledge of the pixon prior, but we simply use the fact that the pixon prior of (**19**) or any sensible prior increases rapidly as the number of pixons are decreased. So at each pseudo-grid point, j, we attempt to increase the local scale until it is no longer possible to fit the data within the noise. In other words at each pseudo-grid point we progressively smooth the current pseudo-image until the GOF criterion is "violated". We then select the largest scale at each point which was acceptable and use these values in the next iteration of step (1).

There is one more practical matter to consider. As with any interative method, there can be convergence problems. With the approach outlined above, we have noticed that if small scales are allowed early in the solution, then these scales become "frozen-in", even if they would have later proved inappropriately small. To solve this problem, we start out the pseudo-image calculation with initially large pixon scales. We then use our pixon-calculator [the code that performs step (2)] to determine new scales. The pixon-calculator, of course, will report that over some of the image the initial scales are fine, but over other parts of the image smaller scales are required. At this point, however, we do not allow the smallest scales requested.

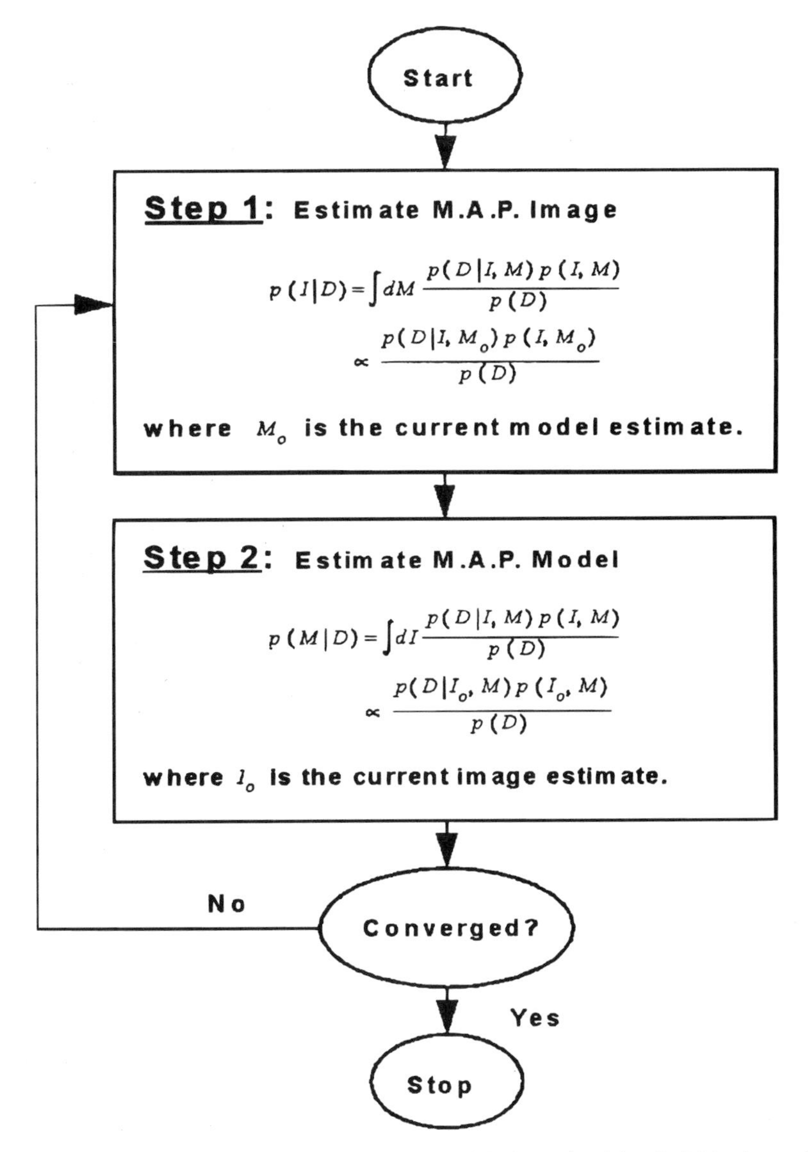

FIGURE 1. Iterative scheme for obtaining the M.A.P. image/model pair. This shows the pixon algorithm as currently implemented. The procedure is divided into two steps: (**1**) GOF optimization step in which the image is optimized while holding the pixon map (local smoothness scales) fixed; (**2**) pixon optimization step in which a new pixon map is calculated while holding the current image estimate fixed. The method alternately iterates between these two steps until convergence is obtained.

Instead, we let the scales get somewhat smaller over the portion of the image for which smaller scales were requested and proceed to step (1) of the next iteration. We repeat this process, letting the smallest scales allowed get smaller and smaller until the method converges. This procedure has proven to be very robust.

APPLICATION OF PIXON-BASED METHODS

Pixon-based reconstruction has now been used by the astronomical community to perform a variety of image reconstruction problems. These include HST spectroscopy,[34] coded mask X-ray satellite imaging,[35] far-IR airborne imaging,[36] IRAS far-IR survey data,[4,7] and mid-IR ground-based imaging.[37] In this section we present a number of applications of pixon-based methods, including adaptive filtering, image restoration, and image reconstruction. The first two examples present noise filtering applications, i.e., (1) temporally adaptive noise filtering of X-ray timing data from the Vela pulsar, and (2) spatially adaptive filtering of a mammogram X-ray phantom. Next we present two examples of image restoration: (1) restoration of a mock data set, i.e., a completely made-up data set in which the underlying image, PSF, and noise frames are manufactured and hence known precisely, and (2) an elliptical pixon-Ybased restoration of Keck 2 μm imaging data for the gravitational lens of FSC 10214 + 4724. Mock data sets are useful in comparisons of image restoration/reconstruction methods since there can be no argument about the goal of the reconstruction. The true, underlying image is known perfectly *a priori*. Furthermore, the noise and all parameters of how the input data was made are completely specified. The second example illustrates the use of non-radially symmetric pixon bases. We shall go into some detail in the discussion of this example principally because most pixon reconstructions presented in previous work have used radially symmetric basis functions and because the science allowed by the reconstruction is quite interesting. Finally, we present three examples of image restoration. The first example is of IRAS 60 μm survey scans of the interacting galaxy M51. The second example is coded Fourier X-ray mask imaging of solar flare data from the hard X-ray telescope aboard the Yohkoh satellite, and the final example is scanned γ-ray collimator imaging data from the OSSE Virgo survey from the Compton Gamma Ray Observatory satellite. The examples of pixon-based applications used in this section have appeared before in references 4, 6, 9, 10, and 11.

Pixon-Based Adaptive Noise Filtering

The present section presents two examples of pixon-based applications that are not directly related to astronomical image restoration/reconstruction. In fact, in both of these examples, no reconstruction is done. Pixon-based methods have been used solely to provide an optimal spatially adaptive noise filter. The first example is presented in FIGURE 2, which displays a sample of timing data from the Vela pulsar, along with the pixon-based estimate of the true underlying signal. The lower curve shows the actual collected data as a function of time (arbitrary units). The upper curve represents the pixon-based best estimate of the true underlying signal. The curve is plotted with an added constant for vertical displacement. The noise in

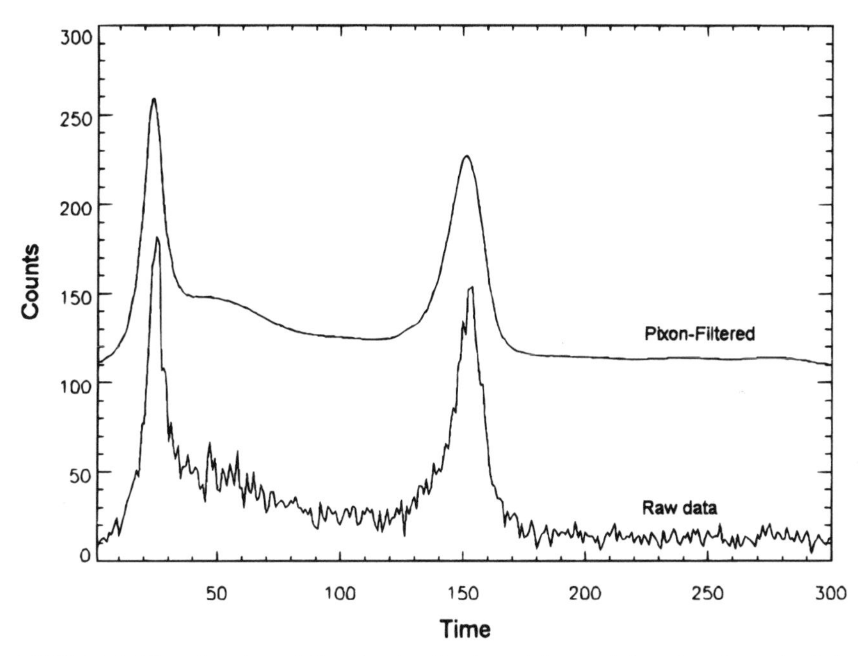

FIGURE 2. Pixon-based spatially adaptive filtering of X-ray timing data from the Vela pulsar. *Lower curve*: Vela pulsar data. *Upper curve*: pixon-filtered data with a constant added to provide a vertical shift. The noise in the data is given by the counting statistics in the signal. The pixon-filtered image displays only those structures calculated to be statistically significant relative to this noise.

the data is due to photon counting statistics in the signal. As can be seen from the illustration, unlike box-car filtered or Fourier-domain filtered data, the pixon-filtered signal retains all of the high frequency information present in the strong pulses. This is because the filtering is temporally adaptive and preserves any statistically justifiable structure by using a broad filter on the low-lying noisy signal and a narrower filter on the stronger sources with just the appropriate widths to preserve the justifiable structure.

The second example is presented in FIGURE 3. This is a mammogram of a standard medical phantom. In this case a fiber of thickness 400 μm has been placed in a piece of material with X-ray absorption properties similar to the human breast. Again, in this example pixon-based methods were used solely to filter the data. No attempt was made to sharpen the image by removing the X-ray beam spread. As can be seen from this example, the pixon filtered image provides vastly superior feature detection contrast and would allow an X-ray technician to scan mammograms much more rapidly and with greater sensitivity to finding features such as cancerous tumors. It is obvious, for example, that even though the fiber present in this image is barely at the detection threshold (as evidenced by the fact that it breaks up into a chain of features due to noise), the pixon-filtered image easily detects the presence of whatever statistically significant structure is present in the image.

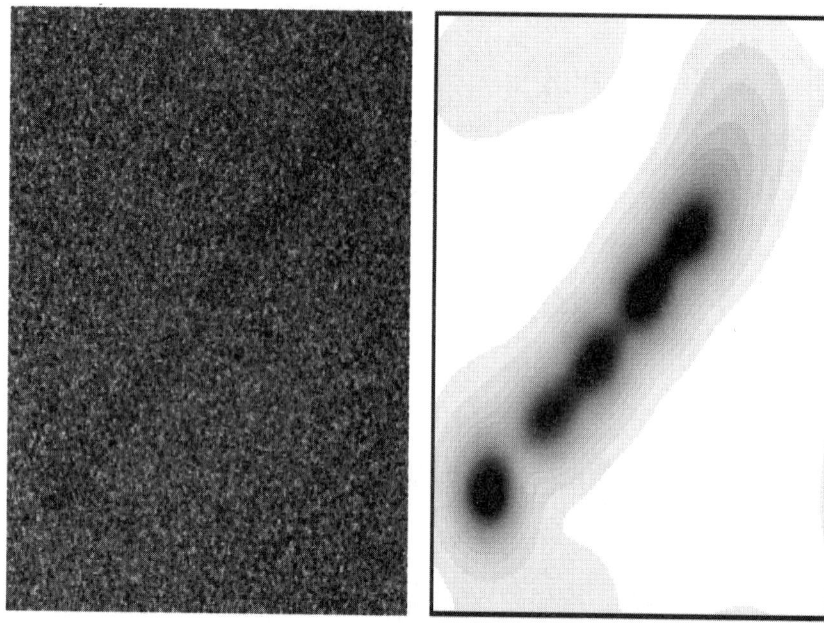

FIGURE 3. Pixon-based spatially adaptive filtering of a standard mammogram phantom. *Left panel*: raw phantom image. *Right panel*: pixon-filtered image. The phantom in this case is a 400 μm thick fiber placed in a piece of material with X-ray absorption properties similar to the human breast. The pixon-filtered image reports only those structures calculated to be statistically significant. Note that in this case the fiber is near the detection threshold—the fiber breaks up into a string of features because in some places the signal is statistically insignificant because of the large noise level. Nonetheless, the pixon-filtered image easily detects the remaining statistically significant structure.

Pixon-Based Image Restoration

We present in this section two examples of pixon-based image restoration. In the first example, we compare pixon-based and MEMSYS 5 restorations of a mock-data set. The ME algorithms chosen for comparison are those embodied in MEMSYS 5, the most current release of the MEMSYS algorithms. The MEMSYS code represents a powerful set of ME algorithms developed by Gull and Skilling.[38] The MEMSYS algorithms probably represent the best commercial software package available for image reconstruction. The MEMSYS reconstruction was performed by Nick Weir of Caltech, a recognized MEMSYS expert, and was supplemented with his multi-channel correlation method which has been shown to enhance the quality of MEMSYS reconstructions.[39,40] The true, noise-free, unblurred image presented in the top row of FIGURE 4 is constructed from a broad, low-level elliptical Gaussian (i.e., a two-dimensional Gaussian with different FWHMs in perpendicular directions), and two additional narrow, radially symmetric Gaussians. One of these narrow Gaussians is added as a peak on top of the low-level Gaussian. The other is subtracted to make a hole. To produce the input

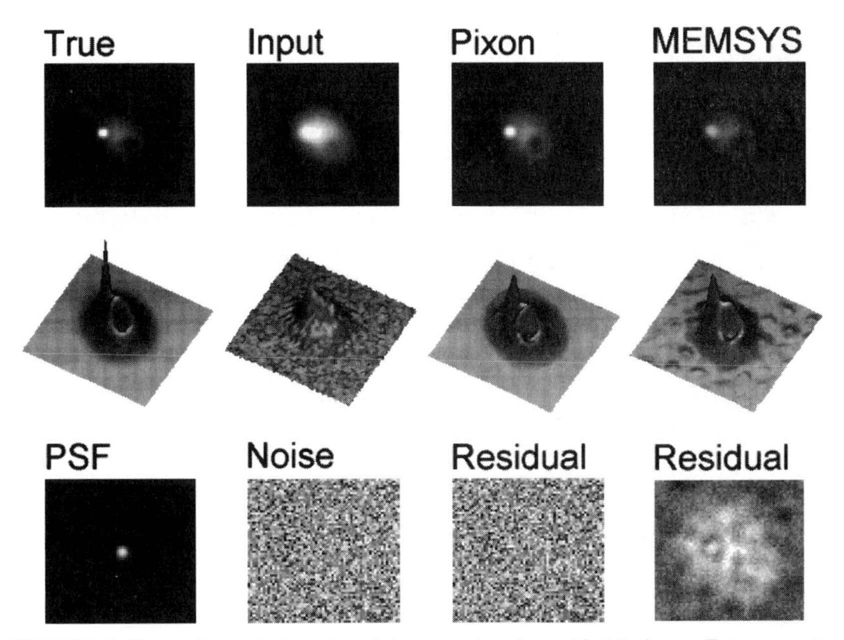

FIGURE 4. Comparison of pixon-based image restoration with Maximum Entropy restoration for a mock data set. Shown is the PSF (Gaussian with FWHM = 6 pixels), the true, underlying image, and the Gaussian noise. The restored image and residuals for radially symmetric, truncated parabolic fuzzy pixons are shown. Also shown are the results and residuals for the multi-channel MEMSYS 5 method. Reproduced from Puetter.[5]

image, the true image was convolved with a Gaussian PSF of FWHM = 6 pixels, then combined with a Gaussian noise realization. The resulting input image is displayed in the top row. The signal-to-noise ratio on the narrow Gaussian spike is roughly 30. The signal-to-noise on the peak of the low level Gaussian is about 20. The signal-to-noise at the bottom of the Gaussian "hole" is 12.

As can be seen, the FPB reconstruction is superior to the multi-channel MEMSYS result. The FPB reconstruction is free of the low-level spurious sources evident in the MEMSYS 5 reconstruction. These false sources are due to the presence of unconstrained degrees-of-freedom in the MEMSYS 5 reconstruction and are superimposed over the entire image, not just in the low signal-to-noise portions of the image. Furthermore, the FPB reconstruction's residuals show no spatially correlated structure, while the MEMSYS 5 reconstruction systematically underestimates the signal, resulting in biased photometry.

In the second example of image restoration, we have applied pixon-based methods to Keck K-band (2.3 μm) imaging data of the gravitational lens system FSC 10214 + 4724 (data presented in reference 41). This object was initially thought to be the most luminous object in the universe, with a luminosity approaching 10^{15} solar luminosities. However, it is now recognized as a gravitational lens with a magnification of roughly 100.[41–43] Hence the luminosity is about 100 times fainter than initially estimated.

In order to make sure that our image description language would not be biased against faint Einstein rings, and is able to describe inherently thin structures, we decided it would be wise to use a set of elliptical pixon shape functions as our language basis. Our elliptical pixon restoration of the Keck K-band image is shown in FIGURE 5. Also displayed in this figure is the Maximum Entropy reconstruction of Graham and Liu.[41] As can be seen from this illustration, our reconstruction reveals the entire Einstein ring. This low-level ring was not detectable by the ME methods employed by Graham and Liu because of the large number and strength of the spurious sources produced by that method. These sources are obvious in the ME result displayed in FIGURE 5.

The science that our better reconstruction allows for these data is quite interesting. First, Graham and Liu[41] concluded on the basis of their inability to see an entire Einstein ring that the size of the emitting source (which has a redshift of 2.3) is 0.25 arcsec or less. They point out that given the lens geometry, if a complete ring was seen, then the size of the emitting region would have to be at least 0.5 arcsec. Hence our first conclusion is that the emitting region of FSC 10214 + 4724 has significant emission out to a size of 0.5 arcsec, roughly twice that deduced by Graham and Liu. Our second result is related to the nature of the gravitational lens. Recent HST images of the Einstein ring at far-UV rest-frame wavelengths (0.25 μm) were obtained by Eisenhardt *et al.*[44] From the narrow nature of the arc, these

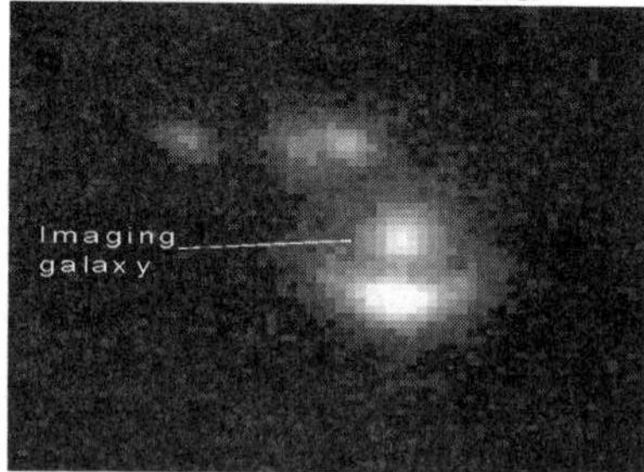

FIGURE 5. Comparison of image reconstruction techniques for the Einstein ring of FSC 10214 + 4724. The elliptical pixon reconstruction used 196 different elliptical pixon kernel functions. Relative to radially symmetric reconstructions, the elliptical pixon reconstruction used roughly 50% fewer DOFs in the image model. Reproduced from Puetter.

authors conclude that the size of the UV source is of order 0.005 arcsec (40 pc), and that the best fit to the shape of the arc requires a contribution to the gravitational lensing potential from the next brightest elliptical galaxy near the main lensing galaxy. Our restored image directly confirms the gravitational contribution of this object to the lensing potential. The K-band Einstein ring is clearly warped in the direction of this source, directly confirming the speculations of Eisenhardt and coworkers. Further discussion of these results will appear elsewhere.

Pixon-Based Image Reconstruction

In this section we present three examples of pixon-based image restoration. The first example is given in FIGURE 6, where we present a reconstructed image from 60 μm IRAS survey scans of the interacting galaxy pair M51. This particular data set is especially useful for comparing the performance of various different image reconstruction techniques since it was chosen as the basis of an image reconstruction contest at the 1990 MaxEnt Workshop.[45] Furthermore, the IRAS data for this object are particularly strenuous for image reconstruction methods. This is because all the interesting structure is on "sub-pixel scales" (IRAS employed relatively large, discrete detectors—1.5 arcmin by 4.75 arcmin at 60 μm) and the position of M51 in

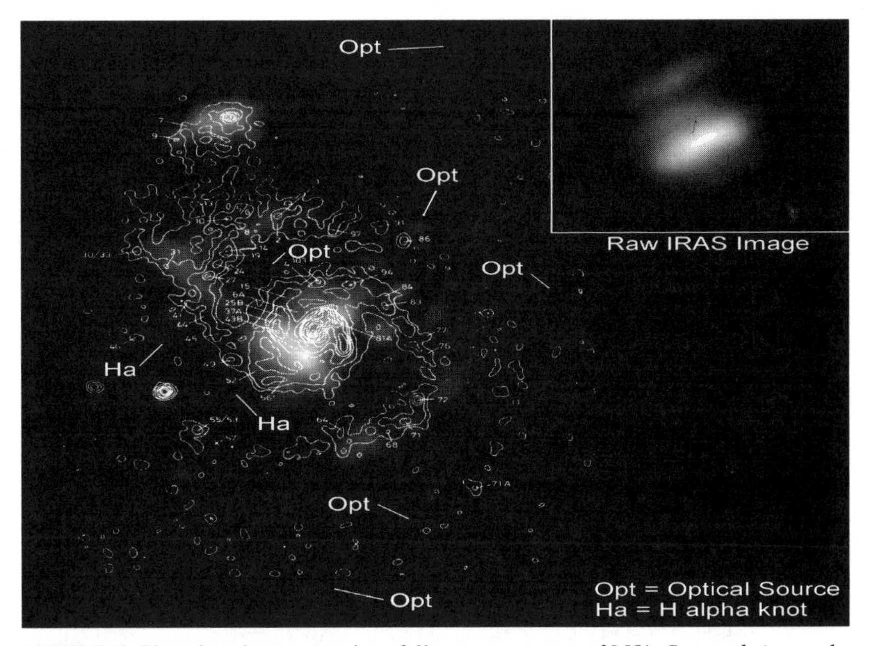

FIGURE 6. Pixon-based reconstruction of 60 μm survey scans of M51. *Gray-scale image*: the pixon-based reconstruction. The faintest sources visible have a formal SNR of roughly 30. *Contours*: the 5 GHz radio flux contours of van der Hulst *et al.*[49] Noted are several of the stronger features in the reconstruction that can be identified with optical sources (labeled "Opt") and *Hα* knots (labeled "Ha"). Figure reproduced from Puetter and Piña.[3]

the sky caused all scan directions to be nearly parallel. This means that reconstructions in the cross-scan direction (i.e., the 4.75 arcmin direction along the detector length) should be significantly more difficult than in the scan direction. In addition, the point source response of the 15 IRAS 60 μm detectors (pixel angular response) is known only to roughly 10% accuracy, and finally the data are irregularly sampled.

In FIGURE 7, our FPB reconstruction is compared to the results from other methods, including the Lucy–Richardson, the Maximum Correlation Method (MCM) (see reference 46), and MEMSYS 3 reconstructions.[47] The winning entry to the MaxEnt 90 image reconstruction contest was produced by Nick Weir of Caltech and is not presented here since quantitative information concerning this solution has not been published—however, see reference 45 for a gray-scale image of this reconstruction. Nonetheless, Weir's solution is qualitatively similar to Bontekoe's solution. Both were made with MEMSYS 3. Weir's solution, however, used a single correlation length channel in the reconstruction. This constrained the minimum correlation length of features in the reconstruction, preventing break-up of the image on smaller size scales. This is probably what resulted in the "winning-edge" for Weir's reconstruction in the MaxEnt 90 contest.[48]

As can be seen from FIGURE 7, our FPB-based reconstruction is superior to those produced by other methods. The Lucy–Richardson and MCM reconstructions fail to significantly reduce image spread in the cross-scan direction, i.e., the rectangular signature of the 1.5 by 4.75 arcmin detectors is still clearly evident, and

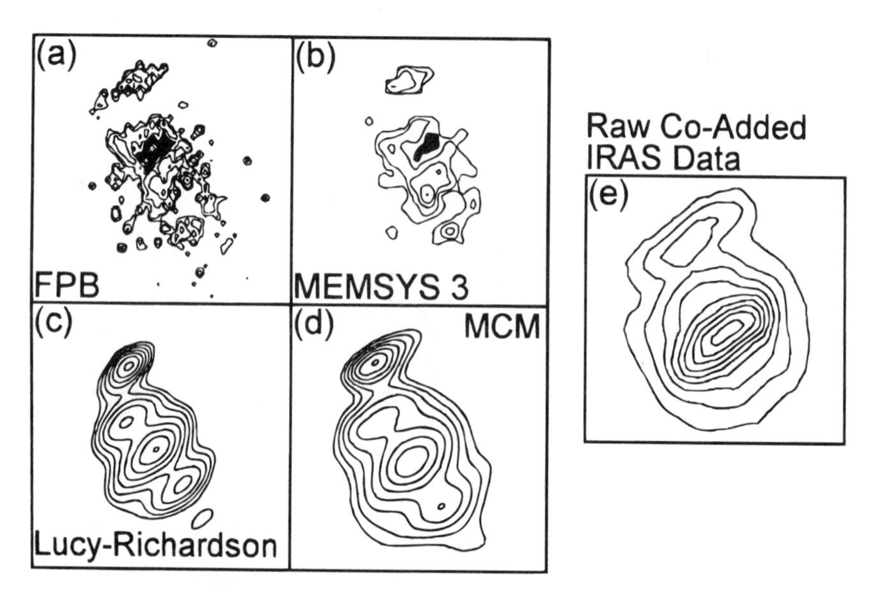

FIGURE 7. Comparison of the performance of various image reconstruction methods for the M51 IRAS survey data. Shown are the results for our pixon-based method (fuzzy, radially symmetric pixons), the Lucy–Richardson method, the "in-house" IPAC MAXIMUM Correlation Method (MCM), and the Maximum Entropy Method (MEMSYS 3). Panel **(b)** is reproduced from Bontekoe *et al.*,[47] by permission of the author, and **(c)** and **(d)** are reproduced from Rice,[46] by permission of the author. This figure first appeared in Puetter and Piña.[2]

fail to reconstruct even gross features such as the "hole" (black region) in the emission north of the nucleus—this hole is clearly evident in optical images of M51. The MEMSYS 3 reconstruction by Bontekoe is significantly better. This image clearly recovers the emission "hole" and resolves the north-east and south-west arms of the galaxy into discrete sources. Nonetheless, the level of detail present in the FPB reconstruction is clearly absent, e.g., the weak source centered in the emission hole (again, this feature corresponds to a known optical source).

To assess the significance of the faint sources present in our FPB reconstruction, our reconstruction in FIGURE 6 is overlaid with the 5 GHz radio contours of van der Hulst *et al.*[49] The radio contours are expected to have significant, although imperfect, correlation with the far-infrared emission seen by IRAS. Hence a comparison of the two maps should provide an excellent test of the reality of structures found in our reconstruction. Also identified in FIGURE 6 are several prominent optical sources and (hydrogen Balmer line emission) knots. As can be seen, the reconstruction indicates excellent correlation with the radio. The central region of the main galaxy and its two brightest arms align remarkably well, and the alignment of the radio emission from the north-east companion and the IRAS emission is excellent. Furthermore, for the most part, whenever there is a source in the reconstruction which is not identifiable with a radio source, it can be identified with either optical or knots. An excellent example is the optical source in the "hole" of emission to the north-east of the nucleus of the primary galaxy or the bright optical source to the north-west of the nucleus (both labeled "Opt" in FIGURE 6). Because of the excellent correlation with the radio, optical, and images, we are quite confident that all the features present in our reconstruction are real. (The faintest sources visible in FIGURE 6 have a formal SNR of 30.)

We next present a sample image reconstruction of coded mask X-ray data from the Hard X-ray Telescope (HXT) on board the Yohkoh spacecraft.[50] FIGURE 8 shows hard X-ray (23–33 keV) images of a solar flare which occurred on 20 August 1992. Since there is no effective method of manufacturing optics with which to focus hard X-ray light, the HXT instrument takes X-ray images by taking pictures through a series of coded masks. The series of coded images is then inverted to yield the underlying source structure. FIGURE 8 shows a time series of three images. Each row of panels shows three different reconstructions. All inversions were performed by T. Metcalf (University of Hawaii). As can be seen, direct linear inversion produces an enormous amount of spurious structure. In addition to hiding low contrast features, the presence of such spurious structure is particularly worrisome since the flux-conserving nature of the algorithm requires that flux placed in spurious sources must come from the true sources, thereby grossly affecting photometry. In this regard, the ME and FPB reconstructions can be seen to be a great improvement over the direct linear inversion. Relative to the pixon-based reconstruction, however, the ME inversion still produces a wealth of spurious emission, resulting in poor photometry and often over-resolves real features. (We know the resolution of the ME image is too high since the quality of the pixon fit is just as good—in fact slightly better for these images—and uses a lower resolution. Hence the ME resolution is unjustified.)

As a final example of pixon-based image reconstruction, we present a comparison of two image reconstruction techniques using 50–150 keV data from the Orient-

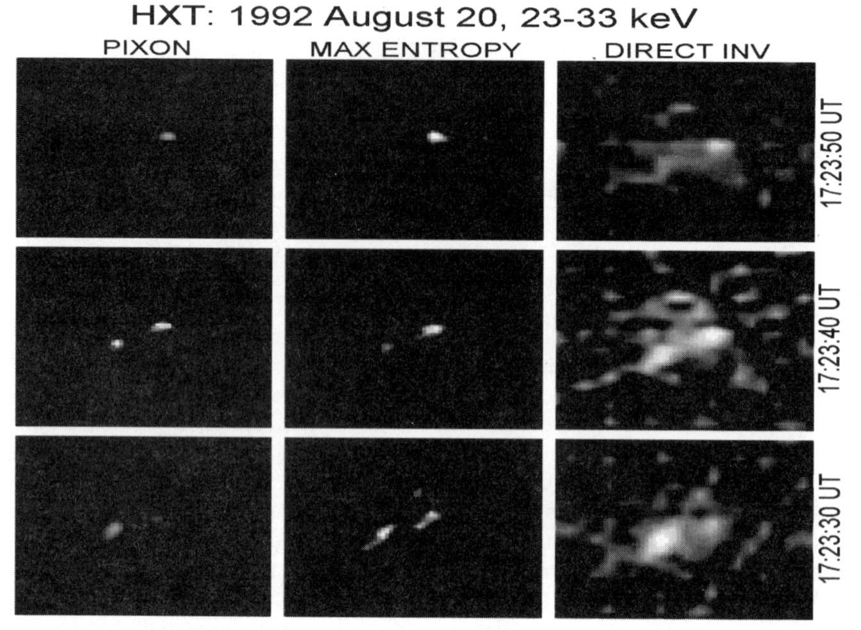

FIGURE 8. Hard X-ray (23–33 keV) images of the 20 August 1992 solar flare. The observations are from the hard X-ray Telescope (HXT) on board the Yohkoh satellite. Each column shows a time series of three image reconstructions. The background patterns which show up in the direct inversion and the ME inversion are artifacts resulting from the sparse UV-plane coverage in the HXT instrument. The pixon-based reconstruction essentially eliminates all of these artifacts. The reconstructions were performed by T. Metcalf of the University of Hawaii. Figure reproduced from Puetter.[6]

ed Scintillation Spectrometer Experiment (OSSE) aboard the Compton Gamma-Ray Observatory (GRO). The OSSE instrument consists of four shielded detectors with a field of view of 3.8° × 11.4° (FWHM). Each detector is mounted on an independent single-axis pointing system which allows for sub-stepping the detector field of view. FIGURE 9 presents a comparison of a pixon-based reconstruction of the OSSE survey data with the Non-Negative Least-Squares (NNLS) method developed at the University of California, Riverside. Both the NNLS and pixon-based reconstructions presented in FIGURE 9 were performed by D. Dixon of UC, Riverside.[51] Details of the UC, Riverside implementation of the pixon-based algorithm are given below. The dark area of the illustration represents all the points for which there is significant exposure time during the scanned observation. Each of the pixels in the reconstructed images in FIGURE 9 has an angular size of 2° × 2°.

As can be seen from FIGURE 9, the NNLS reconstruction produces an image which has the appearance of a random noise field. From this image, it is unclear whether or not there are any real detections of sources. On the other hand, the pixon-based reconstruction clearly finds the two bright sources expected to be seen in the data, i.e., 3C 273 and NGC 4388 (the active galaxy M87 is also in the pixel occupied by NGC 4388, but is not believed to contribute significant 50–150 keV

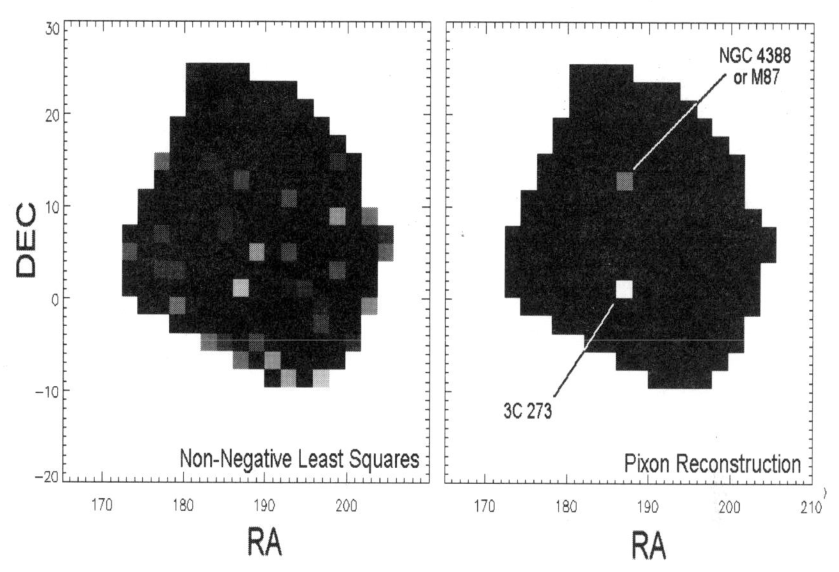

FIGURE 9. Non-Negative Least Squares (NNLS) and pixon-based image reconstructions from the 50–150 keV data of the OSSE Virgo Survey. The reconstructions were performed by D. Dixon of the University of California, Riverside. While the NNLS method gives a result that looks like random noise, Dixon's simulated annealing, pixon-based method clearly reveals the two strong sources expected in these data, i.e., NGC 4388 and 3C 273. It also detects a mild gradient in the γ-ray background. Figure reproduced from Puetter.[6]

emission), and it can be seen that the pixon-based method was very successful in suppressing spurious sources in the reconstruction.

The OSSE data, and γ-ray data in general, present an especially difficult challenge for image reconstruction methods. This is because the signal-to-noise of the collected data is very low. At γ-ray energies there are only very few photons to count. Consequently the standard pixon-based methods developed at UCSD were found to be inadequate—the conjugate gradient method could not find the global minimum. It was for this reason that Dixon (UC, Riverside) decided to use a simulated annealing approach to finding the optimal pixon reconstruction. As previously discussed, simulated annealing is well known for its robust ability to find shallow global minima in the presence of numerous local minima. However, this ability has significant computational costs. To speed the method, Dixon adopted a mean field approach and used only two pixon scales, one appropriate for point sources and one appropriate for the diffuse background. Given the expected nature of the image, this is quite a suitable assumption and gives rise to the excellent reconstruction presented in FIGURE 9. It is clear from the success of this method that pixon-based approaches such as that of Dixon *et al.*[51] have dramatically increased the scientific capability of the OSSE instrument. Current plans at UC, Riverside are to apply these techniques to COMPTEL data as well, and to explore simulated annealing pixon-based approaches with multiple pixon scales.

ACKNOWLEDGMENTS

The author would like to acknowledge the many valuable contributions from his collaborators and colleagues, especially Robert Piña, the co-developer of the pixon concept. The author also thanks N. Weir for insightful conversations and for graciously performing the MEMSYS 5 reconstruction presented in FIGURE 4. The author thanks Romke Bontekoe and Do Kester for providing the M51 IRAS test data set and valuable conversations, Steven Gull and John Skilling for conversations regarding the role of the pixon in Bayesian image reconstruction, and James Graham for making his 2 μm Keck data and ME reconstruction available. The author would also like to thank Tom Metcalf and Dave Dixon for extending pixon-based techniques to X-ray and γ-ray data and for contributing their insights to pixon-based methods. Finally, the author would like to thank Amos Yahil for numerous conversations and insights into pixon-based methods.

REFERENCES

1. PIÑA, R. K. & R. C. PUETTER. 1993. P.A.S.P. **105:** 630.
2. PUETTER, R. C. & R. K. PIÑA. 1993. Proc. S.P.I.E. **1946:** 405.
3. PUETTER, R. C. & R. K. PIÑA. 1993. *In* Maximum Entropy and Bayesian Methods. G. R. Heidbreder, Ed.: 275. Kluwer.
4. PUETTER, R. C. & R. K. PIÑA. 1994. *In* Science with High Spatial Resolution Far-IR Data. S. Tereby & J. Mazzarella, Eds.: 61. JPL.
5. PUETTER, R. C. 1994. Proc. S.P.I.E. **2302:** 112.
6. PUETTER, R. C. 1995. Int. J. Image Sys. Tech. **6:** 314.
7. PUETTER, R. C. 1996. *In* Instrumentation for Large Telescopes. In press.
8. PUETTER, R. C. 1996. *In* Instrumentation for Large Telescopes. In press.
9. PUETTER, R. C. 1996. *In* Instrumentation for Large Telescopes. In press.
10. PUETTER, R. C. 1996. *In* Instrumentation for Large Telescopes. In press.
11. PUETTER, R. C. 1996. *In* Instrumentation for Large Telescopes. In press.
12. KIKUCHI, R. & B. H. SOFFER. *In* Image Analysis and Evaluation: 95. Society of Photographic Scientists and Engineers. Toronto, Canada.
13. BRYAN, R. K. & J. SKILLING. 1980. M.N.R.A.S. **191:** 69.
14. NARAYAN, R. & R. NITYANANDA. 1986. Ann. Rev. Astron. Astrophys. **24:** 127.
15. SKILLING, J. 1989. *In* Maximum Entropy and Bayesian Methods. J. Skilling, Ed.: 45. Kluwer.
16. MACKAY, D. J. C. 1994. *In* Models of Neural Networks III. E. Domany, J. L. van Hemmen & K. Schutten, Eds.: Chapter 6. Springer-Verlag.
17. SIBISIS, S. 1990. *In* Maximum Entropy and Bayesian Methods. J. Skilling, Ed. Kluwer.
18. COHEN, J. G. 1991. *In* Maximum Entropy and Bayesian Methods. W. T. Grady, Jr. & L. Schick, Eds. Kluwer.
19. JAYNES, E. T. 1957. Phys. Rev. **106:** 171.
20. JAYNES, E. T. 1963. *In* Statistical Physics. K. W. Ford, Ed.: 181. Benjamin.
21. HOBSON, A. 1971. Concepts in Statistical Mechanics. Gordon and Breach.
22. SHANNON, C. E. 1948. Bell Syst. Tech. J. **27:** 379.
23. SHANNON, C. E. & W. WEAVER. 1949. The Mathematical Theory of Communication. University of Illinois Press.
24. RABBANI, M. & P. W. JONES. 1991. Digital Image Compression Techniques. SPIE Optical Engineering Press.
25. KHINCHIN, A. I. 1957. Mathematical Foundations of Information Theory. Dover.
26. FEINSTEIN, A. 1958. Foundations of Information Theory. McGraw-Hill.
27. KITTEL, C. 1969. Thermal Physics. Wiley.
28. WICKEN, J. 1987. Philos. Sci. **54:** 176.

29. CHAITEN, G. J. 1982. *In* Encyclopedia of Statistical Science, **1:** 38. Wiley.
30. CHERRY, C. 1978. On Human Communication, 3rd edn. MIT Press.
31. SOLOMONOFF, R. 1964. Inf. Control **7:** 1.
32. KOLMOGOROV, A. N. 1965. Inf. Transmission **1:** 3.
33. CHAITIN, G. J. 1966. J. Ass. Comput. Mach. **13:** 547.
34. DIPLAS, A., E. A. BEAVER, P. R. BLANCO, R. K. PIÑA & R. C. PUETTER. 1993. *In* The Restoration of HST Images and Spectra-II. R. J. Hanisch & R. L. White, Eds.: 272. Space Telescope Science Institute.
35. METCALF, T. R., H. S. HUDSON, T. KOSUGI, R. C. PUETTER & R. K. PIÑA. 1996. Ap. J. **456:** 585.
36. KORESKO, C. D., P. M. HARVEY, D. CURRAN & R. C. PUETTER. 1995. Proc. Airborne Astronomy Symp. on Galactic Ecosystem: From Gas to Stars to Dust, A.S.P. Conf. Series, **73:** 275.
37. SMITH, C. H., D. K. AITKEN, T. J. T. MOORE, P. F. ROCHE, R. C. PUETTER & R. K. PIÑA. 1994. M.N.R.A.S. **273:** 354.
38. GULL, S. F. & J. SKILLING. 1991. MemSys5 Quantified Maximum Entropy User's Manual.
39. WEIR, N. 1991. Proc. ESO/ST-ECF Data Analysis Workshop. P. Grosbo and R. H. Warmels, Eds.: 115. ESO.
40. WEIR, N. 1994. J. Opt. Soc. Am. In press.
41. GRAHAM, J. R. & M. C. LIU. 1995. Ap. J. Lett. **449:** L29.
42. BROADHURST, T. & J. LEHAR. 1995. Ap. J. Lett. **450:** L41.
43. CLOSE, L. M., P. B. HALL, C. T. LIU & E. K. HEGE. 1995. Ap. J. Lett. **452:** L9.
44. EISENHARDT, P. R., L. ARMUS, D. W. HOGG, B. T. SOIFER, G. NEUGEBAUER & M. W. WERNER. 1996. Ap. J. **461:** 72.
45. BONTEKOE, T. R. 1991. *In* Maximum Entropy and Bayesian Methods. W. T. Grady, Jr. and L. H. Schick, Eds. Kluwer.
46. RICE, W. 1993. Ap. J. **105:** 67.
47. BONTEKOE, T. R., D. J. M. KESTER, S. D. PRICE, A. R. W. DE JONGE & P. R. WESSELIEUS. 1991. Astron. Astrophys. **248:** 328.
48. WEIR, N. 1993. Private communication.
49. VAN DER HULST, J. M., R. C. KENNICUTT, P. C. CRANE & A. H. ROTS. 1988. Astron. Astrophys. **195:** 38.
50. KOSUGI, T., K. MAKISHIMA, T. MURAKAMI, T. SAKAO *et al.* 1991. Solar Phys. **136:** 17.
51. DIXON, D. D., O. T. TUMER, J. D. KURFESS, W. R. PURCELL, W. A. WHEATON, R. K. PIÑA & R. C. PUETTER. 1996. Ap. J. Submitted.

Galaxy Classification Using Artificial Neural Networks

S. C. ODEWAHN

Department of Physics and Astronomy
Arizona State University
Tempe, Arizona 85287

INTRODUCTION

Applications of machine-automated reduction procedures and pattern recognition techniques are a key ingredient to the execution of large all-sky digital surveys. The development of objective machine techniques for object detection, measurement, and classification not only allow for a timely reduction of the data, but greatly add to the scientific worth of the final product by imposing a reduction process which is rigorously adhered to. Works by Slezak *et al.*,[1] Heydon-Dumbleton *et al.*,[2] Rhee,[3] and Odewahn *et al.*[4] discuss the use of automated techniques for rudimentary image analysis (i.e., star-galaxy separation), but the area of automated morphological classification of galaxies is in a relative state of infancy. Early investigations have yet to produce a system which can discern the subtle features attainable by experienced human classifiers when assigning a type in the revised Hubble classification system. Using high-quality photometric parameter sets from multiple spectral regions, along with sophisticated classification algorithms, it is only a matter of time before machine classified sets will match the repeatability of the human classifiers. In addition, such techniques will be applied in a more consistent manner to vastly larger data sets than have heretofore been attempted in human visual classification projects.

Thonnat[5] has used a rule-based expert system to derive galaxy classifications from parameters obtained in the image segmentation of PDS density arrays measured on photographic Schmidt plates. Although a well-understood classification system is attained in such a scheme, one must possess either *a priori* knowledge of how the measured image parameter data relate to the galaxy types or a vast amount of examples with which to derive the classification rules. Spiekermann[6] has classified galaxies into five broad morphological classes (E, S0, Sa, Sb, Sc/Ir) using a fuzzy algebra system for digitized IIIa-J Schmidt images. Whitmore[7] presents a classification approach using a principal component analysis (PCA) of a variety of photometric parameters, some of which are based on metric measures, and hence require one to know the galaxy distance. A recent approach along this direction by Han,[8] where PCA is used to characterize the I-band luminosity profile of a galaxy, shows great promise. In recent work, Doi *et al.*[9] have used the surface brightness and image concentration to discriminate between early and late type galaxies. Similarly, Abraham *et al.*[10] use an image asymmetry and concentration parameter to perform a linear separation in a two-dimensional space for galaxy classification in three broad categories. From all these works, it is clear that one wishes to combine as much quantitative information as possible when assigning a Hubble type. An artificial neural network, discussed by Rumelhardt and McClelland,[11] provides an ideal means of combining a large amount of input information to derive a best-guess classification.

ARTIFICIAL NEURAL NETWORKS

Artificial neural networks (ANNs) are systems of weight vectors, whose component values are established through various machine-learning algorithms, which take as input a linear set of pattern inputs and produce as output a numerical pattern encoding a classification. They were designed to simulate groups of biological neurons and their interconnections and to mimic the ability of such systems to learn and generalize. Discussion of the development and practical application of neural networks is presented by Rumelhardt and McClelland.[11] A variety of useful information on this subject may also be found on the internet at the URL address ftp://ftp.sas.com/pub/neural/FAQ.html. In astronomical applications, this technique has been applied with great success to the problem of star-galaxy separation by Odewahn *et al.*[12,13] A neural network classifier developed by Storrie-Lombardi *et al.*[14] is used to assign galaxy types on the basis of the photometric parameters supplied by the catalog of Lauberts and Valentijn.[15] Serra-Ricart *et al.*[16] again use a neural network approach to assign galaxy types based on the ESO-LV data and demonstrate how this approach is superior to a method like PCA which uses only a linear combination of input parameters to perform pattern classification. A similar discussion is presented by Lahav.[17] The application of this method to large surveys of galaxies using photographic Schmidt plate material is described by Odewahn[18] and Naim *et al.*[19]

The neural network literature quoted above contains a wealth of information on the theoretical development of neural networks and the various methods used to establish the network weight values. Summarized here are only the practical aspects of the ANN operation and the basic equations for deriving the network weight values using the method of back-propagation. The design of a typical ANN is schematically illustrated in FIGURE 1. The input information, in this case six photometric parameters known to be correlated with galaxy morphological type,[20] is presented to the network through six input nodes, referred to as the input layer. Each input node is "connected" via an information pathway to nodes in a second layer of nodes referred to as the first hidden layer. In a repeating fashion, one may construct a network using any number of such hidden layers. For the galaxy classification networks discussed here, we have used ANNs with two hidden layers. The final layer of node positions, referred to as the output layer, will contain the numerical output of the network encoding the classification. This system of nodes and interconnections is analogous to the system of neurons and synapses of the brain. Information is conveyed along each connection, processed at each node site, and passed along to nodes further "upstream" in the network. Hence, this type of system is referred to as a feed-forward network. The information processing at each node site is performed by combining all input numerical information from upstream nodes, a_{pj}, in a

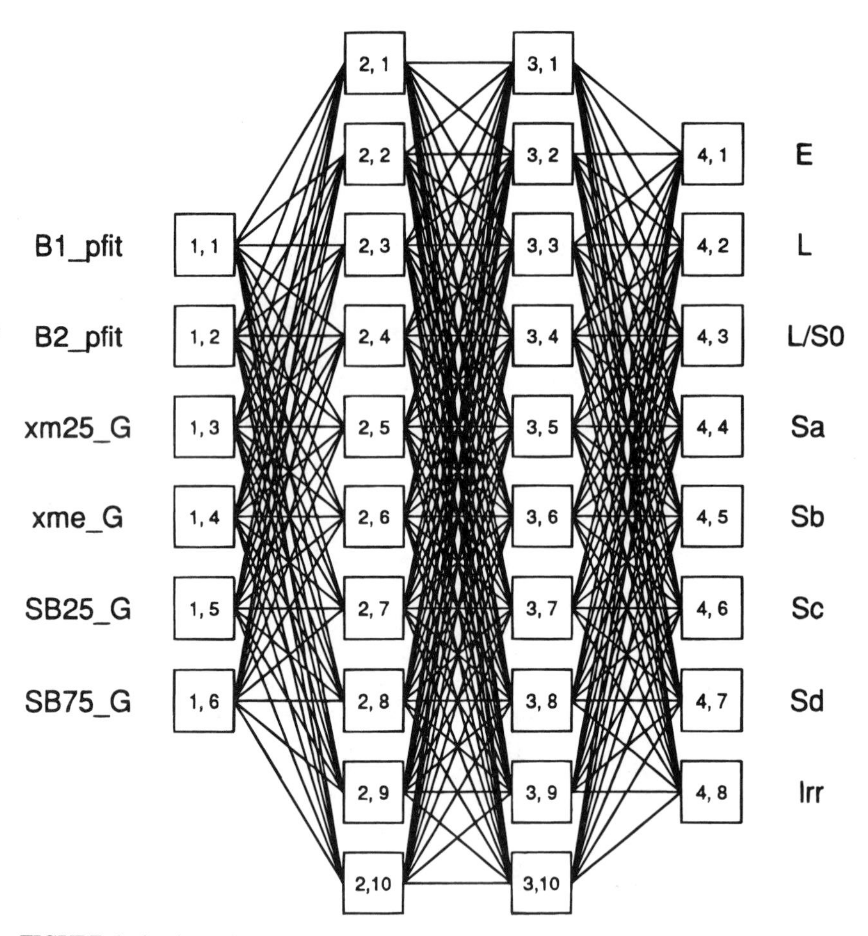

FIGURE 1. A schematic diagram of the typical neural network architecture used to map an input set of six global photometric parameters to an 8-element output vector. Each output node represents two steps in the 16-step scale of the revised Hubble galaxy classification system. To the right of each output node position is printed the mean galaxy type for that output.

weighted average of the form:

$$\beta_i = \sum_j w_{ij} a_{pj} + b_i. \tag{1}$$

The j subscript refers to a summation of all nodes in the previous layer of nodes and the i subscript refers to the node position in the present layer, i.e., the node for which we are computing an output. Each time the ANN produces an output pattern, it has done so by processing information from the input layer. This input information is referred to as the input pattern, and we use the p subscript to indicate input pattern. Hence, if there are five nodes in the previous layer (j), each node (i) in our current layer will contain five weight values (w_{ij}) and a constant term, b_i, which is referred to as the bias. The final nodal output is computed via the activation function, which in the case of this work is a sigmoidal function of the form:

$$a_{pi} = \frac{1}{1 + \exp^{-\beta_i}}. \tag{2}$$

Hence, the information passed from node i in the current layer of consideration, when the network was presented with input pattern p, is denoted as a_{pi}. This numerical value is subsequently passed to the forward layer along the connection lines for further network processing. The activation values computed in the output layer form the numerical output of the ANN and hence encode the classification for input pattern, p.

In order to solve for the weight and bias values of (**1**) for all nodes, one requires a set of input patterns for which the correct classification is known. This set of examples is used is an iterative fashion to establish weight values using a gradient descent algorithm known as back-propagation. For each input training pattern, p, we compute the error function:

$$E_p = \sum_i (t_{pi} - o_{pi})^2, \tag{3}$$

where t_{pi} is the expected value of node i in the output layer given input pattern p, and o_{pi} is the actual numerical value computed for node i. An error of zero would indicate that all the output patterns computed by the ANN perfectly match the expected values and our network is well trained. In brief, back-propagation training is performed by initially assigning random values to the w_{ij} and b_i terms in all nodes. Each time a training pattern is presented to the ANN, the activation for each node, a_{pi}, is computed. After the output layer is computed, we go backwards through the network and compute an error term, δ_{pi}, for each node. This error term is the product of the error function, E, and the derivative of the activation function and hence is a measure of the change in the network output produced by an incremental change in the node weight values. For an output layer node, the error term is computed as

$$\delta_{pi} = (t_{pi} - a_{pi})a_{pi}(1 - a_{pi}) \tag{4}$$

and for a node in a hidden layer we use

$$\delta_{pi} = a_{pi}(1 - a_{pi}) \sum_k \delta_{pk} w_{kj}. \tag{5}$$

In the latter expression, the k subscript indicates a summation over all the nodes in the downstream layer (the layer in the direction of the output layer). The j subscript indicates the weight position in each node. Finally, the δ and a terms for each node are used to compute an incremental change to each weight term via:

$$\Delta w_{ij} = \varepsilon(\delta_{pi} a_{pj}) + \alpha w_{ij}(old). \tag{6}$$

The constant term ε is referred to as the learning rate and determines the absolute size of the weight adjustments during each training iteration. The constant term α is called the momentum. It is applied to the weight change used in the previous training iteration, $w_{ij}(old)$. Both of these constant terms are specified at the start of the training cycle and determine the speed and stability of training.

APPLICATION OF THE TECHNIQUE

Early efforts to build ANN image classifiers for assigning morphological types to galaxies digitized from photographic Schmidt plate material have been presented in Odewahn.[18] More recent work, described in detail in Odewahn *et al.*,[21] makes use of the wider dynamic range and higher angular resolution of images obtained with HST's Wide Field Planetary Camera 2 (WFPC2). These new data allow the determination of the sub-kpc morphology, light-profiles, color-gradients, and morphological types of individual galaxies—as well as the population fractions of different galaxy types—over a wide range of epochs. In view of the many exciting new HST-based studies of galaxy formation and evolution,[22–24] the development of a rigorous morphological classification scheme for the many faint galaxy images expected in deep WFPC2 fields is important to study galaxy formation and evolution in an unbiased manner. This concluding section will review early efforts to develop ANN-based galaxy classifiers for analyzing faint galaxies observed with WFPC2.

To gather a training sample for developing ANN classifiers, Odewahn *et al.*[21] used visual inspection of galaxies from several deep WFPC2 fields to assign morphological types in the revised Hubble classification system. They classified by eye 173 galaxies with $U_{300} \lesssim 26$ mag, 372 in B_{450} ($\lesssim 26.5$) and 542 in V_{606} and I_{814} ($\lesssim 26$), the latter including 173 galaxies in VI from Driver *et al.*,[23] and obtained results mutually consistent within ± 2 (rms) Hubble classes and scale errors $\lesssim 10\%$ on the 16-step scale. As shown in FIGURE 1, each galaxy type on the 16-step scale was converted to an 8-element type code which represents the target pattern, t_p, for the ANN classifier. For the present, we assign a value of 1 to the element representing the proper galaxy class (i.e., element 5 is assigned a value of 1 in the case of an Sb galaxy) and zeroes to the other target vector elements.

Surface photometry was carried out for all galaxies in five different WFPC2 fields using the automated image analysis package MORPHO,[18] resulting in ellipse fits over a range of isophotal levels, elliptically averaged and equivalent radial SB-profiles, and a set of photometric parameters known to show significant dependence with morphological type for local galaxies (see reference 20). Using mean types from the visually-derived type catalogs in $UBVI$, ANN galaxy classifiers were trained using input layer elements composed of six parameters: the surface brightness at the 25% and 75% quartile radii, the mean surface brightness within the effective radius and within an isophotal radius, as well as the slope and intercept of a linear fit to the SB-profile (in either r or $r^{1/4}$ space). A second generation of networks was developed to classify images from several different bandpasses in order to address the uncertainties on our type classification imposed by differing K-corrections over the wide range of redshifts sampled in deep WFPC2 imagery. Images uniformly normalized to a constant peak flux density were used to compute total signal, quartile flux ratios and concentration indices for use as ANN input parameters.

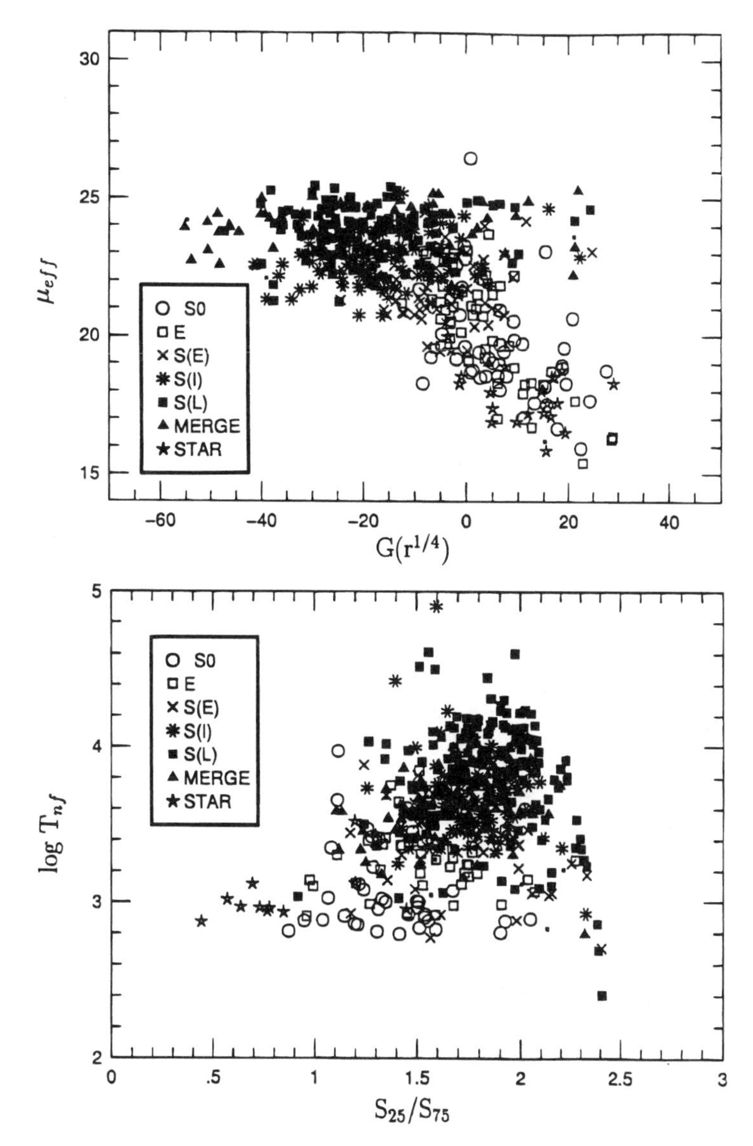

FIGURE 2. Parameter spaces used to separate galaxies by morphological type. *Top panel*: the relationship between surface brightness at the effective radius, μ_{eff}, and the slope, $G(r^{1/4})$, of a linear fit to the radial surface brightness profile in $r^{1/4}$ units. *Lower panel*: the log of the total integrated flux, T_{nf}, as a function of the flux density ratio at the 75% and 25% quartile radii, S_{25}/S_{75}, computed from images normalized to a uniform peak flux density. In both cases, plotted symbols indicate the galaxy types from a weighted mean of independent eyeball estimates by human classifiers. Note the clear separation between early types (E = elliptical, S0 = lenticulars, and S(E) = early spirals) and later types (S(I) = intermediate spirals, S(L) = late spirals). A variety of parameter spaces such as these are combined by the ANN classifiers discussed in the text to estimate morphological types.

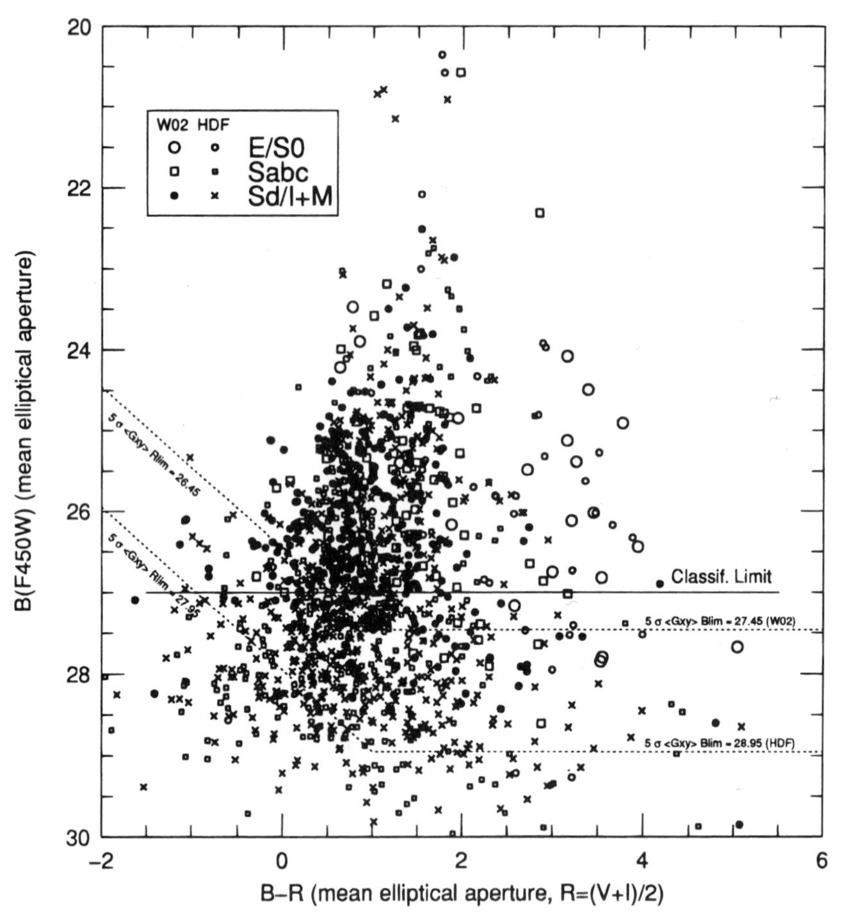

FIGURE 3. The $(B–R)$ versus B_{450} color-magnitude diagram for all classified galaxies in the HDF and 53W002 WFPC2 fields. $(V_{606} + I_{814})/2$ was used as "wide R_{6900}"-band to increase S/N in "$(B–R)$". The B_{450} image is substantially complete for galaxies of average SB at $B_J \lesssim 27.5$ mag (5σ) in the W02 field and $B_J \lesssim 29.0$ mag in the HDF. For $B_J \gtrsim 26.0$ and 27.5 mag, the two fields become increasingly incomplete for the *bluest* objects due to the redder detection limits at $R \sim 26.5$ and 28.0 mag, respectively, as indicated by the two *slanted dashed lines.* As explained in the text, ANN-derived galaxy classes are indicated for E/S0, Sabc, and Sd/Irr + M (M for merger) categories using different symbols. The reddest objects down to $B_J \lesssim 27$ mag are mostly classified as *E*/S0's and Sabc's, and do not increase as rapidly towards the formal completeness limits. The increase of the late-type + merging population galaxies down to the detection and classification limit ($B_J \sim 27$ mag) is quite remarkable and cannot be explained without some form of strong density and/or luminosity evolution.

As can be seen in FIGURE 2, such photometric parameters were chosen on the basis of their high correlation with the morphological types in our mean type catalogs and their ability to separate different type populations. These input vectors, combined with their corresponding output target vectors, were gathered for approximately 500 galaxies and used to train and test two different ANNs, each with different numbers of weights (i.e., learning architectures) for the V_{606} and I_{814} WFPC2 filters, respectively. In practice, we divided the 500-galaxy sample into a training subset and a testing subset. In other words, only a subset of the data set (the training set) was used as target patterns for predicting the weight changes of (**6**). At set intervals in the iterative back-propagation training used to solve for the ANN w_{ij} and b_i terms of (**2**), we correlated the target types with the network-predicted types for the test samples in order to judge ANN performance. Back-propagation training was stopped when type-prediction for the test sample of galaxies failed to improve relative to the type-predictions for the training sample. Although trained on *different* 70% subsets of the existing catalogs, a direct comparison of the types provided by the V_{606} and I_{814}-band classifiers (for $I_{814} \leq 24$ mag) produced a scale difference of only $\lesssim$5% and a scatter of $\lesssim$1.7 steps (rms) on the 16-step revised Hubble system. This level of agreement was found to be quite comparable to that derived from comparisons of visual type estimates from independent human classifiers.

To illustrate the practical application of the WFPC2 ANN-predicted types we plot in FIGURE 3 the total B_{450}-magnitude versus (B–R) color for two faint galaxy fields: the Hubble Deep Fields[25] and the 53W002 field.[22] In this illustration we use different symbols to indicate the ANN-predicted galaxy types in three broad categories: E/S0 (elliptical and lenticular galaxies), Sabc (early-type spiral galaxies), and Sd/Irr + M (late-type spirals irregular and merger systems). Of significance is the clear segregation between the early and the late type galaxies in FIGURE 2 for $B_J \lesssim$ 27 mag, even though *no* color information was used by the ANN classifier. E/S0's are almost without exception the reddest galaxies at any flux level, and Sd/Irr + M's are generally blue, at least down to the formal detection limit. Mid-type spirals (Sabc's) have colors with large dispersion and about as blue as Sd/Irr's, as expected from combining the slightly different colors measured for these populations locally[20] and the K-corrections for galaxies with ongoing star formation at z ~ 1–2.[26] For $B_J \lesssim 27$ mag, where the ANN types should be reliable, this could also be attributed to intrinsic scatter in their colors. For $B_J \gtrsim 27$ mag, the ANN-derived galaxy types have larger classification errors—in addition to possibly larger cosmic scatter in the colors—although E/S0's are still generally redder and have smaller scale lengths, while late-types are generally bluer and have larger r_{hl}. Without further calibration in the rest-frame UV, we do not currently believe that the ANN classifier is reliable for $B_J \gtrsim 27$ mag.

ACKNOWLEDGMENTS

I acknowledge many helpful discussions on the classification and photometry of galaxies with G. de Vaucouleurs, H. C. Corwin, R. Buta and G. Aldering.

REFERENCES

1. SLEZAK, E., A. BIJAOUI & G. MARS. 1988. Astron Astrophys. **201:** 9.
2. HEYDON-DUMBLETON, N. H., C. A. COLLINS & H. T. MACGILLIVRAY. 1989. MNRAS **238:** 379.
3. RHEE, G. 1990. Ph.D. dissert. Univ. of Leiden.

4. ODEWAHN, S. C., E. B. STOCKWELL, R. M. PENNINGTON, R. M. HUMPHREYS & W. A. ZUMACH. 1992. A. J. **103:** 318.
5. THONNAT, M. 1989. The World of Galaxies: 53. Springer-Verlag. New York.
6. SPIEKERMANN, G. 1992. A. J. **103:** 2102.
7. WHITMORE, B. C. 1984. Ap. J. **287:** 61.
8. HAN, M. 1995. Ap. J. **442:** 504.
9. DOI, M., M. FUKUGITA & S. OKAMURA. 1993. MNRAS **264:** 832.
10. ABRAHAM, R. *et al.* 1996. MNRAS **279:** L47.
11. MCCLELLAND, J. L. & D. E. RUMELHARDT. 1988. Explorations in Parallel Distributed Processing. MIT Press. Cambridge, MA.
12. ODEWAHN, S. C., E. B. STOCKWELL, R. M. PENNINGTON, R. M. HUMPHREYS & W. A. ZUMACH. 1992. A. J. **103:** 318.
13. ODEWAHN, S. C., R. M. HUMPHREYS, G. ALDERING & P. THURMES. 1993. PASP **105:** 1354.
14. STORRIE-LOMBARDI, M. C., O. LAHAV, L. J. SODRE & L. J. STORRIE-LOMBARDI. 1992. MNRAS **258:** 8.
15. LAUBERTS, A. & E. A. VALENTIJN. 1989. The Surface Photometry Catalogue of the ESO-Uppsala Galaxies. European Southern Observatory. Garching.
16. SERRA-RICART, M., X. CALBET, L. GARRIDO & V. GAITAN. 1993. A. J. **106:** 1685.
17. LAHAV, O. 1994. Vistas Astron. **38:** 251.
18. ODEWAHN, S. C. 1995. PASP **107:** 770.
19. NAIM, A., O. LAHAV, L. SODRE & M. C. STORRIE-LOMBARDI. 1995. MNRAS **275:** 567.
20. ODEWAHN, S. C. & G. ALDERING. 1995. A. J. **110:** 2009.
21. ODEWAHN, S. C., R. A. WINDHORST, S. P. DRIVER & W. C. KEEL. 1996. Ap. J. **472:** in press.
22. DRIVER, S. P., R. A. WINDHORST, E. J. OSTRANDER, W. C. KEEL, R. E. GRIFFITHS & K. U. RATNATUNGA. 1995. Ap. J. **449:** L23.
23. DRIVER, S. P., R. A. WINDHORST & R. E. GRIFFITHS. 1995. Ap. J. **453:** 48.
24. COWIE, L. L., E. M. HU & A. SONGAILA. 1995. A. J. **110:** 1576.
25. WILLIAMS, R. E. *et al.* 1996. *In* Science with the Hubble Space Telescope-II. P. Benvenuti, F. D. Macchetto & E. J. Schreier, Eds.: 33. STScI.
26. BRUZUAL, A. G. & S. CHARLOT. 1993. Ap. J. **405:** 538.

Atmospheric Turbulence, Speckle, and Adaptive Optics

EREZ RIBAK

Department of Physics
Technion–Israel Institute of Technology
Technion City
Haifa 32000, Israel

The impact of the atmospheric turbulence on astronomical imaging through it is to reduce the final resolution. This process is described through studies of the atmosphere, and through understanding of the imaging process. Speckle imaging is a method to retrieve the image after it has been corrupted. Adaptive optics tries to correct the deleterious effect of the atmosphere before the image is registered.

TURBULENCE AND THE INDEX OF REFRACTION

The index of refraction of air is sensitive to temperature and pressure changes. If light from a star passes through a volume of space where such perturbations occur, it is affected on the spatial and temporal scale of these perturbations. And the perturbations occur because of turbulence which in turn occurs because of temperature gradients on very large scales.

Air temperatures on global and continental scales create winds. These winds tend to be limited within rather thin horizontal layers, and break down inside the layers to turbulence on finer and finer scales. Nighttime astronomy is affected mainly by rather few layers—the boundary layer, near the surface, and higher elevation layers, such as the jet stream, at altitudes up to 15 km above the observatory.[1]

As a planar wave front, arriving from a distant small star, sweeps through the atmosphere, it accumulates fluctuations along its path. It is assumed that these fluctuations, which crinkle the wave front, are weak enough so as not to cause the associated light rays to cross themselves.[2]

The refractive index of air can be written approximately as[1,3]

$$n \cong 1 + 77.6 \times 10^{-6}\left(1 + \frac{0.00752}{\lambda^2}\right)\frac{P}{T} \tag{1}$$

(P is the pressure in millibar, T the temperature in Kelvin, and λ the wavelength in micrometers). The effect of pressure is usually rather insignificant. At sea level and $\lambda = 0.5$ μm, $\Delta n \approx 10^{-6}\Delta T$. The structure function of the refractive index is

$$D_n(\mathbf{r}_1, \mathbf{r}_2) = \langle[n(\mathbf{r}_1) - n(\mathbf{r}_2)]^2\rangle. \tag{2}$$

It can be shown[4] that this is true for one layer crossed, or through many such layers. Moreover, D_n follows D_T, the temperature structure function, and can be shown also to follow the same statistics: when lateral (horizontal) homogeneity and isotropy are assumed, the fluctuations follow the Kolmogorov statistics and the structure function reduces to[2]

$$D_n(\mathbf{r}) = C_n^2 r^{2/3}; \quad l_0 \ll r \ll L_0, \tag{3}$$

where l_0 and L_0 are the inner and outer scales of turbulence. The inertial scale is defined to lie between these two limits (in astronomy: tens of centimeters to tens of meters). Similar effects can be described for the temporal behavior, where the shortest time scales are of the order of milliseconds. The corresponding power spectrum is[1]

$$\Phi_n(\mathbf{k}) = 0.0033\ C_n^2 k^{-11/3}; \quad \frac{2\pi}{L_0} \ll k \ll \frac{2\pi}{l_0}. \tag{4}$$

We see that the atmospheric distortions add up as the aperture grows (3). When the phase deviations of the wave front reach a standard deviation of $\sigma_\phi \approx 1$, the image deterioration becomes significant; this is what is referred to as seeing. The corresponding aperture is defined as Fried's seeing cell size (or coherence width) by

$$r_0^{3/5} = 0.423\left(\frac{2\pi}{\lambda}\right)^2 \int_0^{h_{max}} C_n^2(h)\, dh, \tag{5}$$

where the integration is carried over all atmospheric layers down to the telescope aperture. Using this definition the wave front phase structure function is

$$D_\varphi(r) = 6.88\left(\frac{r}{r_0}\right)^{5/3} \tag{6}$$

and the wave front phase tilt variance (over an aperture of diameter D) is

$$\sigma_\varphi^2(r) = 0.896\left(\frac{D}{r_0}\right)^{5/3}. \tag{7}$$

Since the turbulence is swept in front of the telescope in different layers of altitude h with their corresponding wind speeds $v_{wind}(h)$, the temporal power spectrum of the wave front can be described by[3]

$$F_\phi(f) = 0.0326\left(\frac{2\pi}{\lambda}\right)^2 f^{-8/3} \int_0^{h_{max}} C_n^2(h) v_{wind}^{5/3}(h)\, dh. \tag{8}$$

IMAGING THROUGH TURBULENCE

The angular resolution of astronomical telescopes is severely limited by atmospheric turbulence. Instead of being λ/D, where D is the diameter of the telescope, it is λ/r_0, where r_0 is the lateral correlation length of the wave front. The reduction is by a factor of $D/r_0 \sim 5/0.2 \sim 25$. The reason for this degradation can be shown by using the Fourier depiction of the imaging process (FIG. 1): there is a Fourier relationship between the shape of the astronomical object (hereafter: the star) and

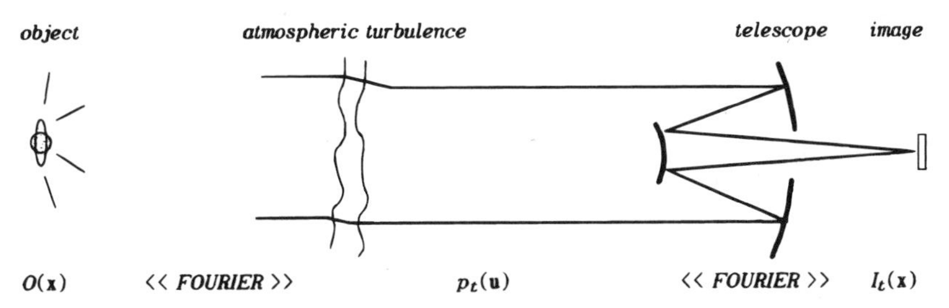

FIGURE 1. Fourier description of the imaging process. The intensity is a power spectrum of the wave front, truncated by the telescope aperture.

the mutual coherence function at the entrance to the atmosphere (the Van Cittert–Zernike theorem).[3] The phases of the wave fronts propagating through the turbulent atmosphere accumulate an additional phase up to the telescope. Another Fourier relationship now exists between the telescope aperture and the image plane. Squaring and adding the fields at this point (the camera), an image of the star is created. This image is a convolution between the star $\mathcal{O}(\mathbf{x})$ and the instantaneous power spectrum of the wave front sampled by the telescope aperture $P_t(\mathbf{x})$,

$$I_t(\mathbf{x}) = \mathcal{O}(\mathbf{x}) * P_t(\mathbf{x}), \tag{9}$$

where $\mathbf{x} = (x, y)$. Thus if one takes a series of short exposures, shorter than the atmospheric response time, then this set of images can be considered a time-frequency graph. Notice the similarity of this graph to time-frequency graphs of audio signals.[5] However, this is a three-dimensional graph, where two axes are the image coordinates (equivalent to the turbulence frequency space coordinates), and the third axis is the time coordinate. In each instantaneous image we can see speckles: diffraction-limited duplicates of the object, copied over many times at different intensities (FIG. 2). Each such speckle represents a significant Fourier component of the atmosphere. Since the telescope samples only a finite section of the

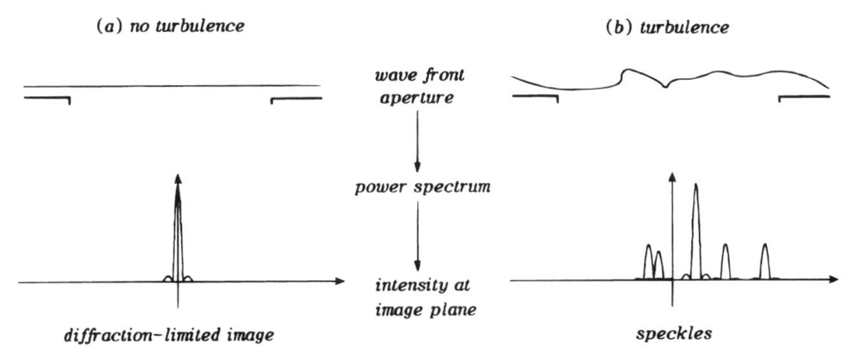

FIGURE 2. Under turbulence, a single image breaks up into many similar speckles. The speckles represent the instantaneous power spectrum of the wave front as sampled by the aperture. The brightest speckle corresponds to the lowest harmonic, which is not necessarily centered.

turbulence, the lowest frequency might have a phase attached to it (corresponding to an average tilt of the wave front). In the Fourier domain such a phase transforms into a shift of the brightest speckle. This is the basis for the shift-and-add method (see below).

Integrating the instantaneous images over time, either in software or by taking a long exposure, one obtains

$$\langle I_t(\mathbf{x})\rangle = O(\mathbf{x}) * \langle P_t(\mathbf{x})\rangle. \tag{10}$$

Unfortunately, the average power spectrum of the wave front does not extend to high frequencies, and the fine details of the image are essentially lost. Speckle imaging methods circumvent this problem by dealing directly with the short-exposure images (**9**), while adaptive optics corrects the perturbed wave fronts before they are integrated as in (**10**).

SPECKLE IMAGING

Labyerie suggested the simplest solution to retrieving the highest frequencies: average the power spectra of the short exposures.[3] Denoting the Fourier transforms by small letters in the $\mathbf{u} = (u, v)$ domain coordinates, and starting from (**9**), we get

$$\langle |i_t(\mathbf{u})|^2\rangle = |o(\mathbf{u})|^2 \cdot \langle |p_t(\mathbf{u})|^2\rangle. \tag{11}$$

The average power spectrum of the atmosphere, $\langle |p_t(\mathbf{u})|^2\rangle$, extends to high frequencies. Moreover, it can be measured by using a point source (an unresolved star), where $|o(\mathbf{u})|^2 = 1$. An inverse transform of (**11**) yields $O(\mathbf{x}) * O(\mathbf{x})$, the autocorrelation of the object. Except for very simple objects (binaries, disks, and other centro-symmetric stars) the autocorrelation is not very useful.[3] Additional methods were proposed to find the missing phase of the Fourier components. Fienup devised a method which bounces between image and Fourier space and adds phases to the spectrum according to constraints such as positivity and limited size of the object.[3] Knox and Thompson suggested using the average cross-spectrum

$$\langle i_t(\mathbf{u})i_t^*(\mathbf{u} + \Delta\mathbf{u})\rangle = o(\mathbf{u})o^*(\mathbf{u} + \Delta\mathbf{u}) \cdot \langle p_t(\mathbf{u})p_t^*(\mathbf{u} + \Delta\mathbf{u})\rangle. \tag{12}$$

At small enough frequency differences $\Delta\mathbf{u} < r_0$, the average *phase* of the cross-spectrum of the turbulence (last term) drops to zero and the object cross-phase follows the intensity cross-phase.[3] This was extended even further using the triple correlation method.[6] The average bispectrum is

$$\langle i_t(\mathbf{u})i_t(\mathbf{v})i_t(-\mathbf{u} - \mathbf{v})\rangle = o(\mathbf{u})o(\mathbf{v})o(-\mathbf{u} - \mathbf{v}) \cdot \langle p_t(\mathbf{u})p_t(\mathbf{v})p_t(-\mathbf{u} - \mathbf{v})\rangle. \tag{13}$$

As in the Knox–Thompson method, the last term is real, and the phases of the object can be found on all possible triangles in the Fourier domain, since they equal the phases of the average bispectrum of the images. This method is very computer intensive.

Efforts to find the missing phase were not limited to the Fourier domain. Bates and Cady[7] proposed the shift-and-add method. Their reasoning was since the brightest speckle looks very much like a diffraction-limited image of the object (FIG. 2). Shifting all the images to center on the brightest speckle and then adding them

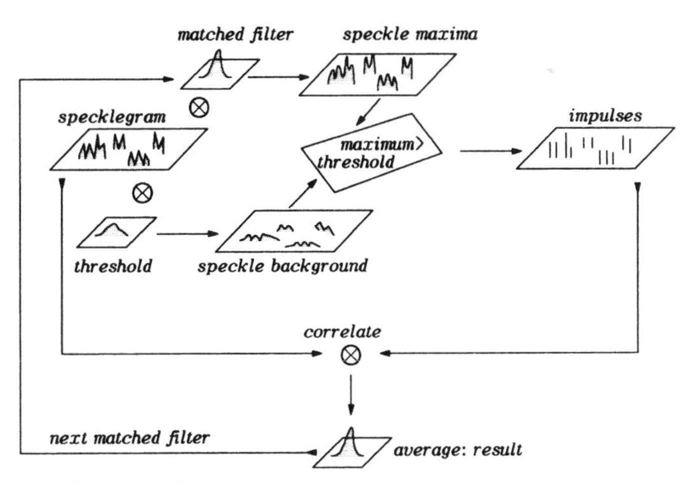

FIGURE 3. Enhanced shift-and-add. Each specklegram is convolved by a matched filter (from the last estimate of the object) and by a low pass filter (from atmospheric estimate). The resulting impulses, which correspond to the powers and locations of the speckles, are correlated with the speckles to yield the object shape.

up will produce a central nice image, albeit with a strong background due to the other speckles. Lynds *et al.*[8] proposed to cross-correlate each image with all the local maxima a_t, which represent the speckle locations. Christou *et al.*[9] improved by giving weights to all the maxima according to the speckle intensities. The estimate of the object, $\hat{o}(\mathbf{x})$, is the inverse transform of the weighted deconvolution:

$$\hat{o}(\mathbf{x}) = F^{-1}\left\{\sum_{t=1}^{n} i_t(\mathbf{u})a_t^*(\mathbf{u}) \bigg/ \sum_{t=1}^{n} |i_t(\mathbf{u})|^2\right\}. \tag{14}$$

Ribak[10] then showed that the best way to locate the speckle positions and intensities is by applying a filter matched to the shape of the expected object; iterations improve the shape of this filter. The general scheme is described in FIGURE 3. His model assumes that the impulses are an unknown point process (due to the random phases of the atmospheric principal Fourier components) to which is added a residual Poisson point process (due to the photon detection).

The bispectrum method and the improved shift-and-add method yield results with a dynamic range of one thousand. The simpler shift-and-add methods are widely used in infrared imaging, where the number of speckles is relatively small. However, most attention has now shifted to the real-time method of adaptive optics.

ADAPTIVE OPTICS

The major disadvantage of speckle imaging is the low signal-to-noise ratio. Because of the short integration times, dictated by the atmosphere, shot noise and read noise (in CCD detectors) limit severely the attainable dynamic range. This limitation is removed if the phases of the atmosphere are measured and corrected in

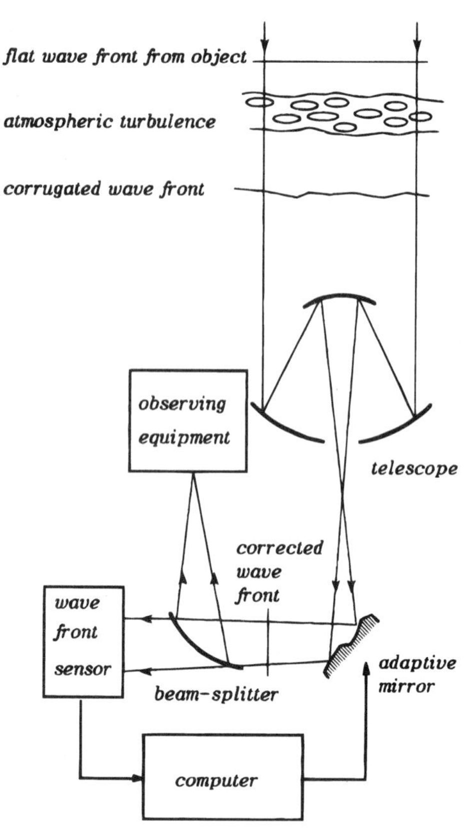

FIGURE 4. Adaptive optics: scheme. The wave front is sensed at a plane conjugate to the aperture *after* its correction by the flexible mirror.

real time, and the integration on the detector can proceed for much longer time. The atmospheric phases are measured by a wave front sensor, and are corrected by a flexible mirror (FIG. 4). The mirror is placed at a plane conjugate to most of the turbulence, while the scientific camera or spectrometer is at the reimaged telescope focus. The sensor only detects residual errors after these have been corrected by the mirror.

The flexible mirrors (FIG. 5) are usually made by a piezoelectric actuator, which translates directly voltages to displacement.[11] The mirror can be made of separate pieces (FIG. 5a) or a continuous thin sheet (FIG. 5b). Another option is to have it made of a thick piezoelectric material (FIG. 5c) where the electrodes are drilled into it. An electrostatic membrane can also serve as a mirror, with a set of electrodes to pull it and an opposite electrode for balance (FIG. 5d). One can also apply moments to bend the mirror (FIG. 5e). We extended this approach to the bimorph mirror (FIG. 5f) where the lever arms become the thickness of the piezoelectric material, much like a bimetallic strip.[11]

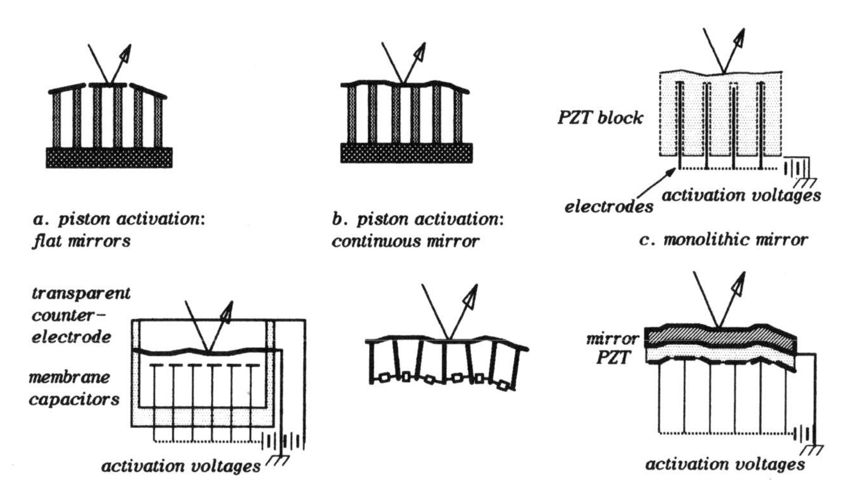

FIGURE 5. Flexible mirrors for adaptive optics. Except for (**d**, **e**), actuation is piezoelectric.

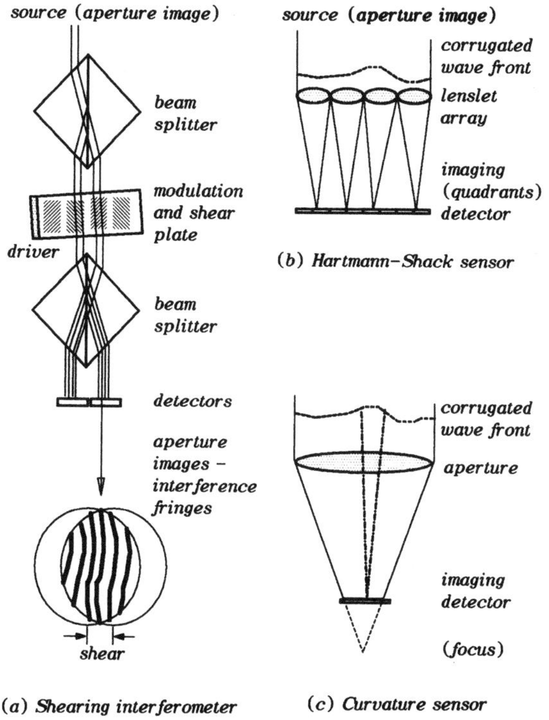

source (aperture image)
beam splitter
modulation and shear plate
driver
beam splitter
detectors
aperture images - interference fringes
shear
(*a*) *Shearing interferometer*

source (aperture image)
corrugated wave front
lenslet array
imaging (quadrants) detector
(*b*) *Hartmann-Shack sensor*

corrugated wave front
aperture
imaging detector
(focus)
(*c*) *Curvature sensor*

FIGURE 6. Wave front sensors. (**a**) Fringes between the sheared apertures are modulated by bulk water waves. Their phase is proportional to the lateral *slope*. (**b**) Focal points move with sub-aperture wave front *slope*. (**c**) More or less light is concentrated at out-of-focus plane according to wave front *Laplacian*.

In adaptive optics, three kinds of wave front sensors are usually used[12] (FIG. 6). A shearing interferometer[13] compares the wave front to its sheared version (FIG. 6a). The result is the derivative of the wave front in the shear direction; an orthogonal measurement is necessary. A Hartmann–Shack sensor splits the wave front to many small pieces by a lenslet array; the average slope over each piece is proportional to the two-dimensional shift of the focal spot (FIG. 6b). A curvature sensor measures the next derivative of the wave front: positive curvature concentrates light into the inside-of-focus image; negative curvature dilutes it (FIG. 6c).

One of the most difficult tasks of adaptive optics is the control loop. The main problems are: (a) a very weak input because of the limited number of photons from the object within each correlation area and time; (b) a very short time to process the input and transfer commands to the mirror; and (3) mismatch between the geometries of the wave front sensor and mirror and between these and the atmospheric turbulence modes. The latter problems are of lesser importance.

A solution to the issue of the weak photon flux is to measure the wave front of the star itself in the visible regime and correct in the near infrared regime. The errors are correlated with the former ones but are reduced by a factor of $\lambda^{5/3}$. When the star is too weak, a reference beacon can be created in the high atmosphere by laser scattering.[14] This is almost always a necessity because of the shortage of true reference stars; even asteroids[15] are not sufficient. A few laser beacons or a laser fringe pattern even allow tomography of the turbulence,[16] with a resultant wide field of view of a few arc-minutes. Some knowledge about high-elevation turbulence can separately be derived from scintillation: the (log of the) intensity pattern is a Laplacian of the high wave front.[17] Another way to partly overcome all these problems is to use our knowledge of the stochastic behavior of the atmosphere.

THE STOCHASTIC ATMOSPHERE

Very early studies of turbulence have indicated that it has a chaotic behavior. Jorgensen[18] proposed that the statistics of the image centroid motion in the focal plane of a telescope could also be chaotic (notice that he did not use the brightest speckle as a trace). He analyzed two data sets and found that indeed it was possible to characterize them using an attractor of dimensions 6.1–6.5 or 5.5–5.7, according to the set. Unfortunately, the number of points he had used (approximately 7000) was quite limited. Still, the finding allows prediction of motion using a neural network.[19]

Does chaos have to be revoked? Would linear prediction suffice? We do know a lot about image motion from Kolmogorov statistics, which were verified experimentally [see, e.g., (**7**)]. The wave front phase structure function (**6**) behaves like $D_\varphi(r) \propto (r/r_0)^{5/3}$ and the spatial power spectrum like $P_\varphi(k) \propto k^{-11/3}$. This reminded us[20] of the fractional Brownian motion (FBm): a Gaussian stochastic process $B_h(t)$ with incremental variance proportional to some power of the time

$$\langle [B_h(T+t) - B_h(T)]^2 \rangle \propto t^{2H}; \quad 0 < H < 1, \tag{15}$$

where H is the Hurst parameter, and $H = 0.5$ represents a classical Brownian motion. From dimensional analysis $P_B(f) \propto f^{-(2H+E)}$, where E is the topological dimension. The fractal dimension of the wave front surface is $F = E + 1 - H$, a FBm with $H = 5/6$ and $F = 13/6$. Since the turbulence is usually moving in front of the telescope faster than internal changes can take place, we call this a Taylor or "frozen flow": a cut in the two-dimensional phase screen. Then $F = 13/6 - 1 = 7/6$; $H = 5/6$, and the spatial power spectrum converts to a temporal one, $P_\varphi(t) \propto t^{-8/3}$.

Simulation and prediction are the two implications from this identification with a fractal. For simulation of wave fronts, various algorithms for fractal creation were tested for this purpose.[20,21] However, a much more important implication is prediction: if the shape of the wave front is known at any time inside the telescope, and if it is indeed fractal, then its values can be predicted for points outside the telescope (or inside the central obscuration). Because this is a three-dimensional process (planar and temporal) a prediction can be given for the wave front for the next time step. The "frozen flow" prediction which assumes that the wave front is just shifted in front of the aperture is thus "thawed" to account for its finer evolution. If we examine, e.g., the normalized temporal correlation of past and future we get

$$\frac{\langle -B_h(-t)B_h(t)\rangle}{\langle [B_h(t)]^2\rangle} = 2^{2H-1} - 1. \tag{16}$$

If $H = 0.5$ then the process is uncorrelated. If $H < 0.5$ the process is anti-persistent and if $H > 0.5$ it is persistent, with positive correlation between past and future without dependence on time. A linear estimator for the phase will in general be a combination of past and neighboring phases

$$\tilde{\varphi}(x, y, t) = \sum_{ijk} r_{ijk}\, \varphi(x + i\Delta x, y + j\Delta y, t - k\Delta t), \tag{17}$$

where the values of the coefficients r_{ijk} are sought. For simplicity, in the temporal prediction case $\tilde{\varphi}(t) = \sum r_i \varphi(t - i\Delta t)$ and we wish to minimize the mean square error

$$\begin{aligned} \langle \varepsilon^2 \rangle &= \langle |\varphi(t) - \tilde{\varphi}(t)|^2 \rangle \\ &= \Gamma(0) + \sum_i^N r_i^2 \Gamma(0) + 2\sum_i^N r_i \Gamma(i\Delta t) + 2\sum_{i>j}^N r_i r_j \Gamma(|i-j|\Delta t), \end{aligned} \tag{18}$$

where $\Gamma(\tau) \equiv \langle \varphi(t)\varphi(t-\tau)\rangle = \Gamma(0) - \frac{1}{2}C\tau^2 H$ is the correlation function and $C\tau^2 H$ is the structure function. Since $\Gamma(0) \to \infty$ for an ideal FBm we use two methods. In the first we require normalization: $\sum r_i = 1$. Minimizing with respect to each r_i we get[20]

TABLE 1. Coefficients for Linear Prediction Using Former Temporal Information

Order	Prediction Coefficients				Residual Error	
N	r_1	r_2	r_3	r_4	$\langle \varepsilon^2 \rangle$	Comments
1	1				$c\Delta t^{5/3}$	Simple time lag
2	1.587 40	−0.587 40			$0.654\,960 c\Delta t^{5/3}$	Two-point prediction
3	1.479 46	−0.344 348	0.153 114		$0.639\,605 c\Delta t^{5/3}$	
4	1.479 42	−0.384 921	0.023 322 7	−0.117 827	$0.630\,725 c\Delta t^{5/3}$	

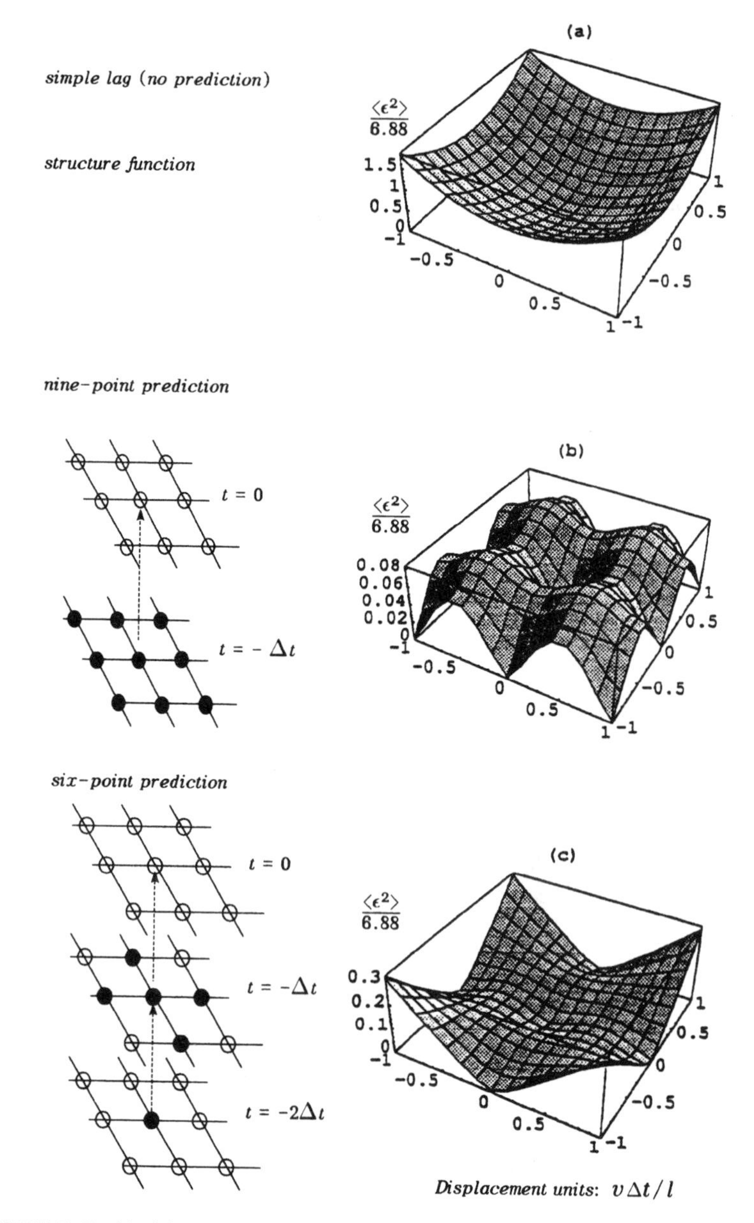

FIGURE 7. Residual (mean square) error across a unit cell without prediction (**a**) equals the structure function. When using past and neighboring information (**b**, **c**) the error drops significantly.

TABLE 1. (In a simple temporal extrapolation $N = 2$, $r_i = \{2, -1\}$.) The improvement drops fast beyond $N = 2$.

In the second method (which can also extend to the spatio-temporal case) we use two models for the spectrum: (a) a cut-on at an outer scale $P_\phi^{(co)}(k) = C_1 k^{-11/3}$, where $k \geq k_0$ (and 0 below), or (b) a Von Kármán spectrum $P_\phi^{(VK)}(k) = C_2(k^2 + k_0^2)^{-11/3}$, where $k \geq 0$. The normalized correlation is $\gamma(r) \equiv \Gamma(r)/\Gamma(0) \simeq 1 - 1.864(k_0 r)^{5/3} + 1.25(k_0 r)^2 + O[k_0 r]^{11/3}$ for the cut-on model and a very close $\gamma(r) \simeq 1 - 1.864(k_0 r)^{5/3} + 1.5(k_0 r)^2 + O[k_0 r]^{11/3}$ for the Von Kármán model. We can also neglect the second-order term: $k_0 r < 10^{-2}$ ($k_0 \sim 0.1$ m^{-1}, $r \sim 0.1$ m). The only parameters in these models are the outer scale k_0 and Fried's parameter r_0 or the wind speed $\mathbf{v}$ (Taylor flow is $\mathbf{r} = \mathbf{v}\Delta t$) which can be estimated from the model, measured, or optimized for.[22]

This method[20] gives for the two-point temporal estimator exactly the same results as derived from the normalization condition above. In FIGURE 7, the drop in the mean square error $\langle \varepsilon^2 \rangle$ is very clear with a nine-point spatio-temporal estimator (same point and its eight neighbors in the former time step, FIG. 7b) or a six-point predictor (same point, one and two steps ago, as well as four neighbors one step ago, FIG. 7c).

Lloyd-Hart and McGuire[23] showed that prediction of real data can be improved even more by using a neural net. This is probably due to their method of using the centroid to track at low signal, instead of the brightest speckle[24,25] (as in shift-and-add, see above). Also, Poisson noise is not accounted for in the model above and can lead to nonlinearities. Aitken and McGaughey[26] tested this possibility by looking at true data and running a rescaled-range analysis. In this analysis a time series x_i is divided into N segments of length d. The cumulative departure from the segment mean is

$$X_n(u) = \frac{1}{d} \sum_{i=nd}^{nd+n} x_i - u \frac{1}{d} \sum_{i=nd}^{nd+n} x_i . \tag{19}$$

The segment range is its deviation from a ramp

$$R_n(d) = [X_n(u)]_{max} - [X_n(u)]_{min} . \tag{20}$$

This range is now scaled by the segment standard deviation $S_n(d)$. The rescaled range is dependent on the sector length d as

$$\frac{R}{S} = cd^J . \tag{21}$$

If $J < 0.5$ the process is anti-persistent, and if $J > 0.5$ it is persistent. It follows from the data that $J \approx 1$ for the wave front itself, being a non-stationary process. The wave front slopes are stationary over a period shorter than $T_p < 0.22$ s, and within this range J converges to the Hurst parameter for this process, $H = 5/6$. This can be explained by the fact that the slopes are measured as an average over the aperture.[26] Hence the process is more persistent than it should be; at even higher frequencies the slopes are indeed anti-persistent. Moreover, the typical time T_p is consistent with the wind speed and the aperture size.[26]

ACKNOWLEDGMENTS

Many thanks are due to Y. Baharav, G. Baum, S. G. Lipson, C. Schwartz and J. Shamir at the Technion; J. R. P. Angel, J. C. Christou, E. K. Hege, and M. Lloyd-Hart at Steward Observatory; E. Gendron and P. Lena at Meudon Observatory; and J. B. Breckinridge at the Jet Propulsion Laboratory.

REFERENCES

1. COULMAN, C. E. 1985. Ann. Rev. Astron. Astrophys. **23:** 19.
2. TATARSKII, V. I. 1971. The Effects of the Turbulent Atmosphere on Wave Propagation. Israel Program for Scientific Translations. Jerusalem.
3. GOODMAN, J. W. 1985. Statistical Optics. Wiley. New York.
4. RODDIER, F. 1981. *In* Progress in Optics XIX. E. Wolf, Ed.: 281. North-Holland. Amsterdam.
5. COHEN, L. 1997. Time-frequency spatial-spatial frequency representations. This volume.
6. WEIGELT, G. P. & B. WIRNITZER. 1983. Opt. Lett. **8:** 389.
7. BATES, R. H. T. & F. W. CADY. 1980. Opt. Comm. **32:** 389.
8. LYNDS, C. R., S. P. WORDEN & J. W. HARVEY. 1976. Astrophys. J. **207:** 174.
9. CHRISTOU, J. C., E. K. HEGE, J. D. FREEMAN & E. RIBAK. 1986. J. Opt. Soc. Am. A **3:** 204.
10. RIBAK, E. 1986. J. Opt. Soc. Am. A **3:** 2069.
11. RIBAK, E. 1993. *In* NATO Advanced Study Institute Summer School on Adaptive Optics for Astronomy. J.-M. Mariotti & D. Alloin, Eds. NATO Advanced Study Series C **423:** 149. Kluwer. Dordrecht.
12. ROUSSET, G. 1993. *In* NATO Advanced Study Institute Summer School on Adaptive Optics for Astronomy. J.-M. Mariotti & D. Alloin, Eds. NATO Advanced Study Series C **423:** 115. Kluwer. Dordrecht.
13. RIBAK, E., E. LEIBOWITZ & E. K. HEGE. 1985. Appl. Optics **24:** 3088; also **24:** 3094.
14. FRIEDMAN, H. W. 1993. *In* NATO Advanced Study Institute Summer School on Adaptive Optics for Astronomy. J.-M. Mariotti & D. Alloin, Eds. NATO Advanced Study Series C **423:** 175. Kluwer. Dordrecht.
15. RIBAK, E. & F. RIGAUT. 1994. Astron. Astrophys. **289:** L47.
16. BAHARAV, Y., E. RIBAK & J. SHAMIR. 1994. Opt. Lett. **14:** 222.
17. RIBAK, E., E. GERSHNIK & M. CHESELKA. 1996. Opt. Lett. **21:** 435.
18. JORGENSEN, M. B., G. J. M. AITKEN & E. K. HEGE. 1991. Opt. Lett. **16:** 84.
19. JORGENSEN, M. B. & G. J. M. AITKEN. 1992. Opt. Lett. **17:** 466.
20. SCHWARTZ, C., E. RIBAK & G. BAUM. 1994. J. Opt. Soc. Am. A **11:** 444.
21. LANE, R. G., A. GLINDEMANN & J. C. DAINTY. 1992. Waves Random Media **2**: 209.
22. GENDRON, E. & P. LENA. 1995. Meudon Observatory, Paris. Private communication.
23. LLOYD-HART, M. & P. MCGUIRE. 1995. *In* Optical Society of America Meeting on Adaptive Optics. F. Merkle, Ed. Munich.
24. CHRISTOU, J. C. 1991. Pub. Astron. Soc. Pac. **103:** 1040.
25. GLINDEMANN, A. 1994. J. Opt. Soc. Am. A **11:** 1370.
26. AITKEN, G. J. M. & D. MCGAUGHEY. 1995. *In* Optical Society of America Meeting on Adaptive Optics. F. Merkle, Ed. Munich.

Dark Matter Image Reconstruction: Parametric and Nonparametric Statistical Arclet Inversion

J. A. TYSON

Bell Laboratories
Lucent Technologies
Murray Hill, New Jersey 07974

INTRODUCTION

Gravitational lens distortion of background galaxies enables calibrated measurements of the distribution of dark matter in the universe and the process of structure formation. This phenomenon is most naturally divided into two broad classes: weak lensing and strong lensing. In weak lensing the gravitational deflection angles are very small and single sources produce single (but distorted) images. In strong lensing, sources appear highly distorted and can form multiple images. Whether a given source is weakly or strongly lensed depends on the impact parameter of the ray: whether its image appears outside or inside the critical "Einstein" radius.

The large number of potential sources in the wide area outside the critical radius of a foreground lens offers the possibility of statistical tomographic reconstruction of the mass distribution in the outer parts of the lens. This weak gravitational lensing provides a direct measure of mass overdensity on large scales (several kpc to tens of Mpc, depending on the distance and compactness of the lens). Through comparison with large N-body simulations for various cosmogonies, this new window on mass in the universe constrains the nature of dark matter. For example, light neutrino hot dark matter clusters strongly at late times on scales larger than a Mpc but free-streams on smaller scales, producing mass overdensities with soft cores of hundreds of kpc. In weak lensing it is necessary to average over the apparent orientation of tens of sources for each resolution element: weak lens statistical inversion of thousands of "arclets" (distorted background galaxies) over a wide field has finite resolution. A number of nonparametric algorithms have been developed for inverting the arclets, and regularized approaches tend to avoid edge effects and systematics near the strong lens regime.

Strong lensing offers the possibility of constraining the mass distribution in the parts of the lens which exceed the critical density for image splitting. In cases where multiple images of a source are created by the lens, the details of the position and distortion of these sub-images are very sensitive to the projected two-dimensional mass distribution in the lens. Parametric models of the lens mass can then be used to first unlens the sub-images to get an image of the source,[1] and then ray trace light from this source past the lens in an iterative fit in the image plane for the lens mass parameters.

In practice, the mass distribution is derived from thousands of weak lensing arclets found in deep multicolor CCD images. Calibration of the mass scale can be done both through simulations and via observation using the same camera/telescope of a mass standard. Realistic simulations of the whole source–lens–atmosphere–detector process, including multiple background galaxy redshift shells, masses for individual cluster galaxies, dark matter lens model, atmospheric seeing, and pixel sampling and sky shot noise, are performed. "Blank" field HST Medium Deep Survey imaging data, together with seeing deconvolved ground-based data, are used to derive the source galaxy angular scales. Strong lensing forms an independent check on weak lensing mass scale calibration. This chapter outlines these nonlinear mass reconstruction techniques which have been used in weak and strong lensing, and then mentions work in progress to combine the two techniques—in lenses which exceed the critical density—into a single nonlinear parametric regularized reconstruction.

THE SOURCES

Since many of the faint galaxies are resolved and are distributed up to high redshifts they may be employed in statistical gravitational lens studies of foreground mass distributions. For mass mapping by statistical gravitational lens inversion, the sources must meet several requirements: The sources must have (1) redshifts large compared with the lens, (2) a number density on the sky sufficient to sample the lens on relevant scales, (3) an intrinsic angular diameter larger than the ratio of seeing FWHM to the magnification of the lens, and (4) other properties (blue color and very low surface brightness) enabling efficient separation of the sources from lens and other foreground objects.

The extreme blue colors of the faint galaxies suggest one is seeing starburst galaxies at redshifts of 1–2, so that their UV excess flux is redshifted into the blue. In a redshift-magnitude plot the trend to redshift ≈ 1 at 25th B magnitude is clear. A typical galaxy seen at $z = 1$ may be a 0.1 L^* galaxy, so a survey at 25th magnitude would cover a wide range in redshifts extending from 1 to 3. For arclet inversion of $z < 0.3$ lenses, the lack of detailed redshift data for each of these sources produces less than a 10% mass scale error.

Galaxies fainter than 26 B magnitude also cover a wide range of angular scales; their average exponential scale lengths are 0.2–0.5 arcsec and typical half-light diameters are one arcsec.[2]

LENS MASS FROM ARCS

Traditional methods of estimating total mass are indirect: mass estimators based on dynamics or X-ray flux maps involve models of orbits or the state of the hot gas, leading to potential systematics in the derived gravitational potential. The path of a photon from a distant source is bent as it passes by the foreground mass (gravitational lens), making the source appear at an altered position.[3] This light-bending is accompanied by another first-order effect: systematic image distortions. If the background source image is resolved then this image stretching is observable. A galaxy of angular size 1 arcsec may appear to be moved by an angle $\beta = 4GM(r)/rc^2$ and distorted into an arc many arcseconds long by sufficient foreground mass $M(r)$ interior to projected radius r. The scattering time is small compared with the period of orbiting test particles, thus avoiding orbit assumptions.

The light deflection is proportional to the mass in the lens and is about 2 arcsec for a typical galaxy.[4] If ϕ is the intrinsic source position in the absence of the lens, the source appears at position θ:

$$\vec{\phi} = \vec{\theta} - \vec{\delta}(\vec{\theta}), \tag{1}$$

$$\vec{\delta}(\vec{\theta}) = (D_{LS}/D_S)\vec{\beta}(\vec{r}),$$

where r is the impact parameter for the ray at the lens, and D_S and D_{LS} are the observer-source and lens-source angular diameter distances. For a thin lens the light bending angle $\vec{\beta}$ is given in terms of the projected two-dimensional mass density Σ:

$$\vec{\beta} = (4G/c^2)\int d^2u\Sigma(\vec{u})(\vec{r} - \vec{u})/|\vec{r} - \vec{u}|^2. \tag{2}$$

We would like to invert this relation and solve for $\Sigma(x, y)$. Since we do not know how the sources were distributed on the sky prior to lensing, this deflection itself is not observable. But its gradient, the shear, is directly observable. In the simple case of a point mass, $\beta = 4GM/rc^2$, and a source exactly behind the mass appears as an "Einstein ring" image of radius $\theta_E = (M/10^{11}\ M_\odot)^{1/2}\ (D/10^6\ \text{kpc})^{-1/2}$ arcsec, where $D = D_L D_S/D_{LS}$. If the lens mass is elliptical in shape or has multiple components this circular ring symmetry is broken, so that in nature circular rings are rare. If the source angle ϕ is less than θ_E then multiple images of the source are formed. Galaxy lenses ($M \approx 10^{12}\ M_\odot$) will produce multiple images with separations ≈ 3 arcsec, while galaxy cluster lenses ($M \approx 10^{14}\ M_\odot$) can produce image separations of 1 arcmin. In the case of an isothermal distribution of gravitationally bound masses of line-of-sight velocity dispersion σ_v, $\beta(r) = 4\pi(\sigma_v/c)^2$, $\theta_E = 29(\sigma_v/10^3\ \text{km s}^{-1})^2 \cdot (D_{LS}/D_S)$ arcsec. Many clusters have measured velocity dispersions $\approx 10^3$ km s^{-1}. More realistic mass distributions, such as a soft core isothermal sphere, may be calculated[5] by substituting the corresponding density $\Sigma(r)$.

TOMOGRAPHIC RECONSTRUCTION OF GALAXY CLUSTER MASS

If the source angle relative to the lens ϕ is larger than θ_E, multiple images of the source will not be formed, but the image of the source will be elongated. The huge mass associated with clusters of galaxies distorts all the background galaxies many arcminutes from the cluster. Foreground galaxy clusters at redshifts 0.2–0.5 with radial velocity dispersions above 700 km s^{-1} have sufficient mass density at significantly distorted background galaxies of redshift greater than 0.4–1. Lensing preserves the surface brightness and spectrum of the source, so that arcs tend to have the very faint surface brightness and blue color of the faint blue galaxies.

INVERSE PROBLEM: MASS FROM IMAGE DISTORTION

This gravitational lens distortion is quantified using the intensity-weighted second moment of the galaxy image orthogonal and along the radius relative to the lens center.[6] A dimensionless scalar alignment T, calculated from these principal axis transformed source ellipticities, is related to the projected mass density clumping and is defined at each point ($\vec{r}$) in the image plane via the (r, θ) principal-axis transformed second moments of the background galaxy image g:

$$T(\vec{r}) = \frac{i_{\theta\theta} - i_{rr}}{i_{\theta\theta} + i_{rr}} = \frac{2\gamma(\vec{r})[1 - \kappa(\vec{r})]}{[1 - \kappa^2(\vec{r}) + \gamma^2(\vec{r})]} \tag{3}$$

where the convergence $\kappa(r) = \Sigma(r)/\Sigma_c$ and the shear[7] $\gamma(r) = [\bar{\Sigma}(r) - \Sigma(r)]/\Sigma_c$, and where Σ_c is the critical surface mass density, related to the distance ratio: $\Sigma_c = c^2/(4\pi GD)$. The distance ratio for a foreground-background pair is[3]

$$D = \frac{(1 - q_o - d_1 d_2)(d_1 - d_2)}{(1 - q_o - d_2)(1 - d_2)(1 + z_{fg})}, \tag{4}$$

where $d_1 = \sqrt{1 + q_o z_{fg}}$ and $d_2 = \sqrt{1 + q_o z_{bg}}$. By averaging over many different galaxies we are averaging D over the distributions of z_{fg} and z_{bg}.

Introducing a galaxy light distribution prior, the tangential second moments are

$$i_{\theta\theta} = M_{20} \sin^2 \phi + M_{02} \cos^2 \phi - 2M_{11} \sin \phi \cos \phi \tag{5}$$

$$i_{rr} = M_{20} \cos^2 \phi + M_{02} \sin^2 \phi + 2M_{11} \sin \phi \cos \phi,$$

where ϕ is the position angle of the vector from the point (x, y) to the background galaxy, relative to the x-axis. The intensity-weighted second moment $M_{lm,\,i}$ of background galaxy g is defined by

$$M_{lm,\,g} = M_{0,\,g}^{-1} \int (\delta x)^l (\delta y)^m W(\delta x, \delta y)[I_g(\delta x, \delta y) - I_0]\, dx\, dy, \tag{6}$$

$$M_{0,\,g} = \int W(\delta x, \delta y)[I_g(\delta x, \delta y) - I_0]\, dx\, dy,$$

where $\delta x = (x - \langle x \rangle_g)$ and similarly for δy, the sky intensity near this background galaxy is given by I_0, and optimal normalized Gaussian weights $W(\delta x, \delta y, I(x, y))$ are calculated from the half-luminosity radius. A randomly placed unlensed population of galaxies randomly oriented will give a net distortion $T(\vec{r})$ of zero at every point in the image plane, while a population of lensed galaxies will give a positive value at the point corresponding to the lens center.

GALAXY-SCALE MASS FROM ARCLETS

How massive is an average galaxy? Multi-band, high spatial resolution Hubble Deep Field data—despite the small size of the field—provide an opportunity to control systematics and bring the signal out of the noise. Systematic image distortion of the distant galaxies about the positions of the brighter galaxies can be compared with simulations and internal tests to give a reliable mass relation. Using full three-dimensional simulations, as well as other HST and ground-based data, one can calibrate the lensing distortion signal and derive a best-fit radial mass relation for the foreground galaxies. The lens distortions for every galaxy in each redshift shell are computed using the galaxies in all the shells in front of it.

Recently, we have used these techniques to obtain calibrated observations of the average mass of galaxies with $22 < I < 25$ in the Hubble Deep Field.[8] The mean mass profile is obtained from the statistical gravitational lens distortion of faint blue background galaxies, based on over 2000 foreground–background pairs. The observed lensing distortion is calibrated via full simulations and other HST and ground-based data on fields with massive clusters in the foreground. All galaxies are modelled as soft core isothermal distributions,[9] but with an outer mass cutoff, giving for the spherical case a surface density distribution outside the core:

$$\Sigma = \Sigma_o(\beta/2r)(1 + r^2/r_{outer}^2)^{-1}, \tag{7}$$

where $\Sigma_o = \sigma_v^2/2G$, σ_v is the line-of-sight velocity dispersion, r_{outer} is the outer mass cutoff, and $\beta = (1 + r_{core}^2/r_{outer}^2)$.

For an isothermal mass distribution, we find a distortion $T(r)$ of 0.06 at a radius of 2 arcsec ($\sim$8 h^{-1} kpc). For such an average galaxy we find an average mass interior to 20 h^{-1} kpc of 6.2×10^{11} h^{-1} $M_\odot$.

CLUSTER MASS PROFILES AND MAPS FROM ARCLETS

The projected mass density about any point in the sky is related to the curl of the source ellipticity about that point. To construct an image of the gravitational lens projected mass distribution, the distortion statistic $T(\vec{r})$ may be computed over a grid of positions as candidate lens centers. In the weak lensing limit it can be shown[7,10] that the tangential alignment T is a measure of the mass contrast:

$$T(\vec{r}) = 2[\Sigma_{av}(<r) - \Sigma(r)]\Sigma_c^{-1}. \tag{8}$$

At any point $\vec{r}$ in the image plane we can sum over the tangential alignment of all source images about that point, creating a continuous scalar distortion statistic D:

$$D(\vec{r}) = \int K(\vec{u})T(\vec{r} - \vec{u})\, d\vec{u}, \tag{9}$$

where the apodization kernel $K(\vec{u})$ weights source images at large radius less, and is generally of the form $K(\vec{u}) = (u^2 + u_0^2)^{-1}$. For D to be simply related to the mass,[11] $K(s)$ must asymptotically approach the power law s^{-2} at large s. Since light bending angles from different mass components add, the distortion D at any point should be related to the mass density contrast, if the mass can be represented as a sum of cylindrically symmetric distributions. The solution to this inverse problem for the contrast of the projected lens mass density Σ is then given by a simple integral of the shear over radius from the lens center. The average projected mass density interior to radius r is given by

$$\bar{\Sigma}(r) = \Sigma_c\, C(r)B(r) \int_r^{r_{max}} T(r)\, d\ln r + \bar{\Sigma}(r, r_{max}), \tag{10}$$

where $C(r)$ is the seeing correction obtained via simulations, $B(r) = (1 - r^2/r_{max}^2)^{-1}$, and $\bar{\Sigma}(r, r_{max})$ is the average density in the annulus between r and r_{max}. For a sufficiently large field $\bar{\Sigma}(r, r_{max})$ is small compared with peak density. A radial plot of the projected surface mass density found using **(10)** is shown in FIGURE 1 for $z_l = 0.3$, $\sigma_v = 1000$ km s^{-1} simulations with two different mass profiles.

The distortion image $D(x, y)$ uniquely locates the lens mass, obtains $M(<r)/r$, and gives its morphological shape on the sky. A useful check of this procedure uses

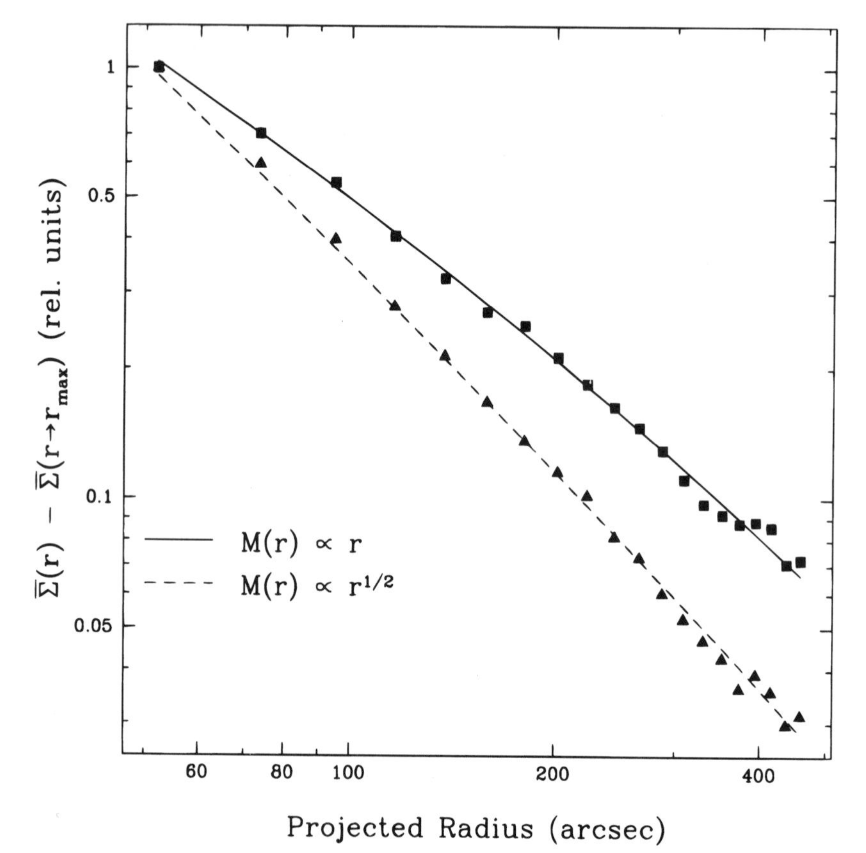

FIGURE 1. A radial plot of the projected mass density contrast obtained by inversion of the arclets in two 29 B mag arcsec^{-2} noisy simulations with $\sigma_v = 1000$ km s^{-1}. *Solid lines*: input mass density.

the mass map (9) as input to a Bayesian search in model mass distribution space, solving for the dark matter mass and core size. For the known source ellipticity distribution, a given source redshift distribution, and a test radial mass distribution model, a maximum likelihood calculation yields the lens M/r (or equivalent velocity dispersion) and core radius. From the inversion of 6000 arclets surrounding the rich $z = 0.18$ cluster A1689, we found a steeper than isothermal profile beyond 300 h^{-1} kpc radius.[12] Some clusters at high redshift are nearly as compact in mass. In rich compact clusters mass appears to trace the cluster red light, on scales greater than 100 h^{-1} kpc, with rest-frame V band mass-to-light ratios of a few hundred h in solar units. The mass core radius is smaller than some observed X-ray core radii in nearby clusters, suggesting that the X-ray gas may be less relaxed dynamically than the dark matter. At least for rich clusters these lens studies appear to confirm the large mass which was implied by virial calculations using velocity dispersions. Comparing the mass morphologies found near clusters with high resolution N-body simulations for various cosmogonies will narrow the candidates for dark matter.

PARAMETRIC MASS MODELS

Several cases of multiple mass clumps on 300 h^{-1} kpc scales have been found in our survey of clusters. However, the resolution of mass maps made from nonparametric arclet inversion is limited by the scale of the smoothing required. The scale of the sub-clumping of the dark matter within clusters of galaxies is also interesting.

In cases where portions of the lens exceed critical density, producing multiple images of some sources, it is possible to obtain higher resolution in the mass map by combining weak lens inversion at large radius with the strong lensing constraints in the inner region. One approach is to use a parametric lens model in which multiple mass components are parametrized by their centroid, mass profile, and ellipticity. An iterative approach, regularized by the weak lens inversion solution for the mass at large radius, is then used to obtain a unique solution for the lens mass map at high resolution.

This has the desirable property that the resulting mass map has high resolution in regions of high mass density. In practice we have found that a reliable way of converging to the solution is to demand that the strong lensed sources, when unlensed by the estimated mass distribution in the lens, reconstruct to an identical single image (see reference 1), and that this image of the source reconstructs to the observed arcs in the image plane when lensed by this same mass. It is necessary to use the lens *mass* distribution, rather than the traditional approach of starting with a two-dimensional lens potential, as noncircular lens potentials can be unphysical in their corresponding mass reconstructions.

With my colleagues Ian Dell'Antonio and Greg Kochanski, we have developed a code for this weak-strong mass reconstruction and applied it to complex lens simulations as well as deep HST imaging of the $z = 0.4$ galaxy cluster CL 0024 + 17. The projected mass density contrast at large radius is obtained from 1200 arclets, excluding the bright long arcs, in a 40 square arcmin field of the KPNO 4-meter CCD. The mass density contrast at intermediate radius (up to 2 arcmin) is obtained from weak lens inversion of 350 arclets found in the HST deep images. Finally, the mass map in the inner region is reconstructed via the procedure outlined above.

LARGE-SCALE DARK MATTER

Larger scale applications of this dark matter mapping may eventually find clumped dark matter unrelated to galaxies or clusters of galaxies. If $\Omega \approx 1$, dark matter will fill the Universe and will likely exist in places where there is no current star formation activity. Larger scale applications of this dark matter mapping may eventually find clumped dark matter unrelated to galaxies or clusters of galaxies.

If we wish to extract only information on the statistics of the foreground mass overdensities, rather than map them, there is a tensor statistic analogous to the scalar two-point correlation function.[13–15] All pairs of galaxies separated by some angle on the sky are summed separately in bins of mutual orientation. This integral tensor statistic, called the orientation correlation function (OCF), is then built from these summed orientation data. The OCF has both parallel and orthogonal orientation components, each a function of angle on the sky. The OCF detects an excess over random for background galaxies to be oriented parallel or orthogonal to one another. In this way much larger areas of the sky may be covered and a smaller amplitude statistical mass fluctuation spectrum may be detected.

The observational challenge in OCF measurement is systematic errors due to variations across the detector of the point spread function, optics distortion, and atmospheric refraction. These systematics can be overcome by chopping and trailing techniques, and calibration in rich star fields. Mosaics of CCDs make such a large-scale search for coherent alignment in the distant faint galaxies particularly attractive, and dark matter on angular scales up to degrees can in principle be studied in this way. Preliminary measurements of the arclet orientation correlation function in random fields are below the CDM prediction. With Gary Bernstein at the University of Michigan, we have built a large area blue-sensitive CCD mosaic camera for the 4-meter Blanco telescope at CTIO. This should enable a more definitive measure of the OCF corresponding to moderate redshift dark matter overdensities.

DARK MATTER TELESCOPE

A definitive measure of the overdensity spectrum of the dark matter on 10 Mpc scales is technically within reach. Deep imaging of hundreds of thousands of distant galaxies at several wavelengths—to get statistical redshifts from their colors—repeated in many different directions, would have the precision to detect 10% of the statistical orientation correlation predicted by standard CDM theory. An accurate OCF measurement in 10 two-degree areas of the sky—to get the cosmic variance—

would provide a direct measure of Ω and the mass fluctuation spectral index. With a dedicated 3- or 4-meter "Dark Matter Telescope", an advanced 8000×8000 pixel mosaic of CCDs covering over half a degree, and deep imaging in four wavelength bands, this project would take five years to complete.

ACKNOWLEDGMENTS

My collaborators in this research are Gary Bernstein, Ian Dell'Antonio, Phil Fischer, Raja Guhathakurta, Greg Kochanski, George Rhee, Jordi Miralda-Escudé, Wes Colley, and Ed Turner.

REFERENCES

1. COLLEY, W. N., J. A. TYSON & E. L. TURNER. 1996. Astrophys. J. Lett. **461**: 83.
2. IM, M., S. CASERTANO, R. E. GRIFFITHS, K. U. RATNATUNGA & J. A. TYSON. 1995. Astrophys. J. **441**: 494.
3. BLANDFORD, R. D. & R. NARAYAN. 1992. Annu. Rev. Astron. Astrophys. **30**: 311.
4. TURNER, E. L., J. P. OSTRIKER & J. R. GOTT. 1984. Astrophys. J. **284**: 1.
5. HINSHAW, G. & L. KRAUSS. 1987. Astrophys. J. **320**: 468.
6. TYSON, J. A., F. VALDES & R. A. WENK. 1990. Astrophys. J. **349**: L1.
7. MIRALDA-ESCUDÉ, J. 1991. Astrophys. J. **370**: 1.
8. DELL'ANTONIO, I. & J. A. TYSON. 1996. Astrophys. J. Lett. **473**: L17.
9. GROSSMAN, S. A. & R. NARAYAN. 1988. Astrophys. J. Lett. **324**: 37.
10. TYSON, J. A., F. VALDES, J. F. JARVIS & A. P. MILLS. 1984. Astrophys. J. Lett. **281**: 59.
11. KAISER, N. & G. SQUIRES. 1993. Astrophys. J. **404**: 441.
12. TYSON, J. A. & P. FISCHER. 1995. Astrophys. J. Lett. **446**: 55.
13. VALDES, F., J. A. TYSON & J. F. JARVIS. 1983. Astrophys. J. **271**: 431.
14. MIRALDA-ESCUDÉ, J. 1991. Astrophys. J. **380**: 1.
15. BLANDFORD, R. D., A. B. SAUST, T. BRAINERD & J. V. VILLUMSEN. 1991. Mon. Not. Roy. Astron. Soc. **251**: 600.

Neural Networks for the Analysis of Stellar Time Series

J. PERDANG[a,b]

Institute of Astronomy
Madingley Road
Cambridge CB3 OHA, United Kingdom

INTRODUCTION

For a broad category of variable stars a detailed theoretical knowledge of the physical mechanisms underlying their variability remains wanting. Prominent instances are the Cataclysmic Variables and the Wolf–Rayet stars. On the other hand, for many variable objects individual observed time series $y(t)$, $t = 0, 1, 2, \ldots,$ show extended gaps; in the specific case of variable Wolf–Rayet stars the cycle length is of the order of days, so that gaps covering large fractions of the cycle become unavoidable. The purpose of this chapter is to develop a method of *reconstructing* such time series on a particular type of Neural Network (NN). Our method has the formal advantage of avoiding assumptions about the character of the variability of the objects under investigation.

The simulations we obtain in this fashion serve the practical purpose of (1) filling in the gaps, and of (2) making predictions on the variability outside the observational range; they are also expected to (3) eliminate at least part of the noise component in the observed signal.

Currently standard methods of interpolation and extrapolation based on local[1,2] or on more global phase space reconstructions[3] rely on a specific assumption on the dynamics responsible for the observed series, namely, on the property of *reducibility to a low order differential system.* In the absence of a precise knowledge of the actual physics of the variability, it becomes impossible to test whether this assumption holds or not. On the other hand, an obvious bonus of this approach is that whenever the assumption applies, the time series supplies direct information on the physical dynamics.

The method adopted in this chapter is predicated on a weaker requirement on the mathematical nature of the time series, namely, essentially on the assumption that a value $y(t)$ at epoch t is determined by the values $y(t')$ at epochs $t' \neq t$ (a property referred to as *determinacy* of the function $y(t)$). It does not attempt to reproduce the dynamics of the actual physics, so that it cannot provide any direct clues on the equations describing the variability of the star. This may be viewed as a substantial drawback as compared to phase space reconstruction methods. It does

[a]Permanent address: Institut d'Astrophysique, 5 Avenue de Cointe, Liège B-4000, Belgium.
[b]Grants of two Royal Society–FNRS European Exchange Fellowships for the years 1994 and 1995 were instrumental for the realization of this work. An FRFC (Belgium) grant also provided partial support.

have the advantage, however, of avoiding a suggestion of a—perhaps—inadequate, or even physically spurious dynamics: in fact, in the presence of insufficient theoretical information on the mechanisms of the variability, there remains a risk that phase space methods when applied to *real*, i.e., noisy observational time series, just reconstruct a *fictitious dynamics* (which under favorable circumstances may be a smoothed version of the true dynamics, but which might as well suggest an inaccurate physical picture of the variability). Within the precision determined by the noise level, such a fictitious dynamics may indeed be perfectly capable of simulating the time series over a limited observational window.

The aim of this work is to illustrate that NNs may perform the task of interpolation and extrapolation in a more direct and entirely *automated* way, without requiring an excessive amount of computation. On the other hand, and equally important, we believe that the NN approach as operated here has the virtue of filtering out the *random* components in the signal (i.e., noise), without forcing the time series to conform to a theoretically biased reference pattern (such as a standard or multiple Fourier series, a phase space of preselected structure, etc.).

ARTIFICIAL NEURAL NETWORKS: INTERNAL DYNAMICS AND ASYMPTOTIC BEHAVIOR

Artificial NNs are schematic models of the brain introduced half a century ago by McCulloch and Pitts[4,5] (see also reference 6). To the extent that they are satisfactory models, they simulate the main properties of the brain, in particular the ability of identifying and *learning* patterns and regularities in (noise corrupted) data, and in applying and *generalizing* the acquired regularities outside the restricted range over which they have been isolated.

The relevance of NNs as efficient automatic devices for analyzing physical data has been appreciated since the middle of the last decade only (presumably as a consequence of the introduction of convenient learning algorithms; cf. references 7–9); in the astronomical literature NNs have made their appearance by the end of the last decade, mainly in the form of computational systems adapted to the purposes of classification or to the technical task of monitoring observations (cf. references 10–12). Although this is not the place for a discussion of anatomical, physiological and biochemical details of the real brain, the general architectural features, and the functioning of artificial NNs, would best be elucidated by an analysis of the brain itself (cf. references 13 and 14; for a more popular account see also reference 15). We believe indeed that a better knowledge of the general properties of the latter may suggest improvements in the design, as well as extensions in the use and applications of the NNs. In fact the Learning Algorithm we propose in this paper is an attempt at duplicating more closely the learning processes taking place in the human brain.

The carriers of information of the brain, the nerve cells or neurons, are simulated in the NN by artificial neurons or nodes (the analogues of the "cells" in a Cellular Automation, CA); exchange of information between real neurons is regulated in the brain by the synaptic junctions or synapses, which play the parts of (adjustable) switches; the latter are modeled by artificial synapses, or links of adjustable transmissions (the analogues of the local evolution rules in a CA). According to a proposal by Hebb,[16] the synapses are modified as a consequence of "learning"; or in the artificial NN the synaptic efficiencies are model parameters which are adjusted in accordance with the information available.

The Internal Network Dynamics

The original model of the brain by McCulloch and Pitts[4] essentially remains the prototype of any artificial NN. It consists in a collection of interacting artificial neurons conforming to the following technical assumptions which describe the *internal dynamics* of the model:

1. The artificial neuron, or node j, $j = 1, 2, \ldots,$ is a two-state automaton whose state $F(t; j)$ at time t is represented by a binary integer ($F(t; j) = 0$: inactive or silent neuron; $F(t; j) = 1$: active or firing neuron).
2. The state $F(t; j)$ of node j at the current timestep t is fully determined by the states $F(t-1; k)$, of nodes k, $k = 1, 2, \ldots,$ respectively at the previous timestep $t-1$; state $F(t; j)$ thus becomes a Boolean function of $F(t-1; k)$; the latter defines a first-order difference equation

$$F(t; j) = R^{(2)}(\Phi(t-1; j)); \tag{1}$$

$$\Phi(t; j) = \sum_{k=1}^{N} J_{jk} F(t; k) - \theta(j). \tag{1a}$$

The function $\Phi(t; j)$ (firing function), defined over $(-\infty, +\infty)$, is an analytic representation of the contribution of the action of nodes k, $k = 1, 2, \ldots,$ on node j (similar to the analytic representation of CA rules); the (nonlinear) function $R^{(2)}(x)$, (synaptic emulation[17]), is required to transform the values of the firing function into permissible node values (analogue of the modulo 2 operation in some CA rules). For further reference we introduce a more general synaptic emulation as follows. If $R(x)$ stands for the unit ramp

$$R(x) = 0 \text{ if } x \leq 0; \quad R(x) = x \text{ if } 0 \leq x \leq 1; \quad R(x) = 1 \text{ if } x \geq 1 \tag{2}$$

then $R^{(z)}(x)$ represents the discrete-valued rounded form of $R(x)$, defined over $\{0, 1/(z-1), 2/(z-1), \ldots, (z-2)/(z-1), 1\}$, (staircase approximation to $R(x)$, with z levels of uniform spacing $1/(z-1)$); in particular $R^{(2)}(x) = 0$ or 1 if $x < 1/2$ or $\geq 1/2$ respectively. The weight coefficients J_{jk}, reals of $(-\infty, +\infty)$, parametrize the synaptic efficiencies. The shift parameter $\theta(j)$ specifies a threshold for the activation of node j.

A convenient visualization of the dynamical behavior of the NN model is provided by a *structure graph* whose vertices are the nodes, and whose oriented arcs represent the artificial synapses; the flow of information follows the oriented arcs. In particular, if the graph shows cycles carrying no outgoing vertices, then these cycles are periodic *attractors*. Nodes of zero out-valency define *fixed points*.

A second more complete visualization of the internal dynamics is provided by a *phase graph* whose vertices are the states of the global network, i.e., the ordered collections of all node states, or points in the phase space of the network; the arcs of the phase space picture the allowed transitions between global network states. Again, the *attractors* and *fixed points* of the NN dynamics are represented by cycles, respectively vertices (without outgoing arcs).

More general NNs are obtained by modifying assumptions (1) and (2).

In the first place, the two-state neurons may be replaced by z-state neurons, z, any integer ≥ 2, so that the node state F then takes any value $0, 1, \ldots, z-1$. It is more convenient to work with a normalized node state $F' = F/(z-1)$ with allowed values $0, 1/(z-1), 2/(z-1), \ldots, (z-2)/(z-1), 1$; in the limiting case $z \to \infty$ the normalized node states densely cover the bounded interval $[0, 1]$. The corresponding synaptic emulation to be used is $R^{(z)}$, or R in the limit $z \to \infty$.

Next, several further technical generalizations lead to a greater flexibility in the dynamics: (a) different delays, of $1, 2, \ldots, d$ timesteps, instead of the single unit delay of the dynamics (**1**) may be adopted;[18] (b) direct nonlinearities may be introduced in the firing function; (c) a variety of different sigmoidal functions may be chosen for the synaptic emulation.[17]

Noise and Phase Transitions

While the original McCulloch–Pitts model is *deterministic*, the actual occurrence of "noise" in the real brain calls for a *probabilistic* modification. A formalism first proposed by Little[19] in the context of binary state neurons consists in substituting to the difference equation (**1**) of the dynamics a probability law

$$P[F(t;j); \Phi(t-1;j)] = \frac{1}{1 + \exp\left[-\beta\sigma(t;j)\Phi(t-1;j)\right]},$$

$$\sigma(t;j) \equiv 2F(t;j) - 1; \tag{3}$$

here $\beta\ (= 1/T)$ is a thermodynamic temperature parameter; σ is a spin variable ($= 1$ or -1, if $F = 1$ or 0 respectively); expression (**3**) gives the probability for node j to take on the binary state value F at the current NN timestep, for a given firing function Φ (**1a**) (at the previous timestep). This stochastic modification of the dynamics can be implemented directly into the original difference equation (**1**) of the McCulloch–Pitts dynamics through the substitution

$$\Phi(t;j) \to \Phi'(t;j) = \Phi(t;j) + \nu(j), \tag{4}$$

where $\nu(j)$ is a stochastic noise term of zero average, whose distribution law is consistent with the probability law (**3**) (cf. reference 13). The NN temperature T is a global measure of the noise level in the network. In the limit $T \to 0$, the network becomes noise-free; we then recover the deterministic McCulloch–Pitts scheme. As T increases, the random contribution to the firing function leads to a more and more unpredictable time–behavior. With the introduction of a temperature, stochastic NNs lend themselves to a systematic analysis in the context of thermodynamics and statistical mechanics (cf. references 13 and 17). The Fermi–Dirac type probability distribution (**3**) indicates in particular that the expression $\sigma\Phi$ is to be interpreted as an energy (namely a spin energy) attached with a given node, so that the stochastic NN is to be interpreted as a spin system. But in the case of particular spin systems (the Ising model in two and higher dimensions) an *order–disorder phase transition* is observed beyond a critical Curie temperature (cf. references 20 and 21). Therefore, in stochastic NNs the counterpart of this phenomenon must occur as well (cf. reference 22), at least in the case of NNs whose architecture, i.e., the NN structure graph, approximates to the lattice structure of these Ising models. With hindsight the existence of these phase transitions is perhaps made even more obvious from a direct inspection of the structure graph and the phase graph, which indicates that a variety of *phases* attached with the attractors of the NN internal dynamics can occur.

If we start with the limit of zero noise–temperature, then the network is deterministic. Given an initial NN state-configuration (states of all nodes are given well-defined values), the dynamics (**1, 1a**) implies that at any later NN time t the state-configuration is uniquely defined. A unique asymptotic limit for t large enough is guaranteed for any initial condition only if all attractors of the dynamics are fixed points, $F_1, F_2, \ldots$, in the NN phase space (i.e., vertices of the phase graph). In the presence of other types of attractor, $C_1, C_2, \ldots$, which are necessarily limit cycles, of periods $P > 1$, in any actually realizable NN (P-cycles of the *phase graph*) there exist initial NN state-configurations producing large time solutions which explore the whole attractor C_1, or the attractor C_2, etc. This alternative can manifestly arise only if the *structure graph* possesses cycles. Therefore, the existence of a unique asymptotic limit is secured for an NN whose architecture and structure graph have no cycles.

Increase now the NN temperature slightly; then the attractors $F_1, F_2, \ldots$ or $C_1, C_2, \ldots$ of the noise-free case transform into *fuzzy* sets: a solution which remained confined to the fixed point F_1, or the limit cycle C_1 at $T = 0$ may now asymptotically explore global states near these attractors; barring special circumstances (illustrated by the one-dimensional Ising model[21]), the average asymptotic behavior (statistical average over a large number of replicas, or time average) will then remain close to the 0-temperature asymptotics. The probabilistic description (**3**) of the transitions of the node states does not alter the structure graph (determined by the set of allowed synaptic weights J_{jk}). However, it does change the phase graph by adding new arcs to the latter: in the probabilistic case ($T > 0$), instead of the unique arc leaving a vertex under the deterministic alternative ($T = 0$), we generally have several leaving arcs, with associated transition probabilities adding up to 1; the addition of arcs is constrained by the consistency requirement with the structure graph. In principle, the modified phase graph may then exhibit cycles, even though the original phase graph, in the absence of noise, had no cycles. The attractor for a given initial global NN state, $A(T)$, depending on the noise temperature, now becomes a set of either connected or disconnected vertices of the phase graph; it is no more restricted to be either a fixed point or a limit cycle. This attractor has the status of a fuzzy set. Typically, at low enough temperatures $A(T)$ will contain a single attractor of the noise-free dynamics, say C_1 $[F_1]$; the asymptotic solution will then be found in the latter set C_1 $[F_1]$ with high probability; in the limit $T \to 0$, $A(T)$ collapses into C_1 $[F_1]$.

While for a small enough noise level the fuzzy attractor $A(T)$ contains a *single* attractor of the noise-free case, the NN architecture may be such that a critical noise level $T_{\rm crit}$ exists such that for $T > T_{\rm crit}$ the attractor $A(T)$ contains *several* noise-free attractors F_i and C_j. At this critical noise temperature we then observe a *phase transition*, namely a "jump" from a relatively ordered phase of the low temperature regime (corresponding to a low level of fluctuations of the global NN state) towards a more disordered phase in the high temperature regime (higher level of fluctuations).

It is clear that such transitions may occur at several critical temperatures. In particular, provided that the structure graph is consistent with sufficiently coupled vertices of the phase graph it may happen that beyond a critical temperature $T_{\rm disorder}$ the fuzzy attractor becomes the collection of *all* vertices of the phase graph (cf. the Ising spin models); in this case, *any* initial global NN state generates the *same* asymptotic global states; i.e., we have a situation of complete disorder.

This qualitative discussion should make it clear that the occurrence and the specific nature of noise-induced phase transitions in NNs are ultimately linked up with the geometry of the structure graph. Provided that the latter has a simple enough geometry, avoiding in particular cycles, we can eliminate phase transitions altogether in the NN behavior.

TASK-ADAPTED NEURAL NETWORKS

The analysis of a physical problem by an NN consists in entering a collection of "questions" (counterpart of the biological *stimuli*); the NN then provides a corresponding set of "answers" (counterpart of the biological *responses*) which are the results of a processing through the internal NN dynamics. Proper answers require that the NN parameters (synaptic weights J, threshold parameters θ) have been adjusted to the specific category of questions we are dealing with: the artificial brain must have previously "learned" the class of questions and answers it is expected to handle.

In analogy with the physiology of the real brain, a particular subset of neurons is specialized in receiving the questions; the latter are referred to as the input nodes; the collection of these nodes form the input layer. A second subset of neurons is specialized in supplying the answers; these are the output nodes which form the output layer. The remaining nodes are the hidden nodes.

More formally, the physical problem is formulated in terms of a finite collection of independent variables $x = (x^1, x^2, \ldots, x^d)$, defined over a space X^d, and a second finite collection of dependent variables $V = (V^1, V^2, \ldots, V^m)$; in an experimental situation the values of V are observed and measured, at given values of x. For our purposes, what is regarded as "dependent" and "independent" variables is largely a matter of the physical setup we are analyzing. The transformation

$$x \mapsto V = V(x) \tag{5}$$

is assumed to be an injection; the variables V are required to depend on x alone (i.e., no other parameter influencing V is allowed to vary within the context of the problem we are dealing with). If we now view the values of the independent variables x as the "questions", and the values of the function $V(x)$ as the corresponding "answers", then the task of *simulating the function* $V(x)$ is formally solved, in principle, on the artificial NN by feeding the values x into the input layer; the corresponding functional values $V(x)$ are then recovered on the output layer.

On a practical level, in order to construct an adequate NN functioning as an actually workable simulation device, several problems have to be resolved.

(1) In the first place the general NN operates as a dynamical system. With x playing essentially a role of initial condition, and $V(x)$ representing essentially the *asymptotic* solution of the dynamics, we have to make sure that

  (a) the asymptotic state is reached in a reasonably short time;
  (b) by virtue of our assumptions on $V(x)$, *any* initial condition x must generate a *unique* asymptotic solution, i.e., $V(x)$ must be a *fixed point* of the internal NN dynamics.

  To be accurate, we mention that the precise NN initial condition and the asymptotic solution refer to the global NN state, and not just to the states of a subset of the nodes (input layer and output layer); it is clear, however, that we can always adopt the convention that the full NN initial state is defined by setting the state function of the hidden and output nodes all equal to 0; the input nodes carry the information x (in a form we have to specify); then the information $V(x)$ is asymptotically carried by the output layer (again in a form to be specified).

(2) In the second place, to conform to an actual observational situation, it is desirable that the NN analysis can take care of observational noise. As mentioned, noise is naturally simulated in the stochastic versions of NNs ((**4**) which holds for z-state nodes). Noise generates *fuzziness* in the network response, in the sense that the transformation (**5**) ceases to be an injection; we have now $x \mapsto V = V(x)$ with probability $p(V;\ x;\ T)$ where the probability distribution of output values V depends on the input values x, as well as on the noise level specified by the NN temperature T. In order to be practically useful the NN must remain discriminatory enough, i.e., the NN must stay in the *ordered phase*; this requires the noise temperature to remain below the Curie temperature.

Perceptrons

Both these requirements are obeyed if we select the subclass of NNs known in the literature as L-Layer Feedforward Non-Recurrent Networks or Perceptrons.

In their most rudimentary form introduced by Rosenblatt in the late 1950s[23,24] (cf. also references 16 and 17), perceptrons are made of two layers of nodes, namely, an input and an output layer. Information can flow "forward" only, i.e., from the first (input) layer to the last (output) layer; no interaction between nodes of the same layer is allowed. More generally, in L-layer perceptrons neurons are arranged in L (≥ 2) layers; the $L - 2$ intermediate hidden layers are the carriers of the hidden nodes; as in the two-layer case, information flows "forward" only, from layer r to layer $r + 1$, $r = 1, 2, \ldots, L - 1$. The physical variables ξ the perceptron is dealing with (input variables $x^1, x^2, \ldots, x^d$; output variables $V^1, V^2, \ldots, V^m$) are *bounded* and *discrete*. We shall indicate by $D(\xi)$ the total number of discrete states of variable ξ. The input layer, carrying $x^1, x^2, \ldots, x^d$, is partitioned into d segments containing $l_1, l_2, \ldots, l_d$ nodes respectively; segment s, of length l_s, encodes the variable x^s, $s = 1, 2, \ldots, d$. In a similar way, the output layer, carrying $V^1, V^2, \ldots, V^m$, is segmented into m parts containing $\lambda_1, \lambda_2, \ldots, \lambda_m$ nodes respectively; segment v, of length λ_v, encodes the variable V^v, $v = 1, 2, \ldots, m$. The layers are labelled by the index $r = 1, 2, \ldots, L-1, L$; layer r contains N_r nodes identified by the ordered pair (r, k), $k = 1, 2, \ldots, N_r$. The state of a node (r, k), at timestep t of the perceptron dynamics, is represented, as above, by a z-ary integer, written here $F(t; r, k) \in \{0, 1, 2, \ldots, z(r, k) - 1\}$; the corresponding normalized state is

$$F'(t; r, k) = F(t; r, k)/[z(r, k) - 1]. \tag{6}$$

In the perceptrons we are concerned with all nodes of a same segment s of the input layer are defined in the same basis, $z(s)$; similarly we express all nodes of the same segment v of the output layer in a same basis, $z_L(v)$; for all nodes of the hidden layers a same basis, z_H, is chosen.

The independent variable x^s, $s = 1, 2, \ldots, d$, carried by segment s of the input layer, is represented by a string of length l_s of $z(s)$-ary digits. Given the precision with which we know x^s, we can estimate the total number of states this variable must possess, $D(x^s)$; therefore

$$D(x^s) \leq z(s)^{l_s} \quad \text{or} \quad l_{s\text{opt}} = \text{int}\,[\ln D(x^s)/\ln z(s)]; \tag{7}$$

the length $l_{s\text{opt}}$ of segment s provides an optimal encoding in the sense that the number of unused states among the potentially available discrete states is minimum (int: integer part). The basis $z(s)$ is a free parameter.

For the encoding of the dependent variable V^v the counterpart of (7) is

$$D(V^v) \leq z_L(v)^{\lambda_v} \quad \text{or} \quad \lambda_{v\text{opt}} = \text{int}\,[\ln D(V^v)/\ln z_L(v)], \tag{8}$$

fixing an optimal choice for the length of segment v; again, $z_L(v)$ is free.

The firing function describing the interactions among the neurons of a perceptron takes the form

$$\Phi(t; r, k) = s(r, k) \sum_{k'=1}^{N_{r-1}} F'(t; r-1, k') J_{r-1k'k} - \theta(r; k); \tag{9}$$

the notations are self-explanatory; the pair of labels (r, k) refers to a single node; $J_{r-1k'k}$ denotes the weight from node k' of layer $r - 1$ to node k of layer r; the summation takes care of the special perceptron architecture. Without loss we restrict the synaptic weights to the interval

$$-1 \leq J_{r-1k'k} \leq +1, \quad r = 2, \ldots, L; \quad k' = 1, 2, \ldots, N_{r-1}; \quad k = 1, 2, \ldots, N_r; \tag{10}$$

these parameters are coded as d_J-digit z_J-ary rationals, or else as reals; the extra parameter $s(r, k) > 0$ rescales the weights J to the physical range; the scaling and the threshold $\theta(r; k)$ are likewise either d_s-digit z_s-ary rationals and d_θ-digit z_θ-ary rationals, respectively, or reals.

It is now quite obvious that with this specification information flows from an input node to an output node in exactly $L - 1$ timesteps; the asymptotic state is thus reached in $L - 1$ timesteps. Moreover, information arriving at an output node cannot leave it, so that the perceptron output corresponds to a fixed point behavior. Hence we satisfy requirements (1a, b).

On the other hand, the phase graph of a perceptron can have no cycles; accordingly phase transitions cannot occur in this class of NNs (requirement 2).

We designate a perceptron by the symbol

$$P[l_1\{z(1)\}, l_2\{z(2)\}, \ldots, l_d\{z(d)\}; N_2, N_3, \ldots, N_{L-1}, \{z_H\}; \lambda_1\{z_L(1)\}, \lambda_2\{z_L(2)\}, \ldots, \lambda_m\{z_L(m)\}] \quad \mathbf{(11)}$$

summarizing the overall architecture, the segmentation of the input and output layers, and the number bases for each type of node (indicated in curly brackets). The behavior of the perceptron is completely defined if besides the symbol (**11**) we specify

(a) the synaptic emulation μ (for our purposes the function R, (**2**), and its variants);
(b) the representation of the synaptic weights, of the scaling, and of the translation parameters (base-z integers, reals).
(c) For the purposes of the learning process of the perceptron we must further specify the level of precision ε to which the data set $V(x)$ is given, i.e., the noise level in the data to be analyzed; in turn, this level fixes the maximum precision to which it remains meaningful to "learn" the data. Moreover, if noise is simulated in the perceptron, a measure of the NN internal noise level must be given (NN internal temperature T).

The Perceptron Internal Dynamics

For perceptrons we can explicitly solve the dynamics, in the sense that we can express the states of the output layer at NN time $(L - 1)$ as a function of the states of the input layer at NN time 0. To this end we associate with each layer $r = 1, 2, \ldots, L$ an N_r-dimensional normalized (row) state vector $\mathbf{F}_r$ whose components are the rescaled node states of that layer

$$\mathbf{F}_r = \left(\frac{F(r, 1)}{z(r, 1) - 1}, \frac{F(r, 2)}{z(r, 2) - 1}, \ldots, \frac{F(r, N_r)}{z(r, N_r) - 1}\right). \quad \mathbf{(12)}$$

We further associate with each layer r N_r-dimensional synaptic-weight column vectors $\mathbf{J}_{rk}$ whose components represent the collection of synaptic weights of that layer r acting on node k in layer $r + 1$

$$\mathbf{J}_{rk} = \begin{pmatrix} J_{r1k} \\ J_{r2k} \\ \cdots \\ J_{rN_rk} \end{pmatrix}, \; r = 1, 2, \ldots, L - 1; \quad k = 1, 2, \ldots, N_{r+1}. \quad \mathbf{(13)}$$

Arranging the collection of the N_{r+1} vectors $\mathbf{J}_{r1}, \mathbf{J}_{r2}, \ldots, \mathbf{J}_{rN_{r+1}}$ in a row we obtain $L - 1$ $N_r \times N_{r+1}$-matrices of synaptic weights attached to the "acting" layers $r = 1, 2, \ldots, L - 1$

$$J_r = (\mathbf{J}_{r1}, \mathbf{J}_{r2}, \ldots, \mathbf{J}_{rN_{r+1}}),\ r = 1, 2, \ldots, L - 1. \tag{14}$$

Likewise we collect together the translation and scaling parameters attached to the "affected" layer ρ in (row) vectors

$$\theta_\rho = (\theta(\rho, 1), \theta(\rho, 2), \ldots, \theta(\rho, N_\rho)), \tag{15}$$

$$\mathbf{s}_\rho = (s(\rho, 1), s(\rho, 2), \ldots, s(\rho, N_\rho)),\ \rho = 2, 3, \ldots, L. \tag{16}$$

We then obtain the dynamics (**1, 1a, 9**) in compact form

$$\mathbf{F}_r(t + 1) = \mu(\mathbf{s}_r \circ (\mathbf{F}_{r-1}(t) \bullet J_{r-1}) - \theta_r),\ r = 2, 3, \ldots, L. \tag{17}$$

In this expression the notation $\mathbf{a} \circ \mathbf{b}$ stands for the term-by-term product of two arrays $\mathbf{a}$, $\mathbf{b}$ of same dimension, producing an array of same dimension whose component $k = 1, 2, \ldots$ is the product of components k of $\mathbf{a}$ and $\mathbf{b}$; the dot $\bullet$ stands as usual for a scalar product; μ is a formal synaptic emulation (in our case R, cf. (**2**), rounded so as to be consistent with the definition of the left-hand side of the equation). On iteration of relation (**17**) we see that if at NN time t the state of the input layer is $\mathbf{F}_1(t)$, then the state of the output layer at NN timestep $t + L - 1$, $\mathbf{F}_L(t + L - 1)$ is given by

$$\begin{aligned}\mathbf{F}_L(t + L - 1) = \mu[\mathbf{s}_L \circ [\mu[\mathbf{s}_{L-1} \circ [\mu[\mathbf{s}_{L-2} \circ [\ldots \\ \ldots(\mu(\mathbf{s}_3 \circ (\mu(\mathbf{s}_2 \circ (\mathbf{F}_1(t) \bullet J_1) - \theta_2) \bullet J_2) - \theta_3)) \bullet \\ \ldots\ldots] \bullet J_{L-2}] - \theta_{L-1}] \bullet J_{L-1}] - \theta_L]].\end{aligned} \tag{18}$$

From this expression the main properties of the $x \mapsto V$ relationship can be read off: *nonlinearity* of this relation is seen to be the direct outcome of the nonlinear operations of the synaptic emulation (truncation and rounding). In an indirect way these nonlinearities are due to the finite number bases z adopted in the data representation; highest nonlinear behavior is forced on this relation through the choice of low enough number bases.

Since observational data tables are given in discrete form, over a grid of discrete values of the independent variables, the dependent variables may undergo jumps from one grid point to a neighbor point at the observationally available level of resolution; on the level of resolution we are working the dependent variables experience *discontinuities*, even though on a finer level of resolution continuity in the independent variables is reestablished. An observed velocity curve $y(t)$ of a variable star, with four or five observations per cycle, is an instance of this occurrence; the velocity change, $y(t + 1) - y(t)$, from one epoch, t, to the next, $t + 1$, is here of the order of $y(t)$ itself, so that the observed behavior appears as discontinuous. In this case we know that the real time behavior of this physical variable is continuous. However, from the available finite (and in practice always short) data series continuity cannot be inferred from the observations alone.

It then seems desirable that an NN designed to simulate experimental relations of type (**5**) should not *a priori* incorporate smoothness of these relations. A lack of smoothness in a function $V(x)$ on a given scale can always be regarded as an effect of nonlinear but smooth behavior on a finer scale. Therefore, our remark on nonlinearity is applicable to non-smoothness as well: we allow and favor non-smooth representations by selecting small number bases for the state functions of the nodes.

Alternatively, in the presence of a large amount of *noise* the NN may serve the purpose of a noise filter. In fact, it is well known that a perceptron can be viewed as a generalized version of a Wiener filter (cf. reference 12). Under such operating conditions it may be useful to force smoothness on the perceptron response. A comment on this point will be made below.

Finally, internal noise is explicitly dealt with in (**18**) through the substitution $\theta \to \theta + \nu$ in the components of the translation vectors (**15**) (ν, a stochastic noise term, cf. (**4**)).

THE NN TREATMENT OF TIME SERIES

The discussion of the previous section refers to arbitrary functional representations (**5**) on a perceptron. In our particular NN treatment of a time series $y(t)$, $t = 0, 1, \ldots, T - 1$, the observational epoch t (not to be confused with the time of the internal NN dynamics) plays the part of the independent variable x; the series itself $y(t)$ is identified, in our simplest scheme, with the dependent variable $V(x)$ (see a comment in the next section on a more general representation). The input layer of the perceptron thus carries a single variable, the epoch t, which for reasons sketched above is represented by a binary integer; the number of nodes N_1 required for this layer is given by (7) ($z(s) = 2$, $l_{sopt} = N_1$, $D = T$).

The output layer, in our simplest scheme, likewise carries a single variable, the time series $y(t)$; we identify the latter with the state function of a single node (so that we need $N_L = 1$ node only in the output layer); from (**8**) the basis z_L of this node is fixed by the number of states of the observational variable y; typically we choose $z_L \approx 125$.

The selection of a single node for the output was motivated by preliminary exploratory numerical experiments with different numbers of output nodes. We find that with more than three output nodes the signal simulation tends to get locked up in an approximation to a devil's staircase function; this unwanted effect is avoided with a single node.

In all of our experiments the NN internal parameters (synaptic weights, scaling and shift coefficients) are represented as reals.

The treatment of time series on an NN as presented here has been suggested by the success of an algebraically similar question of nuclear physics dealt with by Clark.[25-27] Clark simulates the "mass surface" $\delta M(Z, N)$, (dependent variable V, (**5**)), of the nuclei as defined in terms of the atomic number Z and neutron number N (independent variables x), on a perceptron of the class discussed here; after a training in which the NN parameters are adjusted to reproduce the $(Z, N) \mapsto \delta M$ relation for about 1700 nuclides, the calibrated network is able to extrapolate the mass values for nearly 600 nuclei, with an accuracy essentially equal to the learning precision. In conformity with the underlying physics, Clark's network is deterministic.

The analogy we have exploited consists in the following: the *determinacy* of the mass surface, i.e., the actual existence of a map $(Z, N) \mapsto \delta M$, is physically guaranteed by the statistical mechanics of the nucleons; similarly, in the case of stellar time series, the determinacy of $y(t)$ (existence of a map $t \mapsto y(t)$) is guaranteed by the stellar fluid dynamics (plus initial conditions). The epoch in the time series thus plays a part comparable to (Z, N).

The currently popular NN treatment of time series is based on a different approach: at any fixed epoch t, a finite and typically small number d of independent variables, $x^1, x^2, \ldots, x^d$ are selected, which are identified with $y(t-1), y(t-2), \ldots, y(t-d)$; d is independent of the epoch; the value of the time series $y(t)$ at epoch t is then regarded as the single dependent variable ($m = 1$). With these identifications relation (**5**) acquires the status of a *difference equation* of order d. Physically, the latter is an algebraic translation of the physical *assumption* that the signal at time t, $y(t)$, is fully determined by the signal at the d previous timesteps $t-1, t-2, \ldots, t-d$.

In the terminology of the theory of dynamical systems, the phase space representative for the physics underlying the time series is of finite (and small) dimension d, whose possible attractors are therefore *fixed points*, and *tori* of dimension 1 (limit cycles, if $d \geq 2$), 2, ..., or possibly *strange attractors* of low fractal dimension (if $d \geq 3$). The implicit assumption on which this approach hinges may not be guaranteed in the case of time series of the less regular variable stars.

Emulating discretized differential equations of finite order on NNs, this method was proposed by Lapedes and Farber[28] (for further developments see references 29–32). As an interesting illustration we mention that Albano and coworkers[33] carried out NN analyses of the time series $y(t)$ generated by the logistic map and the Hénon map within this framework; for these systems the orders of the difference equations are thus known at the outset ($d = 1$ and $d = 2$ respectively). The experiments show that after training of the deterministic NN carried out on a noise corrupted signal, the trained NN ultimately produces a signal from which noise has been eliminated; the strange attractor can then be reproduced from the simulated signal. In other words, in this specific instance, the NN is found to be able to reconstruct from a noisy signal the "true" dynamics at the origin of the underlying noise-free signal.

However, the success of these experiments should not be overemphasized: with the exact theoretical d value being implemented, the internal dynamics of the perceptron has an algebraic structure consistent with the structure of the algebraic maps; it is therefore clear *a priori* that the result of the iteration *can* be reproduced on the perceptron. Noise is then necessarily eliminated: since noise has infinite dimension, it cannot be reproduced in this perceptron setup (a difference equation of low order d). The reason why in these cases the NN does work as an efficient noise filter is that the main element of the mathematical nature of the underlying dynamics, namely, the dimension d, is exactly known (cf. also comments in reference 34).

We wish to warn that the success of the approach in these artificial instances cannot be extrapolated to the application of the same methodology to real time series of variable stars. As a rule, precise information on the mathematical origin of what constitutes the "signal" is not available (in particular d unknown, possibly unbounded), and the nature and the degree of noise corruption are unknown; in the absence of closer theoretical information it is not clear *a priori* whether a given feature in the observed series is to be interpreted as "essential", i.e., part of the genuine signal, or "accidental" and to be discarded as "noise".

On the other hand, for the purposes of *short term* prediction and interpolation the method is probably acceptable in many cases. In fact, provided only that the signal is smooth enough, it can always be described locally by low order differential or difference equations.

In the astronomical context, the success of Murtagh and coworkers[35] in adapting the perceptron procedure to the (short-range) forecast of *astronomical seeing* is precisely related to this property. More recently, this group has introduced an NN architecture more complex than the perceptron architecture (NNs they refer to as Dynamically Recurrent Networks) which may support a more flexible dynamics;[36] such models suffer from the same implicit restrictive assumptions as the conventional perceptron arrangement, namely, the series $y(t)$ must be defined by a dynamics of low order.

The *sunspot cycle*, which has been a test ground for virtually any method of time series analysis proposed ever since Yule's introduction of autoregressions,[37] has also recently been examined in the framework of the standard perceptron technique;[38] the phase space reconstruction method has been applied to this time series by Kurths and Ruzmaikin.[39]

We now come back to the perceptron method of this chapter. The actual architecture we have adopted for our numerical experiments is that of a perceptron with a *single* hidden layer; according to the remarks made at the beginning of this section, this perceptron is characterized by the symbol

$$P[N_1\{2\};\ N_2\{2\};\ 1\{z_L\}]. \tag{19}$$

Formal discussions of the number of hidden nodes, and the number of hidden layers, required for exact functional approximations $V(x)$ are discussed elsewhere.[40,41] The more pragmatic point of view adopted here is based on the fact that the training set is available with a rather large observational error ε (in the case of irregular stellar variables). Our goal is to obtain a simulation over the training set consistent with the error range.

For completeness we list the input and output normalized state vectors, together with the two synaptic weight matrices

$$\mathbf{F}_1(t) = (F(t;\ 1,\ 1),\ F(t;\ 1,\ 2),\ \ldots,\ F(t;\ 1,\ N_1)),$$

$$\mathbf{F}_3(t) \equiv F_3(t) = F(t;\ 3,\ 1)/[z_L - 1],$$

$$\mathrm{J}_1 = \begin{pmatrix} J_{111} & J_{112} & J_{113} & \cdots & J_{11N_2} \\ J_{121} & J_{122} & J_{123} & \cdots & J_{12N_2} \\ J_{131} & J_{132} & J_{133} & \cdots & J_{13N_2} \\ \cdots & \cdots & \cdots & \cdots & \cdots \\ J_{1N_11} & J_{1N_12} & J_{1N_13} & \cdots & J_{1N_1N_2} \end{pmatrix} = \begin{pmatrix} \mathbf{J}_{11} \\ \mathbf{J}_{12} \\ \mathbf{J}_{13} \\ \cdots \\ \mathbf{J}_{1N_1} \end{pmatrix},$$

$$\mathbf{J}_2 = \begin{pmatrix} J_{211} \\ J_{221} \\ J_{231} \\ \cdots \\ J_{1N_21} \end{pmatrix} = \begin{pmatrix} J_{21} \\ J_{22} \\ J_{23} \\ \cdots \\ J_{1N_2} \end{pmatrix} = \mathbf{J}_2. \quad \mathbf{(20)}$$

Here $\mathbf{J}_{1i}$, $i = 1, 2, \ldots, N_1$ represents the ordered collection (row vector) of the synaptic weights from node (1, i) in the input layer to all nodes in the hidden layer, and the column vector $\mathbf{J}_2$ is the ordered collection of the synaptic weights from all nodes of the hidden layer to the output node. The collection of scaling parameters and translation parameters $\mathbf{s}_2$, θ_2 likewise define N_2 dimensional arrays, while s_3, θ_3 are scalars.

Omitting the dynamical arguments t and $t + L - 1$, **(18)** simplifies to

$$F_3 = R^{(z_L)}\left(s_3\left[R^{(2)}\left(\sum_{i=1}^{N_1} F(1, i)\mathbf{s}_2 \circ \mathbf{J}_{1i} - \theta_2\right)\right] \bullet \mathbf{J}_2 - \theta_3\right). \quad \mathbf{(21)}$$

We observe that the scalar product in this expression involves the N_2-dimensional array $\mathbf{F}_2$ (between square brackets) whose components are 0 or 1, and the array $\mathbf{J}_2$ whose components are reals in $[-1, 1]$. Hence any numerical value α of the argument of the ramp $R^{(z_L)}$ is of the form

$$\alpha = s_3 v - \theta_3 \quad \text{with} \quad v = \sum_{h \in \Lambda} J_{2h} \quad \mathbf{(22)}$$

(Λ, any subset of the labels $\{1, 2, \ldots, N_2\}$). Once the perceptron has been trained, the parameters s_3, θ_3 have fixed values; α then can take at most 2^{N_2} distinct values. Given an observational resolution of the series $y(t)$, the number of states of the output nodes, s_L, is determined; consistency then requires that

$$N_2 > N_{2min} = \ln z_L / \ln 2. \quad \mathbf{(23)}$$

This minimum number of nodes in the hidden layer is adopted in our program when we start the calculations. Relation **(21)** also enables us to analyze the sensitivity of the output to the input. From such a discussion we can derive order of magnitude estimates for the set of internal NN parameters.

THE DYNAMICS OF LEARNING: A DIFFUSION ALGORITHM

It remains to be seen how the Hebbian idea of learning,[16] namely, the progressive modification of the synaptic weights in accordance with the desired output for a given input, can be implemented efficiently in the perceptron **(20)** adapted to the study of time series.

Let $y = y(t)$ be the *actual time series* ($t \in \mathsf{K} \subset \mathsf{T} = \{0, 1, 2, \ldots, T - 1\}$, K, calibration subset). We require the series to be normalized, such as to have $0 < y(t) < 1$ (the bounds 0 and 1 being almost attained at some epochs), and discretized, with a step consistent with the observational precision. Let further $y_P(q) = y_P(t; q)$ be the *simulated time series* on perceptron P whose synaptic weights, scaling and translation parameters have been fixed (values q). The generation of this function on the perceptron automatically secures that it obeys $0 < y_P(t, q) < 1$. As a *measure of discrepancy*, $\delta(q)$, between the simulation $y_P(q)$ and the actual data y

$$\delta(q) \equiv \|y - y_P(q)\|, \tag{24}$$

we choose the following explicit form

$$\|y - y_P(q)\|^2 = \frac{1}{K} \sum_{t \in \mathsf{K}} [y(t) - y_P(t, q)]^2, \tag{24a}$$

with the summation going over the calibration set containing K data points. Let further ε be an estimate of the observational noise level in the data set, formally defined as

$$\varepsilon^2 = \frac{1}{K} \sum_{t \in \mathsf{K}} [y(t) - y_0(t)]^2, \tag{25}$$

where $y_0(t)$ is the ideal noise-free series. The perceptron response, $y_P(q)$, to the calibration epochs K will be a satisfying approximation to the actual time series, y, if we obey the inequality

$$\delta(q) < \varepsilon. \tag{26}$$

Accordingly, given the perceptron architecture and an actual time series, the formal method of training the perceptron to simulate this time series, under condition (**26**), then consists in *minimizing the discrepancy function* with respect to the collection of real-valued NN parameters q defined over a compact set Q of the finite-dimensional parameter space

$$\min_{q \in \mathsf{Q}} \|y - y_P(q)\| = \|y - y_P(q_m)\|. \tag{27}$$

Note that since the perceptron simulation $y_P(t)$ is discrete in the present formalism, the discrepancy function $\delta(q)$ is a step-function of the parameters exhibiting finite discontinuities; moreover, with our specifications of the perceptron the discrepancy obeys

$$\forall_q \in \mathsf{Q} : 0 \leq \delta(q) \leq 1. \tag{28}$$

Manifestly then the very existence of a subset $\mathsf{M} \subset \mathsf{Q}$—or of a class $[\mathsf{M}]$ of disjoint subsets of Q—over which the discrepancy takes its *absolute minimum* value is secured. Besides the absolute minimum, the discrepancy has generally a large number of *relative minima* realized over subsets M′, M″, M‴, ... Provided that there is a minimum set M* (absolute or relative) consistent with condition (**26**), then the time series y is *learnable* by the perceptron P; the training is completed if the array of parameters takes on values $q^* \in \mathsf{M}^*$.

A Dynamical System for Training: Back-Propagation

In a *continuous* formulation (all perceptron variables and parameters represented by reals) the search for a minimum in the discrepancy function (**24**) can be described as the motion of a particle of coordinates $q = (q^1, q^2, \ldots, q^k, \ldots)$ subject to a potential

$$U(q) = \frac{1}{2}\,\delta(q)^2 \tag{29}$$

in a medium Q of high enough "viscosity" η so that the acceleration can be neglected. The particle thus obeys the dynamical system

$$d/d\tau\; q = -\frac{1}{\eta}\,\partial_q\, U(q). \tag{30}$$

The *attractors* of this dynamical system are all *fixed points*, q_F, q'_F, q''_F, ... corresponding to minima of the potential. Hence starting with an initial condition q_o and integrating the dynamics, the asymptotic solution $q(\tau \to \infty) = q_m$ achieves a minimum of the discrepancy (**27**); if consistent with condition (**26**) the training of the perceptron for a given series $y(t)$ is completed. The learning procedure based on these remarks is known as the *Back-propagation Algorithm*[9] (cf. also references 13 and 22). In our case of a perceptron characterized by *discrete* variables and parameters, it is sufficient to discretize the dynamics (**30**) which is transformed into an iterative scheme. If the necessary precautions are taken (steps small enough, etc.), the above conclusion essentially continues to hold. For a given initial condition the long time solution yields a minimal discrepancy.

The natural iterative training algorithm which follows from the above remarks needs no comments. We should observe, however, that the dynamics may get stuck in a secondary minimum which does not satisfy condition (**26**); it is then necessary to restart with a different initial condition.

A Diffusion Dynamics for Learning

One conceivable method which does not share this drawback of getting stuck in an inadequate minimum is a Blind Search Algorithm, in which the configuration space Q is randomly explored, with uniform probability density. At each step, the "best" q value so far encountered is kept; it is clear that formally such a procedure eventually converges to the absolute minimum. On a practical level, however, such a method is inefficient since the parameter space Q, being of high dimension, cannot be sampled properly in any reasonable run (say $< 10^6$ trial steps).

An approach we found useful for the specific learning of time series consists in introducing an auxiliary (homotopic) family of time series, $y_\lambda(t)$, defined by the relation

$$y_\lambda(t) = H(t;\, y(t),\, Y(t);\, \lambda) \equiv (1-\lambda)Y(t) + \lambda y(t),\ \lambda \in [0, 1],$$
$$H(t;\, y(t),\, Y(t);\, 0) \equiv Y(t) \quad \text{and} \quad H(t;\, y(t),\, Y(t);\, 1) \equiv y(t). \tag{31}$$

In this expression $Y(t)$ is a reference time series, $Y(t) \equiv y_0(t)$, which is continuously deformed into the actual time series $y(t) \equiv y_1(t)$ as λ increases from 0 to 1.

The reference series $Y(t)$ is constructed such that we know *a priori* a subset $Q_o \subset Q$ such that for $q_o \in Q_o$ the perceptron response $y_P(t;\, q_o)$ is equal to $Y(t)$ within the precision ε (**26**). We read off from (**21**) that $F_3 = 1/2$ for *any* state of the input nodes, if the NN parameters q are chosen in the set $\mathsf{Q}_o = \mathsf{Q}'_o \cup \mathsf{Q}''_o$, with

$$\mathsf{Q}'_o : \mathbf{J}_{1i} = 0,\ i = 1, 2, \ldots, N_1,\ \theta_2 = 0,\ \theta_3 = -1/2;\ \mathbf{J}_2, \mathbf{s}_2, \mathbf{s}_3 \text{ arbitrary;}$$

$$\mathsf{Q}''_o : \mathbf{J}_2 = 0;\ \theta_3 = -1/2;\ \mathbf{J}_{1i},\ i = 1, 2, \ldots, N_1;\ \theta_2, \mathbf{s}_2, s_3 \text{ arbitrary.} \qquad (\mathbf{32})$$

Therefore we choose as reference series $Y(t) \equiv 1/2$.

Since small enough changes in the parameters generate small (although by construction discontinuous) changes in the NN output (cf. (**21**)) on a given perceptron P, the simulation $y_P(q)$ of y_λ, and the simulation $y_P(q')$ of y'_λ, for $|\lambda - \lambda'|$ small enough, is achieved if q and q' are *close enough* positions in the parameter space Q. The following learning algorithm is based on this remark:

(a) As an initial trial $\tau = 0$, start with $\lambda = 0$ and select $q(0) = q_b \in \mathsf{Q}_o$ (**32**); the parameters left arbitrary in Q_o are fixed by the choice

$$q_b \in \mathsf{Q}'_o \cap \mathsf{Q}''_o \quad \text{and} \quad s_{2h} = 1,\ \theta_{2h} = 0,\ h = 1, 2, \ldots, N_2\,; \qquad (\mathbf{32a})$$

we then have $y_P(q(0)) \equiv Y$.

(b) Increment λ by a small amount $\Delta\lambda$; at any trial $\tau > 1$, rescale s_3 :

$$s_3 \to \frac{\lambda}{\lambda - \Delta\lambda}\, s_3 \,. \qquad (\mathbf{33})$$

(c) Displace the position of the previous trial $q(\tau - 1) \in \mathsf{Q}$ *randomly*, by a small enough amount:

$$q(\tau) = q(\tau - 1) + \Delta q.$$

(d) If

$$\delta_\lambda(q(\tau)) \equiv \|y_P(q(\tau)) - y_\lambda\| < \varepsilon \qquad (\mathbf{34})$$

for $\lambda = 1$, the search is over; if (**34**) is obeyed for $\lambda < 1$ then repeat step (b).

(e) If the inequality (**34**) is not obeyed, then we have two alternatives: if $\delta_\lambda(q(\tau)) < \delta_\lambda(q_b)$, then first reset $q_b = q(\tau)$, and next repeat step (c); or if $\delta_\lambda(q(\tau)) \geq \delta_\lambda(q_b)$, repeat step (c) directly.

In the framework of this algorithm the representative point $q(\tau)$ executes a *random walk* in a small zone of the full parameter space Q, which by construction remains close to a minimum set; we are entitled to expect that this zone is properly sampled if we let this diffusion mechanism operate over an experimentally reasonable number of trials.

We wish to add at this stage a final remark on the perceptron representation of time series. In recent experiments[42] we noticed that the efficiency of the perceptron for simulating stellar variability is substantially increased if we adopt a trial ansatz

$$y(t) = W(t;\, V_o), \quad \text{with} \quad \|y - W(V_o)\| < \varepsilon \qquad (\mathbf{35})$$

where the trial signal W depends on an array of parameters $V_o = (V_o^1, \ldots, V_o^m)$, and the inequality holds for some choice of the parameter values at low enough resolution (ε large enough).

Interpret next the parameters as functions of the epoch t, $V = V(t)$, and adapt the perceptron to simulate $V^j(t)$, $j = 1, 2, \ldots, m$ on the output layer ($N_3 \equiv m$). The learning scheme we adopt follows the pattern described in the diffusion algorithm, with $W(t; V_o)$ now playing the role of the reference function $Y(t)$ in (**31**). Our experiments indicate that with a signal adapted choice of the trial function the perceptron simulation $y(t)$ is not only improved, but high frequency noise is also eliminated to a large degree; when operating in this mode the perceptron seemingly can be used as an efficient noise filter.

NUMERICAL EXPERIMENTS

We briefly report in this section on perceptron simulations of several types of *artificial* time series which bear some similarities with actual observations of stellar variability. The series we have analyzed are typically defined over $T = 200$ or 300 timesteps; the training set has a 10% gap near the middle and a 10% gap at the end of the series. The noise level is set at $\varepsilon = 10$ up to 20%. The training is based on the diffusion algorithm as described above.

Our perceptron program (with single output node) incorporates the following additional features:

1. In order to reduce the number of free parameters all scaling parameters, and all translation parameters of the hidden layer, are set equal among themselves.
2. To keep the number of hidden nodes, N_2, minimal we initially set N_2 equal to the minimum given by condition (**23**); if after a preassigned number of random diffusion steps the "energy" (**29**) has not decreased, then we add one hidden node; several readjustments of N_2 may be required in a run.
3. We substitute to the perceptron simulation $y_P(t; q)$ a best running average, $\langle y_P(t; q)\rangle$; the final simulation is then represented by

$$\langle\langle y_P(t; q)\rangle\rangle = a_o + a_1\langle y_P(t; q)\rangle + a_2\langle y_P(t; q)\rangle^2 + \ldots, \tag{36}$$

   where a_o, a_1, a_2, ... are determined by a least squares fit. With this operation we typically achieve a gain of precision of a few percent.
4. Finally, once acceptable NN parameters have been found, "thermal noise" may be added (according to scheme (**4**)).

Regular Time Series

In a first series of test experiments we have analyzed discretized sine curves given over about 100 cycles, and sampled with a timestep Δt not small in comparison with the period P and incommensurate with the latter (irrational $P/\Delta t$ approximated by sequence of rationals $\{n/m, n'/m', \ldots\}, \ldots$, given by continued fraction expansion). In general the underlying periodicity is not obvious from a mere inspection of these series. However, it can be demonstrated by a Moiré technique consisting in superposing the original series $y(t)$ with the series translated by n cycles, $y(t + nP)$: since $P \approx (m/n)\Delta t$, both series then overlap approximatively.

The perceptron response $y_P(t, q)$ reproduces these series to a precision of better than 10%; missing cycles are correctly recovered, and the Moiré technique supplies again a good estimate of the period. Obviously an elementary Fourier analysis reproduces these results with less numerical labor, so that the real performance of the NN may not be appreciated: the NN technique (in the form operated here) makes use of no mathematical assumption (besides determinacy); a Fourier series representation assumes *regularity* of the series. The fact that the NN method supplies correct results for these elementary signals gives confidence that it should work in the case of complex signals as well; such a confidence is lacking in the case of methods relying on strong *a priori* assumptions on the mathematical nature of the signal.

In a second test series we have analyzed discretized regular signals generated by superpositions of sines of different periods, with a timestep short as compared to the cycle lengths (phase space dimension of signal $d \geq 10$). An instance of such a signal together with its NN approximation is exhibited in FIGURE 1. Just as in the case of simple sines, these regular signals are well approximated; the prediction of the missing parts is made with a precision comparable with the precision of the training. The fact that the NN technique performs well, while no assumptions on the mathematical properties of the time series are made, is indicative that the technique should remain applicable to irregular signals as well.

Irregular Time Series

In a final series of test experiments we have applied the NN analysis to artificial time series associated with two-dimensional CA simulations of the pulsations of plane–parallel stellar atmospheres. These CA models provide rough approximations to the irregular behavior observed in the cooler variable stars (Long Period Variables, etc.).[43] In these experiments the signal $y(t)$ is identified with a space and time average of the local vertical hydrodynamic velocity component, the space average being carried over the whole atmosphere, and the time average being a running average over 10 CA timesteps; accordingly, $y(t)$ roughly mimics an observational velocity curve.

We observe that while the underlying hydrodynamic motions exhibit an extremely complex space–time behavior (cf. the plates in reference 43), presumably to be identified as high-dimensional spatiotemporal chaos, the associated signal $y(t)$ shows a cyclic behavior, with cycles changing in shape and in length. FIGURE 2

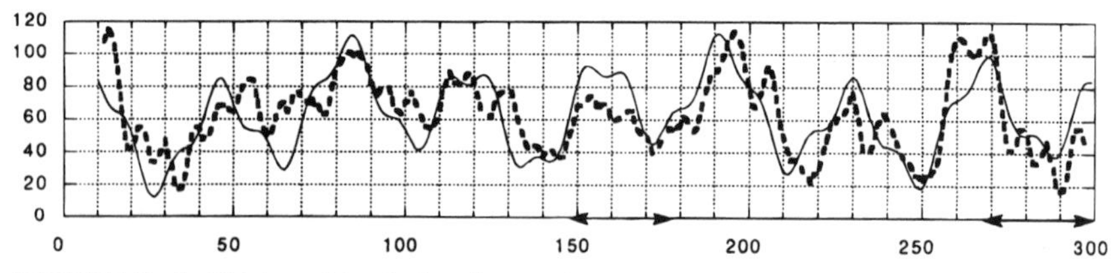

FIGURE 1. Artificial multiperiodic time series (*continuous line*); perceptron simulation (*dashed line*); learning precision 13%; ranges of prediction (*arrows*).

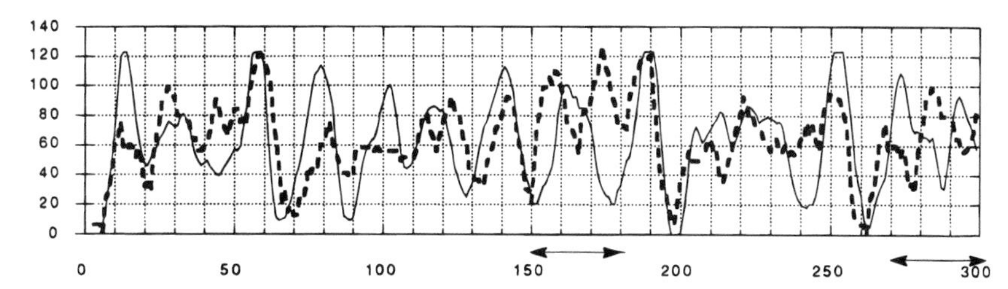

FIGURE 2. CA generated radial velocity curve (*continuous line*) and noisy NN simulation (*dashed line*) (19% noise).

provides a perceptron simulation of an instance of such a signal. We typically observe that for a training precision of about 19%, the precision of the forecast (interpolation + extrapolation) is about 24%, suggesting that the signal contains an element of "randomness" (intrinsic noise, chaos). On the other hand, if we apply the improved procedure adopting as trial signal (**35**) a sine with "parametric" amplitude and period, we typically find perceptron simulations with suppressed noise contributions; for a training precision of about 10% the precision of the forecast remains about 10%. An illustration of this *noise filtering* effect is shown in FIGURE 3.

CONCLUSIONS

The numerical experiments dealing with artificial time series associated with different intrinsic physical dynamics (of dimension ≥ 10 for the regular series, spatio-temporal chaos of high dimension in the CA case) are sufficient to suggest that the framework of NNs in the simplest form discussed in this chapter is a promising practical tool for interpolating and extrapolating gappy time series of variable stars. The class of networks we have adopted for this task are perceptrons $P[N_1\{2\}; N_2\{2\}; N_3\{z_L\}]$ with a single hidden layer, and a single output node ($N_3 = 1$) in the simplest case as discussed in greater detail in this paper; the alternative of two output nodes carrying the "parameters" of a signal-adapted trial function has been briefly sketched as well.

In contrast to NN treatments of time series by other authors, the internal dynamics of our perceptron does not mimic the physical dynamics responsible for

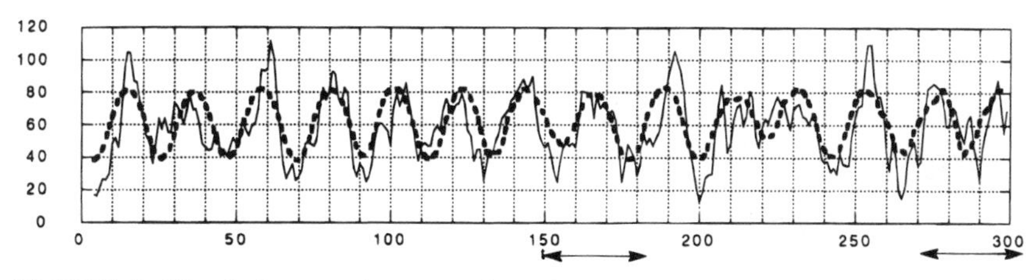

FIGURE 3. CA velocity curve (*continuous line*); signal adapted simulation (*dashed line*); precision: learning 11.7%, prediction 9.9%.

the time series; it merely reduces to a mapping of the epoch on to the stellar observable: $t \mapsto y$.

We have developed a novel *training dynamics* which we found more efficient for our purposes than the conventional Back-propagation Algorithm. Our algorithm consists in a guided diffusion process of the NN parameters: the random walk of the parameters is constrained to remain in a channel leading to the "best" parameter values, the constraint being supplied by a homotopic transformation of a reference signal for which the exact NN parameters are known. In our program the training and the forecast are *fully automated.* We should mention also that internal noise can be included on our perceptron nodes.

We feel that the NN formulation as developed here is not only a technically simple procedure for interpolation and extrapolation of time series, of a performance comparable to the performance of methods directly leaning on an underlying dynamics, but it is also complementary to the latter in several respects:

1. It remains applicable when the data points are too sparse for a meaningful search for a flow reconstruction.
2. It does apply when the actual underlying physical dynamics is high-dimensional, an alternative in principle beyond the theoretical scope of any phase space reconstruction procedure.
3. In its improved version making use of signal adapted trial functions the perceptron can act as an efficient noise filter. The cleaned perceptron signal can then be further analyzed by other methods.

We insist once more that in contrast to phase space methods, the specific NN approach as developed here is not designed to provide direct information on the dynamics of the underlying physics.

ACKNOWLEDGMENTS

I would like to thank Michael Storrie-Lombardi for providing introductory material on neural networks at an early stage of this investigation, and for his encouragement at later stages. I also wish to thank Ted von Hippel for information on the architecture of networks designed for classification purposes, and Robert Buchler for comments on the original version of this paper.

REFERENCES

1. Serre, T. 1992. PhD thesis. Observatoire de Paris. Unpublished.
2. Serre, T., J. R. Buchler & M. J. Goupil. 1991. *In* White Dwarfs. G. Vauclair & E. Sion, Eds.: 175. Kluwer. The Netherlands.
3. Serre, T., Z. Kolláth & J. R. Buchler. 1996. Astron. Astrophys. In press.
4. McCulloch, W. S. & W. Pitts. 1943. Bull. Math. Biophys. **5:** 115.
5. Pitts, W. & W. S. McCulloch. 1947. Bull. Math. Biophys. **9:** 127.
6. Minsky, M. L. & S. Papert. 1969. Perceptrons. An Introduction to Computational Geometry. MIT Press. Cambridge, MA.
7. Rumelhart, D., G. E. Hinton & J. McClelland. 1986. *In* Parallel Distributed Processing, Vol. I: Foundations. D. Rumelhart & J. McClelland, Eds.: 45. MIT Press. Cambridge, MA.

8. RUMELHART, D. & J. MCCLELLAND (Eds.). 1986. Parallel Distributed Processing. MIT Press. Cambridge, MA.
9. RUMELHART, D. E., G. E. HINTON & R. J. WILLIAMS. 1986. Nature **323:** 533.
10. MILLER, A. S. 1993. Vistas Astron. **36:** 141.
11. STORRIE-LOMBARDI, M. C. & O. LAHAV. 1994. *In* Handbook of Brain Theory and Neural Networks. M. A. Arbib, Ed.: 107. MIT Press. Cambridge, MA.
12. LAHAV, O. 1994. Vistas Astron. **38:** 251.
13. PERETTO, P. 1992. The Modelling of Neural Networks. Cambridge University Press. Cambridge, U.K.
14. MÜLLER, B. & D. REINHARDT. 1990. Neural Networks: An Introduction. Springer. Berlin.
15. HINTON, G. E. 1992. Sci. Am. **267:** 145 (see also other articles in this special issue on the brain).
16. HEBB, D. O. 1949. The Organization of Behavior. A Neuropsychological Theory. Wiley. New York.
17. WATKIN, T. L. H., A. RAU & M. BIEHL. 1993. Rev. Mod. Phys. **65**: 499.
18. CAIANIELLO, E. R. 1961. J. Theoret. Biol. **1:** 204.
19. LITTLE, W. A. 1974. Math. Biosci. **19:** 101.
20. LANDAU, L. D. & E. M. LIFSCHITZ. 1966. Statistische Physik. Akademie-Verlag. Berlin.
21. STANLEY, H. E. 1971. Introduction to Phase Transitions and Critical Phenomena. Clarendon. Oxford.
22. SOMPOLINSKY, H. 1988. Physics Today **41:** 70.
23. ROSENBLATT, F. 1958. Psychol. Rev. **62:** 386.
24. ROSENBLATT, F. 1962. Principles of Neurodynamics. Spartan Press. New York.
25. CLARK, J. W. 1988. Physics Rep. **158:** 9.
26. CLARK, J. W. 1992. *In* Proc. Spring College on Many-Body Techniques. M. R. H. Khajehpour, Ed. Plenum Press. New York.
27. GAZULA, S., J. W. CLARK & H. BOHR. 1992. Nucl. Phys. **A540:** 1.
28. LAPEDES, A. & R. FARBER. 1987. Los Alamos Preprint LAUP 87-2662.
29. LANG, K. & G. HINTON. 1988. Tech. Rep. CMU-CS 88-152.
30. WILLIAMS, R. J. & D. ZIPSER. 1989. Connect. Sci. **1:** 87.
31. WAN, E. A. 1993. *In* Time Series Prediction: Forecasting the Future and Understanding the Past. A. S. Weigend & N. A. Gershenfeld, Eds.: 195. Addison–Wesley. Reading, MA.
32. GERSHENFELD, N. A. & A. S. WEIGEND. 1993. *In* Time Series Prediction: Forecasting the Future and Understanding the Past. A. S. Weigend & N. A. Gershenfeld, Eds.: 1. Addison–Wesley. Reading, MA.
33. ALBANO, A. M., A. PASSAMANTE, T. HEDIGER & M. E. FARRELL. 1992. Physica **D58:** 1.
34. RUELLE, D. 1994. Physics Today **47**: 24.
35. AUSSEM, A., F. MURTAGH & M. SARAZIN. 1994. Vistas Astron. **38:** 375.
36. AUSSEM, A., F. MURTAGH & M. SARAZIN. 1995. Dynamical Recurrent Neural Networks—Towards Environmental Time Series Prediction. Preprint.
37. YULE, G. U. 1927. Phil. Trans. Roy. Soc. Lond., Ser. A **226:** 267.
38. MACPHERSON, K. 1994. Vistas Astron. **38:** 341.
39. KURTHS, J. & A. A. RUZMAIKIN. 1990. Solar Phys. **126:** 407.
40. CYBENKO, G. 1989. Math. Control Sig. Syst. **2:** 303.
41. WHITE, H. 1990. Neural Networks **3:** 535.
42. PERDANG, J. 1996. Neural Network Assisted Analysis of the Lightcurves of Cataclysmic Variables. In preparation.
43. PERDANG, J. 1993. *In* Cellular Automata: Prospects in Astrophysical Applications: 341. World Scientific. Singapore.

Index of Contributors